T0235598

The Structures of Mathematical Physics

Steven P. Starkovich

The Structures
of Mathematical Physics

An Introduction

 Springer

Steven P. Starkovich
Department of Physics
Pacific Lutheran University
Tacoma, WA, USA

ISBN 978-3-030-73451-0   ISBN 978-3-030-73449-7   (eBook)
https://doi.org/10.1007/978-3-030-73449-7

This Springer imprint is published by the registered company Springer Nature Switzerland AG
The registered company address is: Gewerbestrasse 11, 6330 Cham, Switzerland

To the memory
of
Fred Cooperstock

Preface

This textbook serves as an introduction to groups, rings, fields, vector and tensor spaces, algebras, topological spaces, differentiable manifolds and Lie groups—mathematical structures which are foundational to modern theoretical physics. It is aimed primarily at undergraduate students in physics and mathematics with no previous background in these topics.

Although the traditional undergraduate course in mathematical methods for physicists is vitally important for a college-level education in physics (a course that is often taken by mathematics and engineering majors as well), too many undergraduate students see mathematical physics as a disconnected set of arbitrary methods, or a bag of manipulative tricks, rather than as being organized around these carefully crafted structures. The first goal of this book is to describe these structures.

Further, students who pursue advanced studies—particularly in physics—are often confronted by a chasm between the methods they learned in school and the structures and methods they find in advanced texts and the professional literature. The second goal of this book is to help bridge that gap.

The motivation for this book is derived from many years of observation of this student experience and its consequences, and the earlier in the student's education we can provide a sense of unity and context to their studies the better. Therefore, the book's principal audience is the undergraduate student in physics or mathematics who is in their second or third year of study; students should not have to wait until they are in graduate school to gain this perspective.

The typical physics student is often dissuaded from pursuing courses in pure mathematics by the seeming lack of an immediate relevance of these courses to physics, and by their emphasis on formal proof. For these students, those doors and the opportunities that lie behind them remain closed.

Rather, what this student needs is a short introduction that articulates the essential concepts and vocabulary in a more expository but nonetheless rigorous fashion, and which does so with an eye on the physics. From atop this kind of "middle ground" a student would then have a secure vantage point from which to survey the possibilities that await. In that spirit, this book seeks to provide a path to that vantage point.

The book is organized around algebraic and topological concepts and structures, rather than methods of solution, and it makes connections to various elements of the undergraduate physics curriculum—a feature that would be out of place in a more formal mathematics text. It is written for the reader who has no formal background in advanced algebra or topology, but those who complete it will be well prepared to delve more deeply into advanced texts and specialized monographs.

Consider a small sample of the things a student encounters in the typical physics curriculum: a multi-variable problem in vector analysis whose domain is three-dimensional Euclidean space; an energy-momentum conservation problem in special relativity whose domain is four-dimensional spacetime; the evaluation of a function in the complex plane via Cauchy's integral formula; a classical mechanics problem that is framed in Hamiltonian phase space; or a Fourier transform of a time signal into an abstract function space (the frequency domain), and then (perhaps after some filtering or analysis) back again into the time domain.

In each of these examples, and the many others encountered over several years of college-level study, there is an underlying mathematical space (a structure) and a collection of tools (operations within a space or between spaces) that are used to define and solve a problem. In a most basic sense, however, there is only one fundamental structure, and there are only three fundamental operations involved; everything else is an elaboration or a specification.

The one fundamental structure is the *set* (or *space*, depending on context). With sets we have equivalence relations, quotient structures and product structures, which are then replicated across groups, rings, fields, modules, linear vector spaces, associative algebras, non-associative algebras and topological spaces.

The three fundamental operations are *composition, taking a limit*, and *mapping*. We can think of a composition—a binary operation where two things combine to yield a third—in the context of algebraic structures. We can think of the process of taking a limit—whereby we assess the continuity of maps (functions) and spaces—in fundamentally topological terms. The mapping concept transcends both algebra and topology, and is the connecting tissue of modern mathematics.

This text is organized accordingly, and consequently the sequence of topics will seem unusual to most physics students and instructors. For example, in a typical mathematical methods text vectors are presented early and groups usually much later. However, vector spaces are rather elaborate algebraic structures compared to groups, so here groups are discussed first. Another consequence is that a single topic may appear in different chapters as different structural aspects are highlighted (e.g., groups generally in Chap. 2, as matrices in Chap. 5 and as manifolds in Chap. 8). At other times we show a fundamental idea (e.g., a map) in different settings.

Chapter 1 is an introduction to sets, relations and mappings and is essential for all that follows. If there is one overriding objective of this chapter, it is to frame the reader's thinking about functions as being maps between sets rather than formulas. Also, although product sets may be familiar to the student, the material here on quotient sets will likely be new.

Chapters 2–5 develop the main algebraic structures of interest. The approach taken in these chapters is to develop the hierarchy of algebraic structures from the bottom up: groups → rings → fields → vector spaces → algebras. One consequence of this approach is that we clearly see how the same set may assume the guise of different algebraic structures, depending on the operations defined on that set.

Chapter 2 introduces continuous groups only in passing and in context with the finite groups discussed earlier in the chapter. We revisit continuous groups at several points later in the text. Antisymmetric groups are introduced here, but symplectic structures are discussed in the context of Hamilton's equations in Chap. 7.

Chapter 3 places the real and complex number fields and the quaternion skew field in context with other rings. I have found this to be an important topic to at least touch upon in my lectures because I have encountered too many students who think of complex numbers as useful contrivances for solving electrical engineering and quantum mechanics problems, and who fail to see complex numbers in a larger algebraic hierarchy. A brief historical account of the development of the complex and quaternion number systems is followed by an introduction to quaternion algebra. The matrix formulation of quaternions appears in the problems in Chap. 5.

Chapter 4 defines a vector space and proceeds to discuss inner products—both bilinear and sesquilinear (Hermitian)—for vectors in real and complex spaces, respectively. For the most part we use the Dirac notation (bras and kets) for vectors. The role of linear functionals (in function spaces) and one-forms (in coordinate spaces) is central to our treatment of the inner product; the higher-order antisymmetric descendants of one-forms (p-forms) are discussed in Chap. 7. Gram-Schmidt orthogonalization is developed along two parallel tracks, with one track for coordinate spaces and another for function spaces, and we include a short account of the defining characteristics of Hilbert spaces. A discussion of sums, products, cosets and quotients of vector spaces rounds out the chapter, with a particular emphasis on the tensor product and tensor spaces. The metric tensor gets special attention; antisymmetric tensors are discussed later in the context of p-forms.

Chapter 5 brings us to the pinnacle of the our algebraic hierarchy, and a good deal of attention is paid to structure constants and associative operator algebras. Lie and Poisson algebras, the vector cross product and Hamilton's equations of classical mechanics appear together in ways that most physics students are unlikely to have seen at this point in their studies. Linear transformations, including unitary and Hermitian operators, are framed both as matrices and as maps between sets. We include a standard account of matrix algebra, eigenvectors and similarity transformations. The chapter closes with a discussion of functions of operators. The exponential mapping will reappear in Chap. 8 in the context of Lie groups.

Chapters 6–8 shift our attention from algebraic to topological and differential structures. Chapter 6 is a survey of general (point set) topology for a reader assumed to have no previous background in the subject. Beyond the standard definitions, this chapter includes an account (with figures) of the meaning of the separation axioms. As important as these axioms are to a mathematician's approach to topology, it is debatable as to whether physicists really must know this. My view is that if a text at this introductory level uses a phrase such as "the space X is a T_2 space," then it owes the reader the courtesy of an explanation as to what that could possibly mean, and whether there are other "T's" we should know about! For us, knowledge of the separation axioms allows us to place metric spaces in their proper topological context. The chapter concludes with a discussion of product and quotient spaces.

After a short review of differentiation and the Jacobian, Chapter 7 introduces the reader to differentiable manifolds and differential forms. These topics are frequently skipped over in the undergraduate curriculum, and yet they are among the most ubiquitous structures in the mathematical physics literature. Therefore, we take some time to develop the subject, but limit this introductory account to $\mathbb{R}^n$. After showing the connection between differential forms and antisymmetric covariant tensors, we explore the properties of the exterior differential operator. Physical or geometric interpretations are given to lower-order p-forms, and the correspondences between exterior calculus and vector calculus in $\mathbb{R}^3$ are then laid out in detail. The application of these ideas to symplectic manifolds is discussed in the context of Hamilton's equations of classical mechanics. A final section discusses the all-important topic of pullback transformations of differential forms.

In Chap. 8 we discuss integration on manifolds, followed by brief accounts of Lie groups and integral transforms. We show (or at least infer) how a Generalized Stokes's Theorem follows directly from the Fundamental Theorem of Calculus. After introducing the concepts of homotopy, simply connected spaces and the winding number, we show how these are relevant to complex analysis. We then use the GST to show the connections between vector integrals in $\mathbb{R}^3$ and the integration of differential forms generally, thereby establishing how our familiar three-dimensional vector calculus is really just a special case of a more comprehensive structure.

The discussion of Lie groups emphasizes their connection to the generators that comprise their corresponding Lie algebras that exist in the tangent space to a manifold. Admittedly, it is pedagogically simpler to introduce Lie groups solely as matrix groups, and there are several excellent introductory accounts available along these lines. However, having by this point developed sufficient background on differentiable manifolds, we can now place Lie groups in their historical context as manifolds that possess group characteristics. Finally, in discussing integral transforms at the close of Chap. 8, we come full circle back to the beginning of the text inasmuch as the fundamental concept underlying an integral transform is that of a map.

The imagined reader of this text has a background that includes single-variable calculus, matrix multiplication, elementary vector algebra, complex numbers and elementary functions, and first-order differential equations. As noted earlier, among college students this is typically someone who is in the middle third of their undergraduate physics or mathematics program; perhaps they are just about to start a mathematical methods course. However, at least as important as a formal background are a modest "mathematical maturity," a willingness to think of familiar things in new ways and an eagerness to expand one's intellectual horizons.

The book is designed for active engagement by the reader. Examples (where the reader is often asked to fill in a few gaps) are woven into the narrative. Problems (many with hints and some with answers) offer both a review and an elaboration of material covered in that chapter. A Guide to Further Study and a list of references are included at the close of each chapter.

Connections to the physics curriculum appear in various places, depending on the topic; sometimes these connections appear in the end-of-chapter problems, but at other times they are part of the narrative. These connections become more frequent in the later chapters.

I have inserted portions of this book's content into several courses, primarily in the upper-division courses in mathematical methods, classical mechanics, electrodynamics and quantum mechanics, as well as in an independent study course. Depending on local circumstances, this book may serve as a text for a standalone seminar course or as an accompanying text for background reading.

Questions or comments may be sent to me at *starkovich@plu.edu*. Suggestions for improvement are especially welcome.

Fulfilling the goals of this text, whose scope is very broad, while keeping it relatively short frequently meant concluding a line of development far sooner than I would have otherwise preferred. Ultimately, though, these tradeoffs will mean the text is more likely to fulfill my wish for this book when placed in the hands of the intended reader—namely, that it might open some new doors for many students, who otherwise may have thought those doors closed to all but a few.

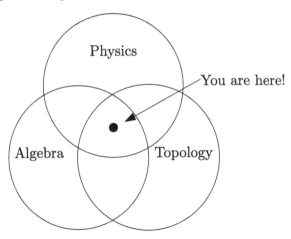

Seattle, WA, USA
February 2021

Steven P. Starkovich

Acknowledgements

Approximately thirty years ago my perspective on mathematical physics began to change. Having previously placed a primary focus on *the path to a solution* to a particular problem, I began to adopt a more comprehensive perspective that respected *the mathematical landscape* in which the problem and its solution reside. I owe this transformation largely to the works of Geroch and Roman [1], and persons who are familiar with their texts may recognize the influence these have had on how I think about mathematical physics. Also important in subsequent years has been the work of Choquet-Bruhat, et al. [2]

The opportunity to now record my thoughts as a text arose from a happenstance conversation with James Overduin (Towson University), and was subsequently nurtured by the steady encouragement and wise counsel of Angela Lahee at Springer. I am deeply grateful to them both.

I have benefited greatly over the years from countless conversations with my colleagues at Pacific Lutheran University, particularly those conversations about how best to meet the academic needs of the ever-evolving college student. Special thanks goes to my colleague, Prof. Bogomil Gerganov at PLU for his enthusiasm for a text like this, and for his words of encouragement for my efforts to write it.

I have been especially fortunate to have spent my career at PLU. Over the years, I have had the pleasure of teaching and mentoring some of the most able, kind and hard-working college students a person could ever hope to meet, and I kept them firmly in mind while composing this text.

Most especially, it is the love and support of my partner, Ruth Williams that made it possible for me to write this book over a span of two years. I am grateful to her for being my partner during this, the most recent leg of our truly excellent adventure.

References

1. Geroch, R.: Mathematical Physics. Chicago Lectures in Physics, Univ. of Chicago Press, Chicago (1985); Roman, P.: Some Modern Mathematics for Physicists and Other Outsiders, 2 Volumes. Pergamon Press, Elmsford, NY (1975).

2.   Choquet-Bruhat, Y., DeWitt-Morette, C., Dillard-Bleick, M.: Analysis, Manifolds and Physics,
     Part I: Basics, 1996 printing. Elsevier, Amsterdam (1982); Choquet-Bruhat, Y., DeWitt-
     Morette, C.: Analysis, Manifolds and Physics, Part II, Revised and Enlarged Edition. Elsevier,
     Amsterdam (2000).

Contents

Glossary of Symbols

Number Systems

$\mathbb{N}$ The set of natural numbers
$\mathbb{Z}$ The set of integers
$\mathbb{Q}$ The set of rational numbers
$\mathbb{R}$ The set of real numbers
$\mathbb{C}$ The set of complex numbers
$\mathbb{H}$ The set of quaternions

Set Theory and General Topology

A	The set A
$x \in A$	The element x belongs to the set A
$A \subset B$	The set A is a subset of (is contained in) the set B
$B \supset A$	The set B is a superset of (contains) the set A
$A = \{x : P(x)\}$	The set A equals the set of all elements x such that $P(x)$ is true
$A \cap B$	The intersection of the set A with the set B
$A \cup B$	The union of the set A with the set B
A^c	The absolute complement of the set A
$B - A$	The complement of the set A relative to the set B
A°	The interior of the set A
∂A	The boundary of the set A
$\bar{A}$	The closure of the set A
A'	The derived set (the set of accumulation points) of the set A
2^A	The power set (the set of all subsets) of the set A
$\emptyset$	The null (empty, void) set; a set containing no elements
$\aleph_0$	Aleph-naught—the cardinal number of a countably infinite set
c	The cardinality of the (uncountable) continuum
$A \times B$	The Cartesian product of the sets A and B

$\prod_{i=1}^{n} A_i$ The product set of the sets A_i for $i = 1$ to n

T_i The ith type of topological space, from the separation axioms

Relations and Maps

$a\mathrm{R}b$	The object a stands in relation R to the object b
$A \prec (\succ)B$	A precedes (follows) B in an order relation
$a; b$	a is equivalent to b (via an equivalence relation)
$\pi = S/R$	π is the quotient set of the set S by the equivalence relation R
$f : A \to B$	The map f maps the set A to the set B
$f : a \mapsto b$	The map f maps the point a to the point b
$f\|A$	The restriction of the map f to the set A
f^{-1}	The inverse (or possibly inverse map) of the map f
p_A	The projection map $p_A : A \times B \to A$
$\mathrm{Ker}(\phi)$	The kernel of the map ϕ
i	The 1-1 insertion map, such as for $i : \mathbb{N} \to \mathbb{Z}$

Basic Algebraic Structures and Groups

$a\square b$	The composition of a and b by the binary operation $\square$
$\Sigma = (S, \square)$	The algebraic structure Σ is the set S with binary operation $\square$
g^{-1}	The inverse of the group element g
C_n	The cyclic group C_n of order n
S_n	The symmetric group S_n of order $n!$
A_n	The alternating group A_n of order $n!/2$
D_n	The dihedral group D_n of order $2n$
$GL(n, \mathbb{R})$	The general linear group of all invertible $n \times n$ real matrices
$GL(n, \mathbb{C})$	The general linear group of all invertible $n \times n$ complex matrices
$G_1 \simeq G_2$	The group G_1 is isomorphic with the group G_2
$G_1 \subset G_2$	The group G_1 is a subgroup of the group G_2
$H \triangleleft G$	H is the invariant subgroup of (is invariant in) the group G
G/H	The quotient group of G by the invariant subgroup H
$G = A \times B$	G is the external direct product of groups A and B
G	G is the internal direct product of groups A and B

Rings

$\Sigma = (S, \oplus, \odot)$	Σ is a ring with binary operations $\oplus$ and e acting in set S
$\oplus$	"o-plus"—referred to as "addition" in the context of rings

$\odot$	"o-dot"—referred to as "multiplication" in the context of rings
$e_\oplus$	The additive identity (the "zero element") of a ring
$a_\oplus^{-1}$	The additive inverse of the element a in a ring
$e_\odot$	The multiplicative identity (the "unit element") of a ring
$a_\odot^{-1}$	The multiplicative inverse of the element a in a ring
$\mathcal{R}(x)$	The ring of all polynomials in x over ring $\mathcal{R}$; also $\mathcal{R}x$
$\mathbb{Z}_n$	The ring of integers modulo n, with characteristic n
J	The ideal (also known as the invariant subgroup) J of a ring $\mathcal{R}$
$\mathcal{R}/J$	The quotient ring of $\mathcal{R}$ by the ideal J
$S = \mathcal{R} \times \mathcal{R}'$	S is the product of the rings $\mathcal{R}$ and $\mathcal{R}'$
$\mathrm{Re}(z), \ \mathrm{Im}(z)$	The real and imaginary parts of a complex number z, resp.

Vector Spaces and Algebras

$\lvert u \rangle, \langle u \rvert$	The ket and bra representations of the vector $\mathbf{u}$, respectively
$\hat{\mathbf{e}}_i$	The standard basis vector for the ith coordinate
$\dim U$	The dimension of the vector space U
$d(u, v)$	The distance function between u and v
$\lVert \mathbf{u} \rVert$	The norm of the vector $\mathbf{u}$
$\langle u \vert v \rangle$	The "bracket" (inner, or scalar, product) of the vectors $\mathbf{u}$ and $\mathbf{v}$
$L(G, \mathbb{C})$	The sesquilinear map $L : G \times G \rightarrow \mathbb{C}$ on complex space G
$L(G, \mathbb{R})$	The biilinear map $L : G \times G \rightarrow \mathbb{R}$ on real space G
$\mathcal{H}$	The Hilbert space $\mathcal{H}$
$U + V$	The sum of vector spaces U and V
$U \oplus V$	The direct sum of vector spaces (or algebras) U and V
$U \otimes V$	The tensor product of vector spaces U and V
g_{ik}	The metric tensor g
X/M	The quotient space X by M
s_{ij}^k	The algebraic structure constants on cyclic indices $i - j - k$
$[A, B]$	The Lie bracket of the vectors, or operators, A and B
$\{A, B\}$	The Poisson bracket of the vectors, or operators, A and B
$\mathcal{A}/\mathcal{U}$	The quotient algebra $\mathcal{A}$ by $\mathcal{U}$
A^T	The transpose of the matrix A
$A^\dagger$	The transpose conjugate of the matrix A
$\det A$	The determinant of the (necessarily square) matrix A
$\mathrm{Adj}\ A$	The adjoint of the matrix A

Differentiable Manifolds

$D\phi(t)$	The Jacobian of the transformation ϕ, with parameter t
$\alpha \wedge \beta$	The wedge product of the p-forms α and β

$\varepsilon^{123...p}_{\alpha\beta\gamma...\pi}$	The Kronecker tensor
$\star\alpha$	The Hodge star operator acting on the p-form α
$d\alpha$	The exterior derivative of the p-form α
$\mathbf{X}_H$	The Hamiltonian vector field in phase space
ϕ_*	The "push-forward" transformation ϕ_* (often just ϕ)
ϕ^*	The pullback transformation ϕ^*
$W(z_0)$	The winding number of the point z_0 *vis-a-vis* a closed curve

Chapter 1
Sets, Relations and Maps

1.1 The Algebra of Sets

The fundamental structure that underlies all of mathematics is the *set*. Informally, a set is a collection of objects, and this definition holds equally well in most formal mathematical contexts. The set concept is a very natural and intuitive idea. Indeed, in everyday conversation we might speak of such things as the set of students in a physics class, the set of universities in the country or the set of teams in our favorite professional sport, but more often than not we talk about these things without ever using the word "set!"

The foundational role of sets in mathematics (and thereby, in mathematical physics) may seem implausible if your only engagement with set theory both began and ended with the Venn diagrams you learned in school. As we will see repeatedly throughout the text, the significance of sets lies partly in the objects that comprise them, but mostly in the operations that are defined within and between sets.

In mathematics we might refer to the set of integers, a set of matrices that meet certain criteria, a set of functions, and so on. A set might be a collection of other sets. Some sets might contain a finite number of objects, while others may be infinitely large. The objects that comprise the set are generally referred to as *elements* of the set, and we often refer to the elements as "points" even when their precise character is perfectly well known.

The operations on sets may be labeled by familiar names (for example, "addition" or "multiplication") with unfamiliar interpretations. We will discuss all of this while developing the various algebraic and topological structures that are the focus of our attention in this book.

© Springer Nature Switzerland AG 2021
S. P. Starkovich, *The Structures of Mathematical Physics*,
https://doi.org/10.1007/978-3-030-73449-7_1

1.1.1 Set Inclusion, Subsets and Set Equality

For notational convenience, we write $a \in A$ to mean "the object a is an element of (belongs to) the set A," and to specify all the elements of a particular set we have at least two options. First, if the set has a relatively small number of elements, we might just choose to list them. In this case we would write $A = \{a_1, a_2, a_3\}$ to mean "the set A consists of the elements a_1, a_2, a_3," or "A is the set whose elements are a_1, a_2, a_3." A shorthand notation is $A = \{a_i, i = 1, 2, 3\}$.

A second and more common notation is to define the elements of a set as being those objects that satisfy a specified condition. For example, if we denote the set of natural numbers by $\mathbb{N}$, then the set of three elements shown above may be written as $A = \{a_i : i \in \mathbb{N}, i < 4\}$, where the colon symbol in this context is to be read as "such that." The full statement reads "A is the set of all a_i such that i is a natural number and is less than 4." The specification that $i \in \mathbb{N}$ identifies the *index set* from which the index i is to be chosen. When a larger "universe" exists from which the elements themselves are chosen, that "universe" is called a *universal set*.

A general structure for statements used to define sets is $S = \{x : P(x)\}$. This is to be read as "the set S is the set of all elements x such that the statement $P(x)$ is true."

Example 1.1 If we take the set of real numbers $\mathbb{R}$ as the universal set, then we might choose to specify the set of all real numbers between (but not including) 0 and 1 as $S = \{x : 0 < x < 1\}$ without specifying $\mathbb{R}$ explicitly. Whenever the context is clear, reference to any set that is implicit in the definition is often omitted.   ▲

A *subset* of a set A is a set whose elements are contained in A. Further, it is *extremely* important to distinguish between an object and the set that contains only that one object. The latter is referred to as a *singleton*, and denotes the fact that the object has met the criteria necessary for inclusion in a particular set.

When denoting a subset some authors use $A \subset B$ to indicate that A is a *proper subset* of B, meaning that A is contained in B but cannot equal B. These authors would then use $A \subseteq B$ to allow for the possibility that $A = B$. In this text, we use $A \subset B$ to accommodate both possibilities. Two sets are equal if they contain the same elements, which implies that each is a subset of the other: $A = B \rightarrow A \subset B$ and $B \subset A$. The converse is also true, and we can write the combined implication using the $\leftrightarrow$ symbol to denote "if and only if," that is, $A = B \leftrightarrow A \subset B$ and $B \subset A$.

Example 1.2 If $A = \{a, b, c\}$ then a list of all the subsets would include at least the sets $\{a\}$, $\{b\}$, $\{c\}$, $\{a, b\}$, $\{a, c\}$, and $\{b, c\}$. We would say, for example, that $\{a, b\}$ is a subset of A, or A *contains* $\{a, b\}$. We would also call A a *superset* of $\{a, b\}$. In addition, the set A itself is contained (trivially) in A, so it too must be included in the list of subsets. The *empty set* (also known as the *null*, or *void*, set) by definition contains *no* elements and is given the symbol $\emptyset$. The null set is (vacuously) a subset of *all* sets and must therefore also be included in the list of subsets.   ▲

Therefore, there are a total of eight subsets of the three-element set A in Example 1.2. A few more examples would suggest that a finite set of n elements would appear to

Fig. 1.1 A Venn diagram showing the "Euler circles" for two sets, A (the "area" on the left) and B (the "area" on the right), that are located within a universal set X and which intersect at $A \cap B$

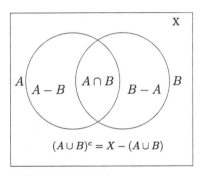

$$(A \cup B)^c = X - (A \cup B)$$

have a total of 2^n subsets. A formal proof of this result is a straightforward application of the method of proof by induction. The set of all subsets of A (in our example, a set with eight elements) is called the *power set* of A and will be symbolized as 2^A.

1.1.2 The Algebra of Sets: Union, Intersection and Complement

The *Venn diagrams* that we encountered early in our mathematics education are useful devices for visualizing the operations associated with the algebra of sets, and they give a certain plausibility to the results. Such "circle diagrams" were used at least as early as 1690 by Gottfried Leibnitz and again later by Leonhard Euler in 1768, after which they became known as "Euler's circles." The 19th-century mathematician John Venn made similar diagrams a prominent part of his book *Symbolic Logic* (1881, 1894), and these illustrations have ever since been associated with his name.[1]

Regardless of their plausibility, however, the results of Venn diagram manipulations do not constitute formal proofs, and the person for whom such methods of proof are both unfamiliar and intriguing can find many excellent introductory guides for further study.[2] With this caveat we nonetheless will use Venn diagrams to assist with our initial understanding of the algebra of sets.

Figure 1.1 provides a pictorial summary of some of the basic relationships in the algebra of sets. Two sets (A and B) are shown to exist within a universal set X. That is, $A \subset X$ and $B \subset X$. That portion of the diagram where the two sets overlap is their *intersection* and represents the set of those elements that belong to *both* A and B, i.e., $A \cap B = \{x : x \in A \text{ and } x \in B\}$.[3] Two sets A and B that do not intersect (i.e., they have no elements in common) are said to be *disjoint*. Their intersection is an example of the null set, which we would write as $A \cap B = \emptyset$.

---

[1]Gottfried Leibnitz (1646–1716); Leonhard Euler (1707–1783); John Venn (1834–1923). See [4], pp. 563-4 and the articles at [9].

[2]For the true beginner of the study of mathematical logic and formal proofs, see [2] and [8].

[3]In the older literature, the intersection is often referred to as the *meet* or *product*.

The *union* of A and B is that set whose elements are in *either A or B*, or both; in set theory, the word "or" is interpreted inclusively.[4] In Fig. 1.1, the union of A and B is that set whose elements lie somewhere within either of the two Euler circles for A and B and is denoted by $A \cup B = \{x : x \in A \text{ or } x \in B\}$.

The set of those elements of A that are *not* also in B is denoted by $A - B = \{x : x \in A \text{ and } x \notin B\}$, and we read this as "A minus B" or the *difference $A - B$*. More formally $A - B$ is called the *complement of B relative to A*. Conversely, the set of those elements of B that are not also in A is denoted by $B - A = \{x : x \in B \text{ and } x \notin A\}$ and is the *complement of A relative to B*.

Set complements are taken with respect to some other set, but when that latter set is the universal set the complement is called the *absolute complement*. It is customary to omit reference to the universal set when the context is clear. Therefore $X - A$ could be written as A^c to denote the complement of A when X is understood to be the universal set to which A is being compared. That is, $A^c = \{x : x \in X, x \notin A\}$ or simply $A^c = \{x : x \notin A\}$. Similarly, those elements of the universal set X in Fig. 1.1 that are in *neither A nor B* (i.e., $\{x : x \notin A \text{ and } x \notin B\}$) are those elements that "lie outside" $A \cup B$ and are therefore the complement of $A \cup B$ relative to X; we would write this as $X - (A \cup B)$ or just $(A \cup B)^c$. One way of writing $A \cup B$ would be to write it as the union of its three distinct parts as shown in Fig. 1.1:

$$A \cup B = (A - B) \cup (A \cap B) \cup (B - A). \tag{1.1}$$

It should be clear that the binary operations of union and intersection are commutative and associative. That is, given three sets A, B and C:

$$A \cup B = B \cup A \qquad A \cap B = B \cap A; \tag{1.2}$$

$$A \cup (B \cup C) = (A \cup B) \cup C \qquad A \cap (B \cap C) = (A \cap B) \cap C. \tag{1.3}$$

In addition, each of these two binary operations is distributive over the other:

$$A \cup (B \cap C) = (A \cup B) \cap (A \cup C) \qquad A \cap (B \cup C) = (A \cap B) \cup (A \cap C). \tag{1.4}$$

Finally, there are two particularly important relationships involving complements called *De Morgan's Laws* which the student should remember:

$$(A \cup B)^c = A^c \cap B^c \qquad (A \cap B)^c = A^c \cup B^c. \tag{1.5}$$

All of these relationships, as well as others that may be quite elaborate, can be shown more formally by employing the "truth tables" of propositional logic (see [8]) or by otherwise keeping careful account of set inclusion.

---

[4]In the older literature, the union is often referred to as the *join* or *sum*.

1.2 Relations Within Sets

Regardless of the precise nature of the elements that comprise a given set, it's important to know how, or whether, the elements of that set are intrinsically related to one another. This is because once those relations are known and described in an unambiguous (though not necessarily unique) manner, it then becomes possible to define algebraic or topological structures on that set.

It is both convenient and customary to adopt the notation $a\mathrm{R}b$ to mean "the object a stands in relation R to the object b" regardless of the specific nature of the objects or the relation. There are two relations, order relations and equivalence relations, that are central to the development of the algebraic structures that underlie much of mathematics generally, and mathematical physics in particular.

1.2.1 Order Relations

In our earliest school days we learn to depict the set of real numbers as a line, with negative numbers to the left and positive numbers to the right. There are subtleties when we invoke geometry to depict the set of real numbers in this way, but implicit in this picture is the idea that some members of the set of real numbers "precede" or "follow" others. Specifically, we use the notation $a < b$ to denote the case where a precedes (is less than) b, and $b > a$ to denote that b follows (is greater than) a. This relation between a and b is unambiguous so long as $a \neq b$, and consequently there is an ordering relation on the set of real numbers.

Definition 1.1 If a consistent and unambiguous relation between elements of a set may be established such that some elements may be said to "precede" and others may be said to "follow," then we can say there is an *order relation* between or among those elements. If such a relation may be established for the entire set, the set is said to be "ordered." The symbols $\prec$ and $\succ$ are used to indicate that some elements of the set precede or follow others, respectively. ∎

How order relations are defined on any particular set depends on the nature of that set. For example, consider the set of five points shown in Fig. 1.2a. We could write $a \prec b \prec c \prec d \prec e$ to mean that a precedes (is to the left of) b, and so forth. Similarly, the set of five sets shown in Fig. 1.2b also may be given an order relation. In terms of subsets we would write $A \subset B \subset C \subset D \subset E$, but we could just as easily write $A \prec B \prec C \prec D \prec E$ if we define "precedes" to mean "is a subset of."

Another example is shown in Fig. 1.2c where $A \prec B \prec C \prec D \prec E$ means the intersections form a chain-like structure. Indeed, sets with order relations like those in Fig. 1.2a–c are called *chains* or *totally-ordered sets*. Further, in many instances it is possible to define a *first element* or a *last element* of ordered sets, or perhaps a *minimum* and *maximum*, even in those contexts where the objects not numbers (see, for example, Problem 1.4). Order relations have a particular relevance to networks.

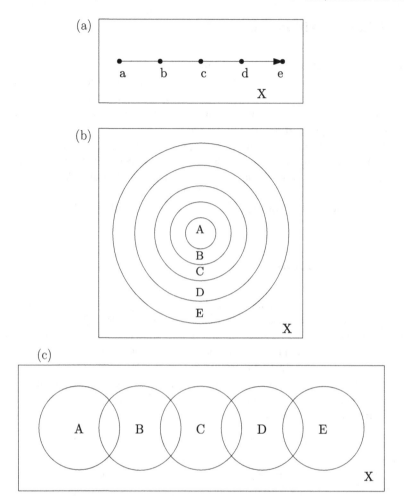

Fig. 1.2 a The set of points $S = \{a, b, c, d, e\} \subset X$ arranged as a chain. **b** The set of sets $S = \{A, B, C, D, E\} \subset X$ arranged as a chain, where "precedes" is defined as "is a subset of." **c** The same sets as in (**b**) rearranged, but still a chain. Here, "precedes" and "follows" are defined in terms of sequential intersections

Next we consider *partially-ordered sets*, or *posets*, wherein one or more order relations are defined among at least some of the elements of the set but perhaps not all. For example, Fig. 1.3a shows a rearrangement of the five points in Fig. 1.2a. For certain, we can write $c \prec d \prec e$, and it seems reasonable to assert $a \prec c$ and also $b \prec c$, if we think of "precedes" in terms of the sequence of points as before. Clearly, however, there is no order relation between the points a and b, nor is there a first element in this poset although a and b both serve as minima. Figure 1.3b, c show the same order relation as in Fig. 1.3a. Posets will appear as *directed sets* when we discuss convergence in a general topological context in Sect. 6.5.

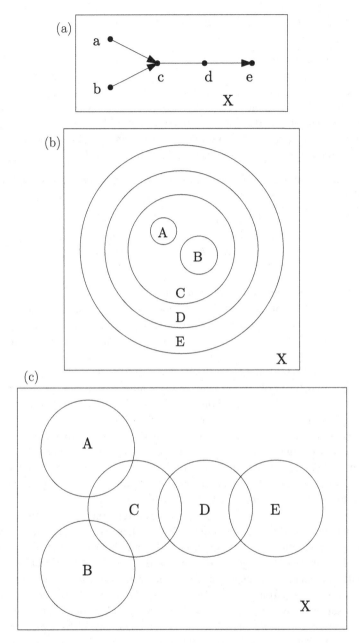

Fig. 1.3   a The set of points $S = \{a, b, c, d, e\} \subset X$ arranged as a partially-ordered set (poset) with two minima (a and b), no first element and a last element (e) which is also a maximum. **b** The set of sets $S = \{A, B, C, D, E\} \subset X$ arranged as a poset. Compare Fig. 1.2b. **c** The same sets as in (**b**) but rearranged as a different poset

1.2.2 Equivalence Relations and Quotient Sets

A second example of a relation between two elements of a set comes from asking whether they are equivalent. This presumes we know what "equivalent" means in any particular context. For example, two integers a and b may be deemed equivalent if they are numerically equal, but we could just as easily define equivalence among the set of integers to mean that two integers are equivalent if they are both even, or both odd, prime, perfect squares, and so on. Such a relation R is called an *equivalence relation*, and the expression $a \simeq b$ means "a is equivalent to b."

Definition 1.2 For objects a, b and c, an *equivalence relation* $\simeq$ is defined to be a relation that is:

$$\text{Reflexive: } a \simeq a;$$
$$\text{Symmetric: } a \simeq b \leftrightarrow b \simeq a; \text{ and}$$
$$\text{Transitive: } a \simeq b \text{ and } b \simeq c \rightarrow a \simeq c.$$

∎

This definition of an equivalence relation is applicable across all types of sets and serves to *partition* a set into distinct subsets. Indeed, it is second nature that we tend to group things based on some shared characteristic, but precisely *how* we group things together (i.e., which characteristic we choose when defining equivalence) says a lot about what our aims might be in partitioning a set.

For example, a person might sort the books in their personal library by defining equivalence among the books as "having the same number of pages," but more likely they would use "same subject area" or "same author" as the defining characteristic because most people organize their books with the aim of being able to find them easily! In elementary particle physics, physicists might choose to classify particles by their electric charge, their spin, or their mass depending on the nature of the experiment or theoretical model under consideration. On the other hand, historians of physics might choose to classify elementary particles by the high-energy accelerator at which they were first observed or by their year of discovery.

As noted above, when we apply an equivalence relation to a set of objects we partition that set into necessarily distinct subsets. These subsets are called *equivalence classes* wherein the elements in each class share the defining characteristic of equivalence (charge, spin, subject area, year of discovery, etc.) for that class. The union of all equivalence classes reconstitutes the original full set.

An important application of equivalence comes when we map (see Sect. 1.3) a set onto its set of equivalence classes. Because all the elements of an equivalence class share the same defining characteristic of that class, any of those elements may be chosen as the "class representative" when referring to that class. For example, consider a set of five objects $S = \{a, b, c, d, e\}$ and assume an equivalence relation R on the set S has been defined in such a way that $a \simeq b \simeq c$ and $d \simeq e$. In this case there are two equivalence classes: $\{a, b, c\}$ and $\{d, e\}$.

Choosing one representative element from each class (say, a and d, or c and e), we use the notation $[a]$ and $[d]$ (or $[c]$ and $[e]$) to represent these two classes. Together, these two equivalence classes form a set called the *quotient set*[5] of S, which we designate as

$$\pi = \frac{S}{R} = \{[a], [d]\}. \tag{1.6}$$

This statement is read as "π is the set of equivalence classes that arises from partitioning the set S by means of the equivalence relation R," a long and cumbersome phrase that is often abbreviated to "π is equal to S modulo[6] R" or "π is the quotient set S by R." In this example, even though the set S has five elements, $\pi = S/R$ has only two elements, the two equivalence classes $[a]$ and $[d]$.

The quotient structure is one of the most powerful and ubiquitous in mathematics. Here it has been applied to sets, but it is applicable across virtually all structures in mathematical physics, including groups (quotient groups), rings (quotient rings), vector spaces (quotient spaces) and others. In every case, the quotient structure involves a map of a set onto its quotient set in a manner that is dictated by the chosen equivalence relation. We can say that *wherever there is an equivalence relation, there is a quotient structure, and vice versa*. We will return to a discussion of equivalence classes, partitions and quotient sets after we discuss maps in the next section.

1.3 Mappings Between Sets

Along with the order and equivalence relations, the mapping concept is central to mathematics and to all that follows in this text.

Definition 1.3 A *map*, or *mapping*, is a rule by which an element of one set is assigned to, or associated with, a *unique* element of another set. If we denote the map as f and the two sets as A and B, then we write $f : A \rightarrow B$ to mean "f maps the set A to the set B." In this arrangement, A is called the *domain* and B is called the *codomain* of f. A subset of B into which some or all of A is mapped is called the *range* of f. ∎

If we wish to refer to the mapping by f of one specific element $a \in A$ to one specific element $b \in B$, then we would write $f : a \mapsto b$. It is in this instance of an element-to-element association that we use the familiar function notation $b = f(a)$, although on occasion the notation $f(A) = B$ is used to refer to the mapping of one set A to another set B. These ideas are illustrated in Fig. 1.4a, b.

In the case of a real-valued function of a single real variable, we often deal with subsets of $\mathbb{R}$ for both the domain and range of the function rather than the entire real line. This is depicted more generally in Fig. 1.4c which shows the *restriction* of f to some $A_0 \subset A$, or $f|A_0$, where the range is $B_0 \subset B$. The restriction of f to A_0 is uniquely defined on A_0 if f is uniquely defined on all of A. However, different maps

---

[5] Also called a *factor set*.

[6] A review of *modular arithmetic* is given in the context of rings in Sect. 3.1.1.

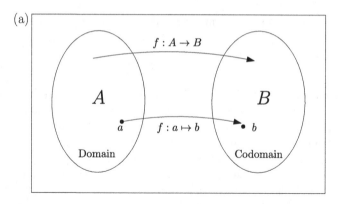

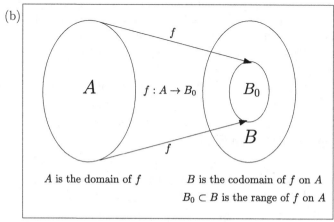

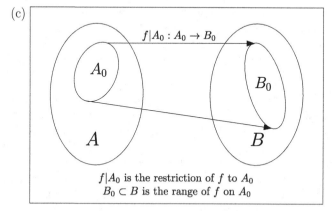

Fig. 1.4 **a** Set-to-set mapping $f : A \to B$, and point-to-point mapping $f : a \mapsto b$ **b** Distinguishing between the codomain B and the range B_0 of a map f. **c** The familiar case of a map being restricted to the subset A_0 of some larger set A

Fig. 1.5 A many-to-one
map is allowed as in (**a**), but
not the one-to-many
assignment shown in (**b**)

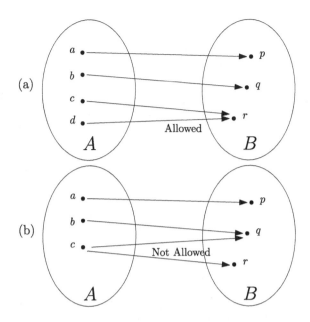

on A may be indistinguishable on A_0, in which case $f(A_0)$ tells us nothing unique
about $f(A)$.

It is important to emphasize the word *unique* in Definition 1.3. A map might
be one-to-one (where one particular element of A is associated with one particular
element of B), or it could be many-to-one (where more than one element of A is
associated with one particular element of B). A map *cannot*, however, be one-to-
many (where one element of A is associated with multiple elements of B). In order
for a rule of assignment or association between set elements to be called a map, it
must be specific and unambiguous in its association of the elements of the domain
to elements of the codomain. Examples of such assignments are shown in Fig. 1.5.

Whenever we draw a graph of a real-valued function $y(x)$ of a real variable x,
we are mapping a set of points $x \in \mathbb{R}$ to a set of points $y = f(x) \in \mathbb{R}$. Figure 1.6
shows two ways of depicting the function $y(x) = x^2$ over a domain and range that
are subsets of the real line. Maps may not always be expressed (or expressible) in
terms of simple functions, so it is important to become comfortable with the more
abstract depiction of maps as discussed in this section.

Given a map $f: A \rightarrow B$, the two sets A and B may be identical, similar or totally
different from each other in the nature of the elements they contain. For example,
the very process of counting the fingers on your hand is a mapping (one-to-one,
in this case)[7] from a subset of the set $\mathbb{N}$ of natural numbers (an abstraction) to a
set of real-life objects S; or $f: \mathbb{N} \rightarrow S$. Examples of maps important in mathemati-

---

[7] A set that can be put in one-to-one correspondence with the set of natural numbers (or a subset of
$\mathbb{N}$) is said to be *denumerable* or *countable*; otherwise it is *non-denumerable* or *uncountable*. See
Sect. 1.3.3.

Fig. 1.6 Two ways of
depicting a map $f: \mathbb{R} \to \mathbb{R}$
for $y(x) = x^2$: **a** abstractly,
and **b** with a formula

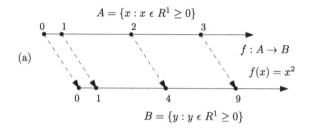

(a)

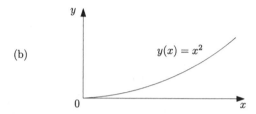

(b)

cal physics include maps that are group-to-group (*homomorphisms*), space-to-space
(*homeomorphisms*), and group-to-space (a focus of homology theory).[8]

1.3.1 Injective, Surjective and Bijective Maps

There is a nomenclature for maps between sets that is used throughout the literature
and advanced works, and it should be committed to memory.

> A map that is one-to-one is called *injective* or *into*.
> A map where the range equals the codomain is called *surjective* or *onto*.
> A map that is both one-to-one and onto is called *bijective*.

The full description of a map requires not only a specification of the rule of
association between set elements, but a clear specification of the domain, codomain
and range of the map as well.

Example 1.3 Consider the map $f: A \to B$, with $f(x) = y = x^2$ for $x \in A$ and $y \in B$. The nature of the map depends on the specific details of the domain, codomain
and range. There are several possibilities:

---

[8]A *homo*morphism is a map that preserves the *same* internal relationships between two algebraic
structures; a *homeo*morphism is a map between two *similar* topological spaces. We discuss these
more fully at appropriate points in the text.

(a) Bijective: $f: A = \{x: x \in \mathbb{R}^1 \geq 0\} \rightarrow B = \{y: y \in \mathbb{R}^1 \geq 0\}$;
(b) Injective but not surjective: $f : A = \{x: x \in \mathbb{R}^1 \geq 0\} \rightarrow B = \{y: y \in \mathbb{R}^1\}$;
(c) Surjective but not injective: $f: A = \{x: x \in \mathbb{R}^1\} \rightarrow B = \{y: y \in \mathbb{R}^1 \geq 0\}$;
(d) Neither injective nor surjective: $f: A = \{x: x \in \mathbb{R}^1\} \rightarrow B = \{y: y \in \mathbb{R}^1\}$.

In other words, the same functional form represents different maps depending on the definitions of A and B.   ▲

These kinds of distinctions must be made for several reasons; chief among them is the necessity to distinguish between *inverses* and *inverse maps*. In one respect, any rule of association between set elements may be reversed to form an inverse simply by "reversing the arrow" in the expression for the rule. That is, if we write the rule as $f: A \rightarrow B$, then we may write the inverse of f as $f^{-1}: B \rightarrow A$ for the rule applied in reverse. However, f^{-1} is not a *map* unless it is defined on B and assigns every element of B to a *unique* element of A. Consequently, a map f must be bijective in order for f^{-1} to be a map.

In Example 1.3, inverse maps do not exist in (b), where f^{-1} is not defined on $y < 0$, nor in (c) where f^{-1} would associate each element of B with two elements of A. Nor, of course, would there be an inverse map in (d). Only in (a), where f is bijective, does the inverse rule f^{-1} constitute a map.

We see this effect whenever we work with trigonometric functions on hand calculators, where it is necessary to define the domain of the angle θ so as to yield a unique value for an inverse function. For example, by defining $\sin\theta$ over the closed interval $-\pi/2 < \theta \leq +\pi/2$, a calculator yields a unique result for $\sin^{-1}\theta$. Another example pertains to the roots and logarithms of complex numbers, where Riemann surfaces are stacked on top of each other so that multiple values appear on different complex planes. Everything is thereby kept nicely one-to-one, and we can define "principal values" to be those that lie within some specified interval.[9]

1.3.2 Continuous Maps

In physics maps typically serve as coordinate transformations, or as transformations between algebraic or topological structures. As such, preference is given most often to bijective maps that have the additional property of being *continuous* or *smooth*.

There are three levels of sophistication in how to think about the continuity of maps. For the sake of illustration we assume that the map is expressible as a real-valued function $y(x)$ of a single real variable x over a well-defined domain and range. First, a continuous function may be naively thought of as one whose graph can be sketched without lifting pen from paper. There is a good deal of wisdom to be found in this intuitive "definition," but we need something a little more precise—and that

---

[9]Most of the standard texts in complex analysis or mathematical methods discuss Riemann surfaces. See the Guide to Further Study at the end of the chapter.

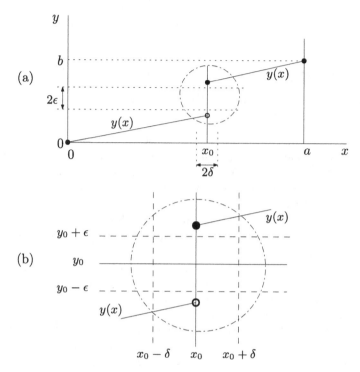

Fig. 1.7 A bijective, but discontinuous, function $y(x)$ shown in (**a**), with detail shown in (**b**)

allows for the possibility that we may not always have a formula—if the level of our work is to advance beyond the most rudimentary applications.

The second level of understanding of continuity comes from the "epsilon-delta (ϵ-δ) formulation" in elementary calculus. Given some function $y(x)$ (see Fig. 1.7), we define the function to be continuous if for *every* ϵ there is a δ such that $\mid y(x) - y(x_0) \mid < \epsilon$ whenever $\mid x - x_0 \mid < \delta$.

Although the function $y(x)$ shown in Fig. 1.7 is bijective, it is clearly discontinuous; there is a "gap" in $y(x)$ that exceeds the chosen ϵ. Formal definitions of continuity usually rely on the convergence of Cauchy sequences, a topic we take up in Sect. 3.4.1.

Implicit in this second definition of continuity is a definition of "distance between points" or "length of an interval" along the two axes (each of which is a subset of the set of real numbers $\mathbb{R}$). The need for such a distance function may not seem particularly burdensome, and in most applications in physics and engineering the notion of distance on a space is second nature. However, as we consider more abstract spaces with less structure, and consider topological spaces in their greatest generality, we will need to define continuity without the benefit of a distance function. As we'll see in Sect. 6.5, continuity and convergence rely fundamentally on the topological properties of the underlying space.

This brings us to the third level of understanding of continuity, for which we generalize the ϵ-δ formulation described above. First, define the two sets $A = \{0 \le x \le a\}$ and $B = \{0 \le y \le b\}$ so that Fig. 1.7 represents a graph of the map $f: A \to B$. Next, in B we observe the open interval $(y_0 - \epsilon, y_0 + \epsilon)$ and think of this as an open subset $V \subset B$.

Definition 1.4 A function $f: A \to B$ is said to be *continuous* if *all* open subsets $V \subset B$ are images of open subsets of A. That is, if U and V are open subsets of A and B, respectively, then f is continuous on A if for *all* open subsets $V \subset B$ we can write $f^{-1}(V) = U$ for some $U \subset A$. ■

By this formulation (and with no definition of the distance between two points or the length of an interval), we can conclude that the function in Fig. 1.7 is discontinuous. All we have done, really, is replace ϵ and δ with the terminology of "open subsets," so all we need for this third formulation of continuity is a definition of what it means for a set or subset to be open. In this regard, the set of real numbers is particularly accommodative since open intervals in $\mathbb{R}$ are also open subsets, but defining open and closed sets is more subtle and may be ambiguous for more general sets and spaces. Indeed, when we return to this topic in our discussion of topology in Chap. 6, we'll see that in some spaces sets may be *both* open *and* closed!

1.3.3 Countable and Uncountable Sets

A set is said to be *countable* or *denumerable* if a one-to-one map (an injection) exists between it and the set (or a subset) of natural numbers $\mathbb{N} = \{1, 2, 3, ...\}$. All finite sets are obviously countable. If the map $f: \mathbb{N} \to A$ is a bijection (so that f^{-1} also is a map), then A is said to be *countably infinite*. If there is no such injection, then A is said to be *uncountable* or *non-denumerable*.

In dealing with infinite sets, it is best to dispense with one's intuition and follow the rules. For example, our common sense would tell us that the set of *all* integers is "larger" than the set $\mathbb{N}$, or that the set of perfect squares is "smaller." We would be wrong on both counts, and Fig. 1.8 shows why this is the case.

These and the other sets shown differ in their "sparseness," but all have the same cardinal number ("size")—a number that is denoted by the symbol $\aleph_0$ (read as

Natural numbers:	1	2	3	4	5	⋯	
All integers:	0	1	-1	2	-2	⋯	Less sparse
All even integers:	0	2	-2	4	-4	⋯	
Even integers > 0:	2	4	6	8	10	⋯	
Perfect squares:	1^2	2^2	3^2	4^2	5^2	⋯	More sparse

Fig. 1.8 Countably infinite sets are those with the same cardinal number ($\aleph_0$) of elements

"aleph naught"). A standard exercise in advanced algebra is to show that the set of all positive rational numbers (among others) is likewise countably infinite. However, a central conclusion that arises from the construction of the real number system (using the concepts of *supremum* and *infimum*—see the discussion immediately preceding Problem 1.5) is that the set of real numbers $\mathbb{R}$ is uncountable and represents a continuum. We leave these topics for the interested reader to explore via the Guide to Further Study.

1.4   Cartesian Products of Sets and Projection Maps

The process of building up new structures from old through a "product" mechanism is in some general sense the opposite of dividing sets into quotient sets via equivalence relations. Central to this process is the *Cartesian product*, which may be applied across all algebraic structures.

Definition 1.5 The *Cartesian product* of two sets A and B is a set whose elements are *ordered pairs*, where the first term of each pair is taken from A and the second term is taken from B. We write the Cartesian product of these two sets as

$$A \times B = \{(x, y) : x \in A \text{ and } y \in B\},$$

an expression that aligns with our mental picture of points in a two-dimensional Cartesian coordinate system. Note that $(p, q) = (r, s)$ only if $p = r$ *and* $q = s$.

 If operations such as addition, multiplication or multiplication by a scalar α are defined on A and B, then these same operations carry over to $A \times B$ as follows:

$$(x_1, y_1) + (x_2, y_2) = (x_1 + x_2, y_1 + y_2),$$
$$(x_1, y_1) \cdot (x_2, y_2) = (x_1 \cdot x_2, y_1 \cdot y_2),$$
$$\alpha(x, y) = (\alpha x, \alpha y).$$

■

When A and B are small finite sets it is straightforward to list the elements of $A \times B$. For example, if $A = \{a, b\}$ and $B = \{c, d\}$, then $A \times B = \{(a, c), (a, d), (b, c), (b, d)\}$. Generally, though, we write the *product set* of the sets $A_1, A_2 ... A_n$ as

$$A_1 \times A_2 \times A_3 \cdots A_n = \prod_{i=1}^{n} A_i = \{(a_1, a_2, a_3, ... a_n) : a_i \in A_i\}. \qquad (1.7)$$

This represents a set whose elements are ordered n-tuples $(a_1, a_2, a_3, ... a_n)$, where $a_i \in A_i$ for each index i. For example, we may think of three-dimensional Cartesian space as a product of three one-dimensional spaces: $\mathbb{R}^3 = \mathbb{R}^1 \times \mathbb{R}^1 \times \mathbb{R}^1$, an expression that is equal to the direct sum (see Sect. 4.5.2) of the same three spaces.

We noted in Sect. 1.3 how a real-valued function of a single real variable such as $y = f(x)$ may be thought of as a map between two copies of $\mathbb{R}^1$ so that $f : \mathbb{R}^1 \to \mathbb{R}^1$. Such a map establishes a set of ordered pairs (x, y) in the Cartesian plane, and the two members of the ordered pair (x, y) are in a relation R to one another as specified by f. We may then write $(x, y) = xRy$, and all such points taken together (the graph of $y(x)$) yield a subset of the Cartesian plane $\mathbb{R}^2$.

Maps may be defined on a product set, and a familiar example is a functional expression for a surface in Cartesian coordinates, such as $z = f(x, y)$. In the language of the present chapter this expression is a map that assigns a specified ordered pair $(x, y) \in \mathbb{R}^2$ to some value $z \in \mathbb{R}$; that is, $f : (x, y) \mapsto z$. This map may (or may not) be bijective, depending on the domain, codomain and range of $f(x, y)$.

Given some product set we can "unwrap it" by extracting its individual pieces using a *projection map*,

$$p_j : \prod_{i=1}^{n} A_i \to A_j. \tag{1.8}$$

For example, given $A \times B$ we could write $p_A : A \times B \to A$ or $p_B : A \times B \to B$, as when the two-dimensional Cartesian plane is mapped onto each axis.

1.5 A Universal Construction for Quotient Sets

The concepts of the quotient set and the projection map may be combined to give one of the more important results in mathematics, namely, the *universal construction of quotients*. We're calling it "universal" because the same quotient structure can be established for virtually all structures—groups, rings, vector spaces and others—as we'll see later in this text. Here, we'll describe the construction for quotient *sets*.

Given a set A, we first identify an equivalence relation R on A. As we discussed in Sect. 1.2.2, the choice of R will be determined largely by the property of A we wish to emphasize. For example, let A be the set of all integers and specify R as the "evenness" property. The resulting partition is the quotient set $\pi = A/R = \{[E], [O]\}$ consisting of two disjoint equivalence classes representing the even and odd integers. Again, there are only *two* elements of the quotient set π.

In essence, we are projecting A onto (i.e., surjectively) $\pi = A/R$. We write this projection as

$$A \xrightarrow{p} A/R.$$

Next, we wish to map π to some set B with a bijection we'll call ϕ'. In our example, this could be the map $\phi' : \pi = \{[E], [O]\} \to \{0, 1\} = B$, if that is the desired target set for B.[10] The combination of these two maps yields

---

[10] $B = \{\text{"apples"}, \text{"oranges"}\}$ works, too!.

Fig. 1.9 A universal
construction for quotient sets

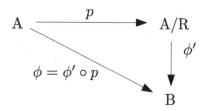

$$A \xrightarrow{p} A/R \xrightarrow{\phi'} B.$$

The map that takes us directly from A to B is a composition of two maps: the surjective projection map p followed by a bijection ϕ'. This composition is a surjection and is written as $\phi = \phi' \circ p$, an expression which is to be read right-to-left. The usual way of depicting this arrangement is shown in Fig. 1.9. Closed mapping diagrams like Fig. 1.9 are said to *commute*.[11]

The projection map p is sufficiently important that it is often referred to in the literature as the *canonical map*[12] from a set to a corresponding quotient set. The nature of the map ϕ' depends on the the properties of ϕ, and consequently ϕ' is called the *induced map*,[13] i.e., induced by ϕ. The equivalence relation R is called the *kernel* of the map $\phi: A \to B$ and is denoted by $R = \text{Ker}(\phi)$. The precise interpretation of $\text{Ker}(\phi)$ will depend on context, and we will encounter it again as we discuss other algebraic structures.

Finally, we can imagine that B might be a proper subset of a set B' and that the map ϕ is not from A to B but from A to B'. In this circumstance, the procedure outlined above is still correct, except as a last step we would need to apply the one-to-one *insertion map* $i : B \to B'$ as a final step. The full succession of maps is now

$$A \xrightarrow{p} A/R \xrightarrow{\phi'} B \xrightarrow{i} B'.$$

All of this may be summarized by writing $\phi : A \to B'$, where $\phi = i \circ \phi' \circ p$. In our example, B' might be the set of all non-negative integers,[14] and i maps the set $B = \{0, 1\}$ into $B' = \{0, 1, 2, ...\}$.

---

[11] This use of the word "commute" is not to be confused with its more common usage in algebra where it refers to the reversibility of an algebraic operation.

[12] Also called the *natural* projection.

[13] Also called a *canonical bijection*.

[14] Or, correspondingly with the earlier footnote, all types of fruit.

Problems

1.1 Let $A = \{1, 2, 3\}$, $B = \{2, 3, 4\}$ and $C = \{5, 6\}$. Find: (a) $A \cup B$; (b) $A \cap (B \cup C)$; (c) $(B \cup C) - A$; (d) $B \cup (C - A)$.

1.2 A "truth table" for $A \cup B$ may be written as

$x \in A$	$x \in B$	$x \in A \cup B$
T	T	T
T	F	T
F	T	T
F	F	F

where the first row states that if $x \in A$ and $x \in B$, then $x \in (A \cup B)$. Construct a truth table for the following: (a) $A \cap B$; (b) $(A \cup B) \cap C$; (c) $(A \cup B)^c$; (d) $(A \cap B)^c$.

The next two problems assume you have had no previous background in formal logic or methods of proof. Consequently, you are asked to use either Venn diagrams or truth tables. In at least a few cases you may wish to use both. If you have this more formal background, then you may use the methods you learned there.

1.3 Verify Eqs. 1.1–1.4.

1.4 Verify De Morgan's Laws (Eq. 1.5).

Before considering the next several problems, we wish to expand on the material in Sect. 1.2.1 regarding order relations. First, consider the set of points shown below and which are ordered left-to-right. In (i), the point a is both a minimum and a first element, and the point d is both a maximum and a last element. In (ii), the points a and b are minima, and the points d and e are maxima, but there is no first or last element.

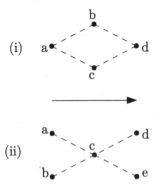

Next we consider the concepts of *supremum* and *infimum*. The supremum of a set A is written as $\sup(A)$ and is often called the "least upper bound" of the set. The infimum, denoted as $\inf(A)$, is often called the "greatest lower bound." Neither $\sup(A)$ nor $\inf(A)$ is necessarily an element of A. For example, the set $A = \{x : 0 < x < 1\}$ as a subset of the real numbers $\mathbb{R}$ has $\inf(A) = 0$ and $\sup(A) = 1$ even though neither belongs to A. However, if $A = \{x : 0 \le x \le 1\}$, then $\inf(A)$ and $\sup(A)$ take on the same values but are now elements of A.

With this background, we can now consider a few problems regarding order and equivalence relations.

1.5 Let $a, b \in A = \{2, 3, 4, 5, 6, 7\}$.

(a) Order the set of all rational numbers b/a for $b > a$. Identify any of the following elements that exist: first, last, minimum and maximum.

(b) What are some of the ways in which you might partition A? What is the corresponding equivalence relation?

1.6 Consider the set of all rational numbers $\mathbb{Q}$. Now imagine "cutting" (*Hint*: this is called the *Dedekind cut*) $\mathbb{Q}$ into two subsets (call them L (for "left") and R (for "right")) so that the number x is a supremum of L and an infimum of R, but x belongs to neither L nor R. Is $x \in \mathbb{Q}$? How do we describe x? Give an example.

1.7 The set of four points arranged in a loop as shown below contains two posets, $\{a, b, d\}$ and $\{a, c, d\}$, where we move left-to-right across the loop.

(a) How many posets are in the network of n loops arranged as shown in (i), where the motion is again left-to-right?

(b) How many posets are in the network of n triangles as shown in (ii), where "vertical" motions along an edge of a triangle *are* allowed?

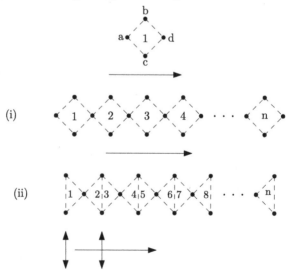

(c) How many posets are there in (ii) if the vertical motions in (b) are *not* allowed (*Hint*: compare with part (a))?

(d) Examine how the number of posets that you found in (b) and (c) depends on whether the two ends of the network of triangles consist of two edges (as shown in (ii)), two vertices, or one vertex and one edge.

1.8 Examine the extent to which equivalence relations exist for the following:

(a) The chemical elements in the periodic table;
(b) The planets of the solar system;
(c) The fundamental particles: leptons, mesons and baryons.

1.9 Draw sketches of the appropriate axes to reflect the domain and range in each of the four cases in Example 1.3. Include a sketch of the function as well.

1.10 Repeat Example 1.3 for (a) $y(x) = x^3$; (b) $x^3 - 4x^2$; (c) $y(x) = e^x$; (d) $y(x) = \sin x$.

1.11 Let $x, n \in \mathbb{N}$. Identify the following maps as being either bijective, injective only, surjective only, or neither surjective nor injective.

(a) $x \mapsto x + n$;
(b) $x \mapsto 3x + 2$;
(c) $x \mapsto x/2$.

1.12 If $f: A \to B$ and $g: B \to C$ are bijections, then $gf: A \to C$ also is a bijection (gf is a composition of the two maps that is read right-to-left). Show that $(gf)^{-1} = f^{-1}g^{-1}$. Make a sketch showing A, B, C, f and g that illustrates this result.

1.13 The function $f(x) = |x|$ is continuous at $x = 0$. Given some $\epsilon > 0$, find the maximum value for δ such that $|f(x) - 0| < \epsilon$ when $|x - 0| < \delta$?   (*Ans.* $\delta \leq \epsilon$)

1.14 Given $x \in \mathbb{R}$, let

$$f(x) = \frac{x^2 - 2}{x - 2} \quad \text{for} \quad x \neq 2;$$

$$f(x) = 4 \quad \text{for} \quad x = 2.$$

(a) Is $f(x)$ continuous at $x = 2$? Justify your answer in two ways: (i) graphically, and (ii) by applying the ϵ-δ formulation.

(b) Refer to *l'Hopital's rule* in your calculus text and find $f(x)$ as $x \to 2$. After you have reviewed the derivation of this rule, can you relate it to (ii) in part (a)?

1.15 Let $a \in A = \{1, 2, 3\}$ and $b \in B = \{4, 5, 6\}$.

(a) Find the elements of $A \times B$;
(b) Apply the equivalence relation $R =$ "b/a is an integer" and form the quotient set $\pi = (A \times B)/R$ whose elements are "Y" (yes) and "N" (no). List the elements of $A \times B$ assigned to each equivalence class in π;
(c) Define a set S (your choice!) and describe the induced map $\phi' : \pi \to S$;
(d) Write (or describe in words, if necessary) the map that goes directly from $A \times B$ for your chosen set S.

1.16 Consider the following four sets in the two-dimensional plane:

$$U_1 = \{x : 1 \le x \le 3\};$$
$$U_2 = \{y : 1 \le y \le 4\};$$
$$V_1 = \{x : 2 \le x \le 4\};$$
$$V_2 = \{y : 0 \le y \le 2\}.$$

(a) Draw graphically correct Venn diagrams (rectangles) for $U_1 \times U_2$ and $V_1 \times V_2$. Drawn correctly, the two rectangles intersect in a set we'll call S.
(b) From your drawing, show that $S = (U_1 \times U_2) \cap (V_1 \times V_2) = (U_1 \cap V_1) \times (U_2 \cap V_2)$. This result (an identity) plays an important role in describing product topologies (Sect. 6.6). The qualitative answer to this problem is shown in Fig. 6.11.

Guide to Further Study
Somewhere in the course of one's studies, it is important to gain at least a basic understanding of the history and philosophy of one's chosen field of specialization. Unfortunately, some of the older works in the history and philosophy of mathematics fell prey to the "great men" approach, a malady that once plagued much of the historical writing in science and mathematics all too often and for far too long.

Things have changed greatly in recent years, but even among the older works there are a few standouts worth considering whose focus is on the mathematics. The comprehensive three-volume set by Kline [6] is one of those, as is the much shorter monograph by Eves [3], whose Chap. 9 gives a clear and succinct introduction to philosophical matters. The historical writings of Grattan-Guinness are always a good bet; [4] is one example. A very accessible account of the broad themes of 20th century mathematics, particularly as they relate to physics, was written at the turn of the century by Atiyah [1] and is highly recommended. I have found the online source cited in [9] to be a treasure trove of information regarding the origins of mathematical terms and the stories behind them.

One of the premises for this text is our desire to provide a rationale for the essential ideas in mathematical physics without getting "bogged down" in proofs. Still, proofs are what ultimately matter in mathematics, and there is a wide range of choices from which to choose if you wish to learn more about methods of proof. I have found the work by Cupillari [2] to be a good starting point. The same holds true for studies in mathematical logic. In my opinion there are few introductory works better than that of Stolyar [8], but that's because I found the early and heavy emphasis on truth tables to be particularly helpful.

The content in the first chapter of this text is standard material and is found in many places. In writing this chapter, I chose what amounts to a distillation of the first two chapters in Roman [7], but advanced algebra texts carry it as well. We'll refer to these latter works in the context of other topics in later chapters. Terminology in some areas of mathematics changes over time, and this has been true for set theory; we have mentioned a few such cases in the footnotes in this chapter. The same holds true for notation. With these caveats, I highly recommend the classic work on set theory by Halmos [5]. The word "naive" in the title is meant to be inviting.

References

1. Atiyah, M.: Mathematics in the 20th century. Bull. London Math. Soc. **34**, 1–15 (2002)
2. Cupillari, A.: The Nuts and Bolts of Proofs. Wadsworth, Belmont, CA (1989)
3. Eves, H.: Foundations and Fundamental Concepts of Mathematics, 3rd edn. PWS-Kent, Boston (1990); Dover, Mineola, NY (1997)
4. Grattan-Guinness, I.: The Norton History of the Mathematical Sciences, first, American edn. W.W. Norton & Company, New York (1998)
5. Halmos, P.R.: Naive Set Theory. Van Nostrand, Princeton, NJ (1960)
6. Kline, M.: Mathematical Thought from Ancient to Modern Times, published in 3 Volumes. Oxford University Press, Oxford (1990)
7. Roman, P.: Some Modern Mathematics for Physicists and Other Outsiders, 2 Volumes. Pergamon Press, Elmsford, NY (1975)
8. Stolyar, A.A.: Introduction to Elementary Mathematical Logic. MIT Press, Cambridge (1970); Dover reprint, New York (1983)
9. Venn Diagram; Euler's Circles, Euler's Diagram. Earliest Known Uses of Some of the Words of Mathematics. http://jeff560.tripod.com/mathword.html. Accessed most recently in November, 2020

Chapter 2
Groups

2.1 Groupoids, Semigroups and Monoids

Consider a set S with a single binary operation $\Box$ (read as "box"). This generic symbol as well as others will assume different meanings in different contexts; in some instances it might represent ordinary addition, while at other times it might represent matrix multiplication, and so forth.

A *composition* is a binary operation that combines two elements of a set to yield a third element.[1] For example, given a set $S = \{a, b, c\}$ the $\Box$ operation might combine a and b to yield c. We would write this as $a \Box b = c$ in the same way we would express the addition of two numbers or the multiplication of two matrices.

Another way of thinking about this draws upon our earlier discussion of maps and the Cartesian product of sets. From this perspective we would write $\Box\colon S \times S \to S$, where $\Box$ is now a map whose domain is the set of all ordered pairs in the product set $S \times S$, with both elements in the ordered pair drawn from S. The range of $\Box$ is likewise contained in S, although $\Box$ may not necessarily be an onto map. We would write the composition of a and b to yield c as $\Box\colon (a, b) \mapsto c$, although reversing the order of a and b might yield a different result. Both perspectives of composition will be used in this text, and in either case it is important to emphasize the *closure* property whereby the third element c (in our example) is also a member of S.

Definition 2.1 Taken together, a set S and the binary operation $\Box$ that insures closure in S form an *algebraic structure* or *algebraic system* denoted as $\Sigma = (S, \Box)$. When discussing algebraic structures, it is essential that both the underlying set S and the binary operation $\Box$ be specified. A set without the binary operation is just a set, and different binary operations on a given set yield different structures. ■

With no further requirements or conditions imposed, the elemental structure described in Definition 2.1 is one definition of a *groupoid*. Although groupoids are foundational for all subsequent algebraic structures, they are rarely encountered

---

[1] We also speak of the composition of maps within or between sets and spaces.

© Springer Nature Switzerland AG 2021
S. P. Starkovich, *The Structures of Mathematical Physics*,
https://doi.org/10.1007/978-3-030-73449-7_2

in physical applications in their barest form. The process for constructing more elaborate algebraic structures proceeds from groupoids by adding, in a step-by-step manner, more features to the binary operation $\Box$ or to the set S.

Definition 2.2 Starting with a groupoid $\Sigma = (S, \Box)$, the next step is to require $\Box$ to have an associative property defined as $a\Box(b\Box c) = (a\Box b)\Box c$ for all $a, b, c \in S$. This combination of "groupoid + associativity" is called a *semigroup*. ∎

You may reasonably ask why associativity is chosen as an essential ingredient at this stage rather than, say, commutativity. The predisposition to the commutative property is driven by our experience with ordinary arithmetic, but it has been faced and overcome in some notable historical instances such as in Hamilton's development of *quaternions*.[2] Further, non-commutative operations are quite common in everyday experience (as in, say, finite rotations of objects around different axes) although they are nonetheless associative. This experience needs to be respected. In physics the non-commutativity of various operations is a feature of great significance in quantum mechanics. Hence, commutativity turns out to be a rather special property. In the event $\Box$ *is* commutative it is said to be *abelian*[3].

The next property to consider is whether S contains an element e such that $\Box: (e, a) \mapsto a$, or $e\Box a = a$, for all $a \in S$. This e is called an *identity element*[4] of the set S.

Definition 2.3 Given a semigroup $\Sigma = (S, \Box)$, the inclusion of an identity element $e \in S$ yields a structure Σ called a *monoid*. We may think of a monoid as a "semigroup + identity." ∎

A survey of the mathematics literature would suggest that the interest of most algebraists begins with semigroups and monoids. The interest of physicists, however, begins primarily with groups to which we devote the remainder of this chapter.

2.2 Groups

Consider a monoid with the additional feature that for each element $a \in S$ there is also an element $a^{-1} \in S$ such that $\Box: (a, a^{-1}) \mapsto e$, or $a\Box a^{-1} = e$. We call a^{-1} the *inverse* of a, and the resulting structure Σ is an *algebraic group*.

Definition 2.4 An algebraic structure $\Sigma = (S, \Box)$ is a *group* if:

---

[2] See [7] for a biographical "mathematical appreciation" of the Irish mathematical physicist William Rowan Hamilton, (1805–1865) and his work on this topic. We will define quaternions as part of our discussion of fields in Sect. 3.4.

[3] Neils Henrik Abel, (1802–1829), a Norwegian algebraist and analyst.

[4] If $\Box: (e, a) \mapsto a$, then e is called a left identity. If $\Box: (a, e) \mapsto a$ then e is a right identity. For groups they are the same, and a similar situation holds for left and right inverses. See Sect. 2.2.

1. Σ is closed under the operation of $\Box$;
2. $\Box$ is associative for all combinations of elements of S;
3. There is an identity element e in S such that $e\Box a = a\Box e = a$ for all $a \in S$;
4. There is an inverse element a^{-1} in S such that $a\Box a^{-1} = a^{-1}\Box a = e$ for all $a \in S$.

We call these four criteria the *group axioms*. An important part of this definition is the stipulation that the left identity is the same as the right identity. The same stipulation applies to the left and right inverses; they are the same. ∎

We can summarize the hierarchy of structures we've defined thus far as follows:

Semigroup = Groupoid + Associativity
Monoid = Semigroup + Identity
Group = Monoid + Inverse.

It is difficult to identify a branch of physics where groups do *not* play a significant role. Groups are central to the formulation of conservation laws and symmetries, they underlie the special functions of mathematical physics, they can help determine the solutions to differential equations, they are prominent in the kinematics and dynamics of elementary particle physics, and much more. The list of topics where groups not only are descriptive but also predictive of natural phenomena is long.

Not surprisingly, the literature on "group theory in physics" is vast. Our purpose in this chapter is to describe the essential mathematical characteristics of groups. As the text proceeds we will encounter some applications that students will recognize from their coursework in physics.

It is worth verifying that the identity element $e \in S$ is unique, and that for each element $g \in S$ its inverse g^{-1} is unique. First, we will simplify the notation by omitting the $\Box$ symbol and write ab for what is formally $a\Box b$. We take care to distinguish ab from ba so as not to assume commutativity of $\Box$. Our method of proof is proof-by-contradiction (a standard technique in mathematics), and the proofs shown here are found in all introductory accounts of group theory.

To show the identity element e is unique, assume there are two such identity elements e and f. Then for any group element g we have both $eg = ge = g$ and $fg = gf = g$. But we also would have $ef = f$ (where e is the identity), and the same $ef = e$ (where f is the identity). Therefore, $e = f$ and the identity is unique.

To show the inverse to g is unique, assume there are two such inverses h and k. Then for any group element g we have both $gh = hg = e$ and $gk = kg = e$. This leads to $gh = gk = e$. Next, compose this latter formulation with h to give $h(gh) = h(gk)$. By associativity we have $(hg)h = (hg)k$, or $eh = ek$, or $h = k$. Therefore, the inverse to g is unique, and we denote it by g^{-1}.

For abstract groups to be of practical use they need to be expressed in some tangible way through what is called a *realization* of the group. In physics, one of the most common realizations (particularly for continuous groups—see Sect. 2.3.5) arises from defining a bijective map between the abstract group elements and a set of $n \times n$ matrices (see Chap. 5) so that the set S of matrices are the elements of group $\Sigma = (S, \square)$ and $\square$ is matrix multiplication.

In this case, the realization is called a *faithful representation* of the abstract group — "faithful" because of the bijective nature of the map. The number of elements in the group (i.e., the number of matrices in the set S) is the *order* of the group, and the parameter n is the *dimension* of the representation. It often happens that the same abstract group will have representations of different dimension, although not all would be faithful representations.

Example 2.1 Let Σ be the set of natural numbers under the operation of ordinary addition; that is, $S = \mathbb{N} = \{1, 2, 3, \ldots\}$ and $\square = +$. Although $\mathbb{N}$ is closed under $+$ and $+$ is associative, there is neither an identity element nor an inverse in $\mathbb{N}$; therefore, $\Sigma = (\mathbb{N}, +)$ is not a group. Expanding S to include zero adds an identity element, but not until we expand our set to include negative integers do we have a group $\Sigma = (\mathbb{Z}, +)$. Further, Σ is a group under $\square = +$ when we replace $\mathbb{Z}$ with $\mathbb{Q}$, $\mathbb{R}$ or $\mathbb{C}$, the sets of rational, real or complex numbers, respectively.   ▲

Although the set of integers $\mathbb{Z}$ is a group under addition, it is *not* a group under ordinary multiplication because not every element of $\mathbb{Z}$ has a multiplicative inverse in $\mathbb{Z}$. This illustrates the point made earlier in this section, namely, that an algebraic structure is defined by specifying *both* the set *and* the rule of composition.

Expanding $S = \mathbb{Z}$ so as to include all rational numbers $\mathbb{Q}$ would at first glance seem to provide inverses, but this is still not a group because $\mathbb{Q}$ includes zero, and division by zero is not defined. However, the set of *non-zero* rational numbers *is* a group under ordinary multiplication, as are the sets of non-zero real and non-zero complex numbers. All the groups in the examples discussed above are *abelian groups*, and many of the examples to follow will be as well.

2.3   Some Noteworthy Groups

We wish to examine the cyclic, symmetric, alternating and dihedral groups. These are finite discrete groups in the sense that they contain a finite number of group elements with discrete steps from one group element to the next. These groups are usually referred to as *point groups*. We also introduce one-parameter continuous groups which typically are represented by sets of $n \times n$ matrices whose entries are continuous functions.

2.3.1   Cyclic Groups

Consider the set $S = \{e, a\}$ and four examples. First, let $e = +1$, $a = -1$ and let $\square$ be ordinary multiplication. Second, imagine two objects on a table. Let e represent the initial configuration, a be the state where the objects have switched places, and let $\square$ be the act of switching the objects. Third, consider a collection of points in three-dimensional space with a coordinate system defined. Let e represent the initial distribution of the points in space, a be the state where the coordinates of all points are inverted though the origin and let $\square$ be the process of spatial inversion. Fourth, define the $\square$ operator as matrix multiplication, and let

$$e = \begin{pmatrix} 1 & 0 \\ 0 & 1 \end{pmatrix} \quad \text{and} \quad a = \begin{pmatrix} -1 & 0 \\ 0 & -1 \end{pmatrix}.$$

These four examples are superficially very different from one another, but they all exhibit the same abstract group structure; they are four different realizations of the same abstract group. The group operations are $e\square e = e$, $e\square a = a\square e = a$ and $a\square a = e$. These can be conveniently summarized in a *group multiplication table*,

$$C_2 = \begin{array}{c|cc} \square & e & a \\ \hline e & e & a \\ a & a & e \end{array}, \tag{2.1}$$

where the leftmost column represents the lefthand element in the composition and the topmost row represents the righthand element.

This group, designated as C_2, is the cyclic group of order 2 whose underlying set is $S = \{e, a\}$. From the table we can see that the four group axioms are satisfied: (i) S is closed under $\square$; (ii) $\square$ is associative (for example: $a(ae) = (aa)e$) for all combinations of elements of S; (iii) there is an identity element e; and (iv) an inverse exists for each element, namely $a^{-1} = a$, and e is always its own inverse. As an additional feature, the group also happens to be abelian.

In a cyclic group, each group element is cyclically permuted by repeated application of the binary composition $\square$. In the case of C_2, with just e and a as elements, we can write this cyclical pattern variously as $a\square e = a$ followed by $a\square a = e$, or as $e \rightarrow a \rightarrow e$. The arrow notation $e \rightarrow a$ is to be read as "e becomes a," "e is replaced by a," or "the cyclic permutation takes (changes) e into a."

We can also write C_2 as a "two-cycle": $(e\, a)$ or $(a^0\, a^1)$. The latter notation emphasizes the fact that all elements of a cyclic group arise from repeated composition of only one element with itself — hence the exponent on a. Generally speaking, if all the elements of a group arise from the composition of elements that lie solely within a subset of the group, then the elements of that subset are said to be the *generators*[5] for that group. For C_2, we also have $a^0 = a^2 = e$.

This cycle notation has the advantage of being more compact than a group multiplication table, and the cycle notation makes a table's construction trivial as each row and column of the table must preserve that same cyclic order. For example, the

---

[5]The concept of a group generator is central in establishing the connection between a continuous group and its corresponding algebra. We will return to a discussion of generators in the context of *Lie groups* at the end of Chap. 8.

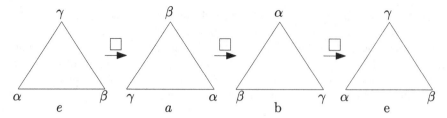

Fig. 2.1 Successive application of a $\theta = 2\pi/3$ CCW rotation in C_3

three-cycle of the cyclic group of order three, designated as C_3 and whose three group elements are $S = \{e, a, b\}$, may be written as $(e\ a\ b) = (a^0\ a^1\ a^2)$, and the corresponding group multiplication table is

$$
C_3 =
\begin{array}{c|ccc}
\Box & e & a & b \\
\hline
e & e & a & b \\
a & a & b & e \\
b & b & e & a
\end{array}, \tag{2.2}
$$

from which it is easy to see, for example, that a and b are each other's inverses.

Example 2.2 The group C_3 describes the symmetries associated with the rotation of an equilateral triangle about an axis that passes through its center and is perpendicular to the plane of the triangle. The $\Box$ operator represents a counterclockwise (CCW) rotation by an angle $\theta = 2\pi/3$; each application of $\Box$—each rotation by θ—yields a triangle that is indistinguishable from the triangle before the rotation.

What is important in the cycle notation is the cyclic order within the cycle, not the cycle's "starting point." Consequently, the CCW rotation shown in Fig. 2.1 may be written in three equivalent ways: $(\alpha\gamma\beta) = (\gamma\beta\alpha) = (\beta\alpha\gamma)$. The cycle for the clockwise (CW) rotation by $\theta = 2\pi/3$ would be written as $(\alpha\beta\gamma)$ or either of the other two equivalent expressions.   ▲

The pattern illustrated for C_2 and C_3 continues for higher-order cyclic groups, and in general we may define C_n, the cyclic group of order n, by its *n-cycle* $C_n = (a^0\ a^1\ a^2 \ldots a^{n-1})$, where $a^0 = a^n = e$. The generator of C_n is the single element a, and the four criteria for a group are met. The group C_n is an abelian group. In geometric terms we may think of C_n as the group associated with an object which has a single n-fold rotational axis of symmetry—rotation around that axis by an angle $\theta = 2\pi/n$ leaves the object invariant.

2.3.2 Symmetric Groups

In defining the cyclic groups C_n, we imposed the condition that the permutation of the elements maintain a cyclical order. Relaxing that condition and allowing the

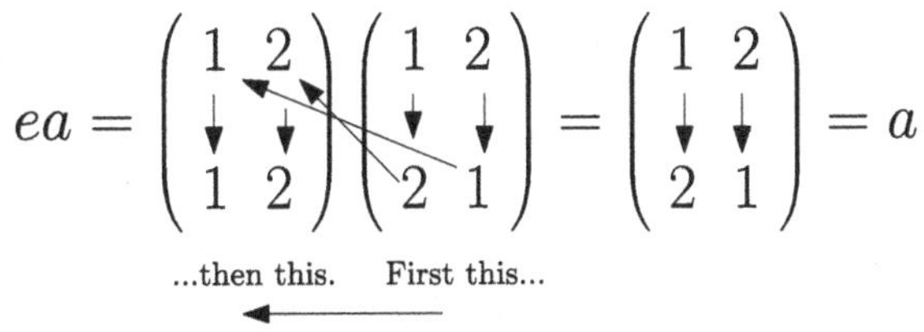

Fig. 2.2 The composition
ea in the group C_2

$$ea = \begin{pmatrix} 1 & 2 \\ \downarrow & \downarrow \\ 1 & 2 \end{pmatrix}\begin{pmatrix} 1 & 2 \\ \downarrow & \downarrow \\ 2 & 1 \end{pmatrix} = \begin{pmatrix} 1 & 2 \\ \downarrow & \downarrow \\ 2 & 1 \end{pmatrix} = a$$

...then this.   First this...

permutations to be non-cyclical yields the symmetric group S_n, which consists of all possible permutations of the n elements in the underlying set.

Consequently, there are $n!$ elements in S_n, whereas there are only n elements in C_n. Each element in the symmetric group S_n is represented by a matrix with two rows and n columns, where the top row specifies the initial state and the bottom row specifies the permuted state.

Example 2.3 Consider S_2 and denote the underlying set as $\{1, 2\}$ rather than $\{e, a\}$. The result is a group whose structure is the same[6] as C_2:

$$e = \begin{pmatrix} 1 & 2 \\ 1 & 2 \end{pmatrix} \quad \text{and} \quad a = \begin{pmatrix} 1 & 2 \\ 2 & 1 \end{pmatrix}.$$

Reading down each column, from the top row to the bottom row, the element e is the identity element of S_2, where $1 \to 1$ and $2 \to 2$ ("1 becomes 1, and 2 becomes 2"). Similarly, the element a represents the permutation where $1 \to 2$ and $2 \to 1$.   ▲

A composition of the two elements in S_2 is not a matrix multiplication but a succession of permutations that is read and implemented right-to-left.[7] For the composition ea, Fig. 2.2 shows the sequence $2 \to 1 \to 1$ and $1 \to 2 \to 2$. The end result is $2 \to 1$ and $1 \to 2$, or $ea = a$.

Figure 2.3 shows a list of the six elements of S_3 and a geometric interpretation of each element in terms of the symmetries associated with an equilateral triangle.[8] The (6×6) multiplication table for S_3 is left as an exercise.[9]

Example 2.4 An example of a composition among the elements of S_3 is

$$\alpha\delta = \begin{pmatrix} 1 & 2 & 3 \\ 3 & 1 & 2 \end{pmatrix}\begin{pmatrix} 1 & 2 & 3 \\ 3 & 2 & 1 \end{pmatrix} = \begin{pmatrix} 1 & 2 & 3 \\ 2 & 1 & 3 \end{pmatrix} = \zeta.$$

---

[6]Here, "the same" means that S_2 is *isomorphic* to C_2, meaning that there is a bijection between them. We'll define this and other morphisms in Sect. 2.4.

[7]The right-to-left convention is far and away the most common, as in [5]. Still, occasionally you'll see left-to-right. You should always check the conventions in use when comparing texts.

[8]The rightmost column in Fig. 2.3 refers to the group D_3 discussed below.

[9]When constructing these tables, we adopt the convention of placing the lefthand factor in the composition in the leftmost column and placing the righthand factor in the topmost row.

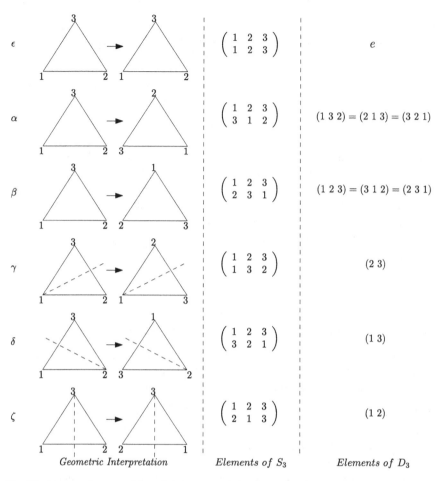

Fig. 2.3 The six elements of S_3 and their geometric interpretations in terms of the symmetries of an equilateral triangle. The identity is ϵ, while the elements α and β correspond to CCW and CW rotations, respectively, by an angle $\theta = 2\pi/3$ around a central axis perpendicular to the plane of the triangle. The elements γ, δ and ζ correspond to reflections across the respective dashed lines. The rightmost column labels these elements as denoted in the group D_3

Separately, the inverses of α and β are found by reversing the directions of the cycles of permutation. For example, α is the permutation $1 \to 3, 3 \to 2$ and $2 \to 1$, so α^{-1} is $1 \to 2, 2 \to 3$ and $3 \to 1$, which we see is just β. Thus, $\alpha\beta = \beta\alpha = \epsilon$. Also, each of the two-cycles (γ, δ and ζ) is its own inverse.

However,

$$\delta\beta = \begin{pmatrix} 1 & 2 & 3 \\ 3 & 2 & 1 \end{pmatrix} \begin{pmatrix} 1 & 2 & 3 \\ 2 & 3 & 1 \end{pmatrix} = \begin{pmatrix} 1 & 2 & 3 \\ 2 & 1 & 3 \end{pmatrix} = \zeta,$$

whereas

$$\beta\delta = \begin{pmatrix} 1\ 2\ 3 \\ 2\ 3\ 1 \end{pmatrix} \begin{pmatrix} 1\ 2\ 3 \\ 3\ 2\ 1 \end{pmatrix} = \begin{pmatrix} 1\ 2\ 3 \\ 1\ 3\ 2 \end{pmatrix} = \gamma.$$

Consequently, S_3 is our first example of a *non-abelian* group. ▲

It is often convenient to write an element of S_n as a composition of lower-order cycles. For example, one of the $6! = 720$ elements of the group S_6 is

$$a = \begin{pmatrix} 1\ 2\ 3\ 4\ 5\ 6 \\ 4\ 6\ 3\ 5\ 1\ 2 \end{pmatrix} = (1\ 4\ 5)(2\ 6)(3).$$

Here, the element a is written as a composition of a three-cycle, a two-cycle and a one-cycle. Because there are no overlapping terms in these three lower-order cycles they can be implemented in any order, i.e., they are abelian with respect to one another even though S_6 may be shown to be a non-abelian group.

2.3.3 Alternating Groups

The number of permutations from the standard order 1-2-3-···-n necessary to create any given element of S_n is either even or odd, an attribute known as the permutation's *parity*. As we will see, those elements of S_n that arise from an even number of permutations form a subgroup[10] of S_n called the *alternating group*, denoted as A_n. For small values of n it is often feasible to determine the parity by inspection, but for larger n a systematic method is needed.

Given a sequence of integers, the number of inversions[11] is the number of integers smaller than the first integer in the sequence. We find the *total* number of inversions from the standard order by progressively finding the number of inversions in the sequence and each subsequence and then taking their sum.

That is, after finding the number of inversions in the full sequence, we then truncate the first integer and find the number of inversions in the remaining subsequence, and continue in this fashion until the sequence is exhausted. Whether that total number of inversions is even or odd determines the parity. Although the number of permutations rarely equals the number of inversions, they have the same parity.

Example 2.5 Consider the element $\delta \in S_3$:

$$\delta = \begin{pmatrix} 1\ 2\ 3 \\ 3\ 2\ 1 \end{pmatrix}.$$

---

[10] A subgroup is a subset of a group that itself satisfies the group axioms; not all subsets are subgroups. See Sect. 2.5.

[11] These inversions are not to be confused with the inverses of the elements of S_n.

The standard order is 1-2-3. There are two inversions in the sequence 3-2-1 ($2 < 3$ and $1 < 3$) and there is one inversion in the subsequence 2-1. The total number of inversions is 3, so the number of permutations in the standard order necessary to create the element δ is likewise going to be odd. By inspection we see that δ contains just a single (odd number) permutation, so while the number of inversions is *not* equal to the number of permutations they have the same parity.

Although the inspection method is sufficient for S_3, it becomes much harder for, say, S_6. Consider the element $a \in S_6$, where

$$a = \begin{pmatrix} 1\ 2\ 3\ 4\ 5\ 6 \\ 4\ 6\ 3\ 5\ 1\ 2 \end{pmatrix},$$

that we mentioned earlier. When we tabulate the number of inversions in the sequence 4-6-3-5-1-2 and each subsequence we find the total to be 11:

$$4 - 6 - 3 - 5 - 1 - 2 = 3 \text{ inversions}$$
$$6 - 3 - 5 - 1 - 2 = 4 \text{ inversions}$$
$$3 - 5 - 1 - 2 = 2 \text{ inversions}$$
$$5 - 1 - 2 = 2 \text{ inversions}$$
$$1 - 2 = 0 \text{ inversions.}$$

The total number of inversions is an odd number, and therefore so is the number of permutations. The element $a \in S_6$ has odd parity, so $a \notin A_6$.   ▲

We still need to show that the set A_n (consisting only of the even parity elements of S_n) satisfies the group axioms of closure, associativity, identity and inverse. The inverse operation simply reverses the permutations, and the identity element makes no permutations (even parity). Associativity is an attribute of all elements of S_n. Closure within A_n is established by considering the parity of the composition of two permutations, p_1 and p_2, each with its own parity, and noting that a permutation and its inverse have the same parity. The first line of Table 2.1

establishes closure (two even permutations combine to give an even permutation), so A_n is a group. We leave it as an exercise to show that the order of A_n is $n!/2$.

Table 2.1 Parity of compositions

p_1	p_2	$p_1 p_2$
Even	Even	Even
Even	Odd	Odd
Odd	Even	Odd
Odd	Odd	Even

Fig. 2.4 A rectangle has a D_2 symmetry. The C_2 axis is through C and perpendicular to the page. The lines l_1 and l_2 are the $n = 2$ two-fold axes perpendicular to the C_2 axis

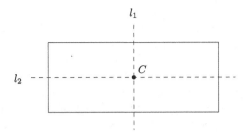

2.3.4 Dihedral Groups

The dihedral[12] group D_n may be thought of in geometric terms as a group of rotations and reflections that describes the symmetries associated with the regular polygons. More precisely, it is group that (like C_n) has a single n-fold axis of symmetry, but which also has an additional set of n two-fold rotation axes (that is, rotations through an angle $\theta = \pi$) that are at right angles to the C_n axis. The rotations around each of these two-fold axes also may be thought of as a reflection across the perpendicular plane containing it. Consequently, the group D_n contains $2n$ elements.

Example 2.6 The elements of the group D_2 may be illustrated by their effect on the plane rectangle in Fig. 2.4.

The axis through C and perpendicular to the plane of the page is the rotational axis of symmetry as it would be described in C_2, while the lines l_1 and l_2 represent the system of two-fold axes perpendicular to the C_2 rotation axis. Equivalently, the two-fold rotations around l_1 and l_2 may be thought of as reflections across the planes containing each line and perpendicular to the plane of the rectangle. Let the four elements of D_2 be $\{e, a, b, c\}$ where e is the identity, a is the $\theta = \pi$ rotation around (reflection across) l_1, b is the $\theta = \pi$ rotation around (reflection across) l_2, and c is the rotation around the C_n axis by an angle $\theta = 2\pi/n = \pi$. We leave it as an exercise for the reader to show that the group multiplication table for D_2 is

$$
D_2 = \quad
\begin{array}{c|cccc}
\square & e & a & b & c \\
\hline
e & e & a & b & c \\
a & a & e & c & b \\
b & b & c & e & a \\
c & c & b & a & e \\
\end{array}.
$$

▲

The details of D_3 were illustrated in Fig. 2.3. There, the three-fold rotational axis of symmetry is through the center of the triangle and perpendicular to the plane

---

[12]The word *dihedral*, derived from the Greek *di-*, twice, and *hedra*, seat, means "to be bounded by two planes, or two plane faces." A dihedral angle may be thought of as the angle between two intersecting planes.

of the page (as in C_3), and the system of three two-fold axes perpendicular to the C_3 axis are the three dashed lines in Fig. 2.3 (or, equivalently, they mark the three planes perpendicular to the plane of triangle across which the triangle is reflected). For $n > 2$, the D_n group is non-abelian.

Applications of the point groups are prominent in chemistry and solid state physics, particularly in the study of the transitions between molecular states and the electromechanical properties of crystals. If you are interested in learning more about point groups, please consult the references mentioned in the Guide to Further Study at the end of this chapter.

2.3.5 Continuous Groups

We saw earlier that the algebraic structure $\Sigma = (\mathbb{R}, +)$ is a group, and because $\mathbb{R}$ is continuous we can say that Σ is a *continuous group* under addition. An example of this one-parameter group is the one-dimensional translation of an object through space. Such translations are closed (the sum of two translations in $\mathbb{R}$ is a translation), associative, there is an identity element (whereby there is no translation), and there is an inverse for each translation (a translation in the opposite direction).

Example 2.7 A rotational example of a one-parameter continuous group would be the rotation of a circular disc about its central axis perpendicular to the plane of the disc (see Fig. 2.5).

The infinite set of rotations is closed (the sum of two rotations around the given axis is a rotation), associative, there is an identity element (no rotation), and there is an inverse for each rotation (a rotation in the opposite direction).   ▲

Because the elements of continuous groups are often expressed as matrices, such groups are usually referred to as *matrix groups*. Given how matrix multiplication is

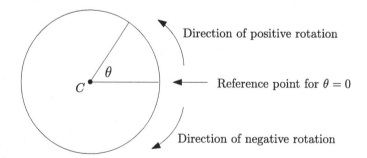

Fig. 2.5 A set of rotations forms a one-parameter continuous group under rotation about the axis through C and perpendicular to the disc. Contrast this with the C_n point group where symmetry is preserved via discrete steps

defined,[13] the requirement that the group axioms be satisfied means that all elements of a matrix group will be $n \times n$ square matrices with inverses for each matrix (i.e., the matrices must be invertible, or non-singular, meaning they have a non-zero determinant). In addition, the set of matrices must include an identity matrix and must be closed under matrix multiplication. When these conditions are met, Σ is referred to as the *general linear group of all invertible $n \times n$ matrices* and is denoted as $GL(n, \mathbb{R})$ or $GL(n, \mathbb{C})$ depending on whether the matrix entries are real or complex functions, respectively.

The groups $GL(n, \mathbb{R})$ and $GL(n, \mathbb{C})$ are very broadly defined, and the applications of matrix groups to physics usually involve placing other conditions on the entries in the matrices. The resulting subgroups (see Sect. 2.5) of the general linear group find wide use in the application of group theory to physical problems. It is here, for example, that we encounter unitary and orthogonal groups, and groups associated with the names Dirac, Lorentz, Poincare and others. We will return to a discussion of continuous groups at various points in the text, most especially in Chaps. 5 and 8.

2.4 Morphisms, and a Glance at Algebraic Topology and Categories

Figure 2.6 shows two algebraic structures Σ and Σ', and for the moment let them both be groups. Let ϕ be a bijective map $\phi \colon \Sigma \to \Sigma'$, thereby insuring the existence of an inverse map $\phi^{-1} \colon \Sigma' \to \Sigma$. We can write the map between individual group elements of Σ and Σ' as $\phi \colon a \mapsto a'$, as $\phi(a) = a'$, or even more simply as $\phi a = a'$ provided we remember that ϕ is a map while a is a group element. A similar notation may be used for ϕ^{-1} in mapping elements of Σ' to elements of Σ. Further, within each group is a map ψ (for Σ) and ψ' (for Σ') where, for example, $\psi \colon a \mapsto b$ and $\psi' \colon a' \mapsto b'$.

A map from one set to another is called a *morphism*, and a structure-preserving map is called a *homomorphism*, although this term is sometimes used in other contexts as well.[14] For groups a homomorphism ϕ is defined as a map where $\phi(a \square b) = \phi(a) \square \phi(b)$. In this example, with ϕ being bijective, we see that the rules of composition internal to Σ correspond in a one-to-one fashion to those in Σ'.

If we follow the sequence of maps in Fig. 2.6 from a' to a (via ϕ^{-1}), then from a to b (via ψ) and finally from b to b' (via ϕ), we see that the combination[15] $\phi\psi\phi^{-1}$ maps a' to b', thereby giving the same result as ψ' acting on a'. Therefore, $\phi\psi\phi^{-1} = \psi'$. This relationship is a direct consequence of ϕ being bijective.

---

[13]I have assumed that you are familiar with the method of multiplying matrices. We review a few properties of matrices in Sects. 5.5.1 and 5.5.2.

[14]We use "morphism" to be synonymous with "map" or "function." See the discussion around Table 2.2 for various uses of the word "homomorphism.".

[15]Remember to read this composition of maps right-to-left!.

Fig. 2.6 A bijective
morphism ϕ, where the
internal rules of composition
in Σ are mirrored by those in
Σ'. As a structure-preserving
morphism, ϕ is called a
homomorphism

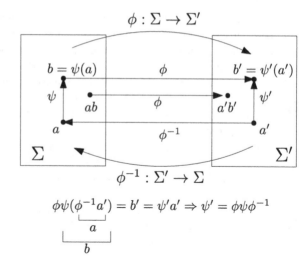

$$\phi : \Sigma \to \Sigma'$$

$$\phi^{-1} : \Sigma' \to \Sigma$$

$$\phi\psi(\phi^{-1}a') = b' = \psi'a' \Rightarrow \psi' = \phi\psi\phi^{-1}$$

Table 2.2 Types of morphisms between groups

One-to-one (1–1)	Injective	Monomorphism
Onto	Surjective	Epimorphism
(1-1) and onto	Bijective	Isomorphism

The relationship $\phi\psi\phi^{-1} = \psi'$ appears in matrix algebra (e.g., $ABA^{-1} = B'$)
as a *similarity transformation* that preserves certain matrix properties (in this case,
between B and B'). However, this relationship is a general result that depends only on
ϕ being bijective (or, in the case of matrix algebra, the matrix A being non-singular)
and is not specific to matrices. Indeed, although we have assumed here that Σ and
Σ' are groups, we did not assume a matrix representation for either.

In the context of groups we use the terms *monomorphism, epimorphism* and *isomorphism* for the injective, surjective and bijective maps of set theory, respectively.[16]
This is summarized in Table 2.2. We also speak of an *endomorphism* to refer to a
map of a set onto itself. An isomorphism of a set onto itself (i.e., an "isomorphic
endomorphism") is called an *automorphism*.

At this point we wish to throw some passing glances in the direction of two
branches of mathematics: algebraic topology and category theory. We assumed in
the discussion of Fig. 2.6 that Σ and Σ' are groups. However, it may happen that they
are not groups, nor may they even be the same type of mathematical structure. For

---

[16]Terminology varies among sources. For example, "homomorphism" is often used in lieu of "epimorphism," as in the case for quotient groups (Sect. 2.7 and Fig. 2.7), where a group is mapped to
its quotient group via a canonical map (an epimorphism). Because it preserves the group operation,
many sources call this a "homomorphism that is not necessarily one-to-one." If that weren't bad
enough, some authors use the word "morphism" to be the homomorphism we have defined here.
Our terminology preserves all the distinctions; they can always be blurred later.

example, in the event that Σ is a group and Σ' is a topological space, we are essentially mapping a group structure onto a space and converting topological problems that we may wish to solve into algebraic problems. This is the essence of the field of mathematics known as *algebraic topology*. We then say (perhaps too loosely) that "the space has a group structure," or that "the space is a group." However, there is no requirement that a given topological space be a group; it depends on the space, the group, and whether there exists a structure-preserving map between them.

Category theory is a framework for exploring and developing new structures or correspondences between structures. A *category* consists of two things: (i) a set of *objects*, such as a set of sets or a set of groups, and (ii) a set of morphisms (known as *arrows*) on that set. These are supplemented by maps between categories that preserve internal morphisms; these maps are called *functors*.

In this way, we speak of the "category of sets" or the "category of groups," and so on for other algebraic and topological structures and the functors between them. The precise nature of the morphisms and functors will, of course, depend on the structures under consideration. For example, the study of algebraic topology involves the use of functors between an algebraic category and a topological category.

For physicists, categories provide a taxonomy for organizing the various mathematical structures of interest. If you aspire to seek new applications of mathematics to physics, it may be best to approach this endeavor from the perspective of category theory. For the present, we may think of each new structure and its related morphisms that we introduce in this text as an introduction to a new category.[17]

2.5 Subgroups

First and foremost, a subgroup is a group. Unlike subsets whose elements may be chosen randomly from among the elements of the full set, the elements of a subgroup must be chosen so as to satisfy the group axioms under the same rule of composition that applies to the full group. Consequently, for a given group $\Sigma = (S, \square)$, the subgroup $\Sigma_0 = (S_0, \square)$ with $S_0 \subset S$ must include the (unique)[18] identity element of Σ. All inverses of the elements of S_0 must be contained in S_0 as well. In this way, Σ_0 will be closed under $\square$. As for notation, we write $\Sigma_0 \subset \Sigma$ to denote that Σ_0 is a subgroup of Σ, although some authors use the "$<$" symbol.

We can readily identify subgroups from among the groups we have discussed. For example, the set of real numbers $\mathbb{R}$ is a group under addition, and the sub*sets* $\mathbb{Q}$ and $\mathbb{Z}$ are sub*groups* under addition. Similarly, the non-zero real numbers are a group under multiplication, and the non-zero rational numbers are a subgroup.

---

[17] See [3] for a treatment of the structures of mathematical physics that develops categories from the outset. This reference will be very accessible to the reader who completes the present text.

[18] Multiple subgroups of a given group must each contain the identity element of S. Consequently, subgroups do not partition a group into distinct sets. Such partitioning is very important, however, and is discussed in Sects. 2.6 and 2.7.

An example of a subgroup of the complex numbers $\mathbb{C}$ under multiplication is the set of those complex numbers that lie on the unit circle centered on the origin in the complex plane. These complex numbers may be written as $z = e^{i\theta}$, and they form a group under multiplication as may be seen by considering the four group axioms:

(i) closure: $z_1 z_2 = e^{i(\theta_1 + \theta_2)} = e^{i\theta} = z$;
(ii) associativity: $z_1(z_2 z_3) = e^{i[\theta_1 + (\theta_2 + \theta_3)]} = e^{i[(\theta_1 + \theta_2) + \theta_3]} = (z_1 z_2)z_3$;
(iii) identity: $1 \cdot z = e^{i0}e^{i\theta} = e^{i[0+\theta]} = e^{i\theta} = z$;
(iv) inverse: $zz^{-1} = e^{i\theta}e^{-i\theta} = e^{i0} = 1$.

Example 2.8 Consider again the group $S_3 = \{e, (12), (13), (23), (123), (321)\} = \{\epsilon, \zeta, \delta, \gamma, \beta, \alpha\}$ (see Fig. 2.3).[19] It is straightforward to verify that the group axioms are satisfied for each of the following four sets of elements from S_3: $A = \{e, (12)\}$, $B = \{e, (13)\}$, $C = \{e, (23)\}$ and $H = \{e, (123), (321)\}$.

The first three subgroups are isomorphic to C_2, while H is isomorphic to C_3. When we include the two trivial subgroups of S_3, namely $\{e\}$ and S_3 itself, we see there are six subgroups. It is worth observing that the union of subgroups is not necessarily a subgroup. For example, the union $A \cup B \cup C$ yields a set that does not satisfy the group axioms.   ▲

There are many possible subgroups among the general linear matrix groups $GL(n, \mathbb{R})$ and $GL(n, \mathbb{C})$ whose entries are continuous functions. By imposing certain conditions on these functions, or representing a given abstract group with matrices of different dimension n, it is possible to form subgroups that are of particular importance in physics. We will explore this topic further in Chaps. 5 and 8.

2.6 Classes and Invariant Subgroups

As noted earlier, subgroups do not partition a group into distinct sets of group elements. We saw this explicitly in the case of S_3, but it is true generally since each subgroup must share at least the unique identity element of the full group.

However, group partitions are important for many advanced applications in physics, so we need to explore this. Recall that equivalence relations serve to partition sets (see Sect. 1.2.2), and the same holds for groups. If we can identify an equivalence relation among the elements of a group, then we will have a means of partitioning that group.

---

[19] S_3 is both large enough to be illustrative of important concepts and small enough for manageable hand calculations. Hence our repeated use of it in examples.

Two ways of partitioning a group are to form either the *classes* or the *cosets* of the group. An important distinction between them is that classes are a characteristic of the full group, while cosets are always defined with respect to particular subgroups within the group. Further, in the course of defining classes we will discover how to define and identify the *invariant subgroups*[20] of a given group.

Definition 2.5 Consider two elements of a group G denoted as a and b, and an arbitrary third element g. If a and b are related to each other via a similarity transformation such that $b = gag^{-1}$ for all $g \in G$, then we say that a and b form a conjugate pair. The set of all group elements that are conjugate to each other define a *conjugate class*, which is usually just referred to as a *class*. Further, conjugation is transitive, reflexive and symmetric, and therefore *conjugation defines an equivalence relation that partitions a group into distinct classes* (see Problem 2.13).   ∎

Example 2.9 We can see how this works by taking a step-by-step approach to identifying the classes of the group $S_3 = \{e, (12), (13), (23), (123), (321)\}$ $= \{\epsilon, \zeta, \delta, \gamma, \beta, \alpha\}$. Starting with the defining relationship $b = gag^{-1}$, first let $a = \gamma$ and then let g be each of the other elements in S_3. No matter the choice for g, the result for b will be either γ, δ or ζ. The same holds true if a is δ or ζ; the result for b is always either γ, δ or ζ. We conclude, therefore, that these three elements of S_3 form a class.

Second, if a is either α or β, then a similarity transformation with each of the other elements of S_3 always returns b as being either α or β. Therefore, α and β form another class of S_3. Finally, if $a = \epsilon$, then a similarity transformation with each $g \in S_3$ gives $b = \epsilon$, so the identity element ϵ is in a class by itself.

Therefore, S_3 has three conjugate classes: $K_1 = \{\epsilon\}$, $K_2 = \{\gamma, \delta, \zeta\}$ and $K_3 = \{\alpha, \beta\}$. Seen another way, the n-cycles of S_3 form classes for $n = 1, 2$ or 3.   ▲

Definition 2.6 A subgroup $G_0 \subset G$ that is closed under conjugation with all $g \in G$ is called an *invariant subgroup* of G. We often say that "G_0 is invariant in G," which we write as $G_0 \lhd G$.   ∎

Example 2.10 We showed in Example 2.8 that $A = \{\epsilon, \zeta\}$, $B = \{\epsilon, \delta\}$, $C = \{\epsilon, \gamma\}$ and $H = \{\epsilon, \beta, \alpha\}$ are the four non-trivial subgroups of S_3. None of the first three is closed under conjugation with each of the other elements of S_3. For example, $\gamma\zeta\gamma^{-1} = \delta \notin A$. We might be tempted to check whether a union of the first three subgroups is closed, but as we noted earlier the union of these subgroups is not even a subgroup, let alone an invariant subgroup.

That leaves $H = \{\epsilon, \beta, \alpha\}$ among the non-trivial subgroups of S_3, and a straightforward calculation (using some of the results from previous examples and problems) shows that H is indeed closed under conjugation with each $g \in S_3$. Therefore H is an invariant subgroup of S_3.   ▲

We can generalize this argument to any group G with invariant subgroup H. Given two (generally different) elements $h_i, h_j \in H$ and $g \in G$, the elements of H satisfy

---

[20] A alternative to the name *invariant* subgroup is *normal* subgroup.

$gh_ig^{-1} = h_j$, or $gh_i = h_jg$. This last expression is often used as the definition of an invariant subgroup. One consequence is that *any* subgroup of an abelian group will be an invariant subgroup; if we let $h_i = h_j$ in an abelian group G, then the expression $gh_i = h_ig$ is satisfied for any element h_i of any subgroup in G, and for all $g \in G$.

Finally, the two trivial subgroups of S_3 ($\{e\}$ and S_3 itself) are invariant subgroups, also said to be trivial. Groups that contain only the two trivial invariant subgroups are called *simple groups*, while those groups that contain only *non-abelian* invariant subgroups are called *semi-simple groups*. Neither of these descriptions applies to our example of S_3, which contains a non-trivial *abelian* invariant subgroup.

2.7   Cosets and Quotient Groups

Definition 2.7 The second method for partitioning a group G is to start with a non-trivial subgroup G_0 and then compose its elements with each of the other $g \in G$ that are *not* in G_0. The results are called the *cosets* associated with G_0, and taken together G_0 and its cosets partition the full group G. ∎

Suppose our group is $G = \{e, g_1, g_2, g_3, g_4\}$, and one of the subgroups is $A = \{e, g_1, g_2, g_3\}$. That leaves g_4 outside the subgroup. Now let g_4 act on each element in A. The result is $M_A = \{g_4e, g_4g_1, g_4g_2, g_4g_3\}$ which we call the *left coset* of A because we acted upon A from the left. The *right coset* would be defined accordingly.[21]

Example 2.11 Recall $S_3 = \{e, (12), (13), (23), (123), (321)\} = \{\epsilon, \zeta, \delta, \gamma, \beta, \alpha\}$. We can use this group to illustrate all the essential features of cosets. From our earlier discussion, the four subgroups are $A = \{e, (12)\} = \{\epsilon, \zeta\}$, $B = \{e, (13)\} = \{\epsilon, \delta\}$, $C = \{e, (23)\} = \{\epsilon, \gamma\}$ and $H = \{e, (123), (321)\} = \{\epsilon, \beta, \alpha\}$.

Consider the cosets of the subgroup A. We operate on the elements of A with each of the four elements not in A and generate the following cosets:

$$\delta\epsilon = \delta; \quad \delta\zeta = \beta \quad \rightarrow \quad M_{A1} = \{\delta, \beta\};$$

$$\gamma\epsilon = \gamma; \quad \gamma\zeta = \alpha \quad \rightarrow \quad M_{A2} = \{\gamma, \alpha\};$$

$$\beta\epsilon = \beta; \quad \beta\zeta = \delta \quad \rightarrow \quad M_{A3} = M_{A1} = \{\delta, \beta\};$$

$$\alpha\epsilon = \alpha; \quad \alpha\zeta = \gamma \quad \rightarrow \quad M_{A4} = M_{A2} = \{\gamma, \alpha\}.$$

Although there appear to be four cosets (one for each element not in A), only two of these cosets are distinct. Taken together with the subgroup A, these cosets serve to partition S_3 into three distinct subsets:

---

[21] Hence, an invariant subgroup H is a subgroup whose left and right cosets are equal by virtue of the relation $gh_i = h_jg$ for all $h_{i,j} \in H$ and $g \in G$.

$$A \cap M_{A1} \cap M_{A2} = \emptyset; \quad A \cup M_{A1} \cup M_{A2} = S_3.$$

The cosets for B and C may be determined in similar fashion to yield:

$$B = \{\epsilon, \delta\}; \quad M_{B1} = \{\zeta, \alpha\}; \quad M_{B2} = \{\gamma, \beta\};$$

$$C = \{\epsilon, \gamma\}; \quad M_{C1} = \{\zeta, \beta\}; \quad M_{C2} = \{\delta, \alpha\},$$

where of the four cosets in each case only two are distinct, and where each coset is of order $n = 2$ (has two elements). ▲

When finding the cosets in Example 2.11 we see that the order of the subgroup (e.g., $n_A = 2$) divides the order $n_G = 6$ of the full group S_3. This implies that groups of prime order (e.g., C_3, C_5) have only trivial subgroups (the identity, and the group itself). This is a general result and is the substance of *Lagrange's Theorem* as that term is used in group theory.

Following this same procedure, we would expect to find up to three cosets of the (invariant) subgroup $H = \{\epsilon, \beta, \alpha\}$. In fact, those three cosets of S_3 are identical, and we have

$$H = \{\epsilon, \beta, \alpha\} \quad \text{and} \quad M_H = \{\zeta, \delta, \gamma\}, \tag{2.3}$$

a determination we leave as an exercise.

These results for S_3 motivate the following definition:

Definition 2.8 An invariant subgroup H and its coset M_H serve to partition G, and together they form the *quotient group of G by H*. We write this quotient group as:

$$G/H = \{[H], [M_H]\}.$$

A quotient group is often referred to as *factor group* or a *coset group*. Only invariant subgroups can partition a group into a quotient group (Problem 2.17). ■

In general, there may be more than one coset for a given $H \lhd G$, and the order of G/H is $n_{G/H} = n_G/n_H$. We will limit our examples to those cases where, like S_3, the quotient group has order $n_{G/H} = n_{S_3}/n_H = 6/3 = 2$.

There are at least two things we need to consider regarding Definition 2.8. First, calling G/H a quotient *group* does not automatically make it a group! We need to show that the elements of the set G/H satisfy the group axioms. Second, we wish to further explore the nature of the equivalence relation that leads to the partition of G into the quotient group G/H. Recalling our discussion of quotient sets in Sect. 1.2.2 and how an equivalence relation provides a rule for partitioning a set into distinct equivalence classes so as to emphasize one feature of the set over other features, we want to get a sense of the nature of the "rule" (the feature being highlighted) inherent in the equivalence relation that gives us this partition.

First, regarding the group axioms, recall that $h_{i,j} \in H$ and $m_{i,j} \in M_H$. Because H is a subgroup we have $h_i h_j \in H$, but we can also think of the object $[H]$ as

Fig. 2.7 A universal
construction for quotient
groups

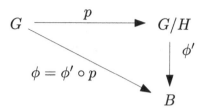

representative of any and all $h_i \in H$. Therefore, we abbreviate the precise expression $h_i h_j \in H$ as $HH = H$, dropping the bracket signs.[22] In addition, we have $m_i h_i \in M_H$, or $MH = M$. This follows directly from the definition of the coset of H. Therefore, we can already see that H (or, more precisely, $[H]$) plays the role of an identity element in G/H.

We also need to consider the composition of two elements $m_i h_i$ and $m_j h_j$ of $[M]$. Using the associative property and the definition of the coset of H, we may write

$$(m_i h_i)(m_j h_j) = m_i(h_i m_j)h_j = m_i(m_j h_k)h_j = m_i m_j(h_k h_i) = (m_i m_j)h_l, \quad (2.4)$$

where we have been careful to distinguish among different possible elements of H. This tells us that the composition of two coset elements is in the coset of their composition. More succinctly we have $(MH)(MH) = (MM)H$. This establishes closure within G/H under coset multiplication. Finally, the existence of an inverse follows directly from Eq. 2.4 by letting $m_j = m_i^{-1}$, and we find $(M^{-1}H)(MH) = H$ as expected since H has already been shown to serve as an identity in G/H.

It follows from Eq. 2.3 that $m_i m_j \in H$, or $MM = H$, for S_3. The results of the various coset multiplications for S_3 now may be summarized in a group multiplication table:

$$
G/H = \quad
\begin{array}{c|cc}
\square & H & M_H \\
\hline
H & H & M_H \\
M_H & M_H & H
\end{array}
\quad \xrightarrow{\phi'} \quad C_2 =
\begin{array}{c|cc}
\square & e & a \\
\hline
e & e & a \\
a & a & e
\end{array}, \quad (2.5)
$$

where ϕ' is an isomorphism from the two elements of G/H to the two elements of the cyclic group C_2 discussed in Sect. 2.3.1.

With regard to the nature of the equivalence relation in G/H, we first observe that the universal construction for quotient sets discussed in Sect. 1.5 is replicated here for quotient groups. Starting with a group G (S_3 in our example), we find an invariant subgroup H and its associated coset M_H. Next we apply the canonical map (an endomorphism) $p: G \to G/H = \{[H], [M_H]\}$ followed by the isomorphism $\phi': G/H \to C_2$. The resulting structure is shown in Fig. 2.7, where $B = C_2$ in our particular example. This is the same fundamental structure as was shown in Fig. 1.9 for the universal construction of quotient sets.

As an example of the potentially confusing terminology noted earlier (see Table 2.2 and the nearby footnote), it is common to refer to the canonical map p

---

[22]Expressions like this represent what is called *coset multiplication*.

as a homomorphism in this context because it preserves the group operation, even though it is not one-to-one. Under the action of ϕ', the equivalence class $[H]$ is mapped to the identity element e of C_2.

Consequently, when we consider the composite effect of $\phi = \phi' \circ p$ on the group G we see that ϕ maps all elements of the full group G that belong to H to the identity element e ($\phi: h_i \mapsto e$, or $\phi(h_i) = e$). The invariant subgroup H is the *kernel* of ϕ acting on G, or $H = \text{Ker}(\phi)$. All the remaining elements of G (those in M_H) are mapped by ϕ to the element a in C_2. Thus, the equivalence relation that underlies the partition of G into the quotient group G/H is the identification of all $h_i \in H$ with an identity element. It is sometimes said that H and M_H "absorb" their respective sets of equivalent group elements.

Another example is the quotient group S_n/A_n, where the alternating group A_n is a kernel of the symmetric group S_n (Sects. 2.3.2 and 2.3.3) and serves to partition S_n into two equivalence classes by the parity of the permutation. Consequently, the quotient group S_n/A_n is isomorphic to C_2. Still another example involves the partition of C_4. We explore these examples in the end-of-chapter problems.

2.8 Group Products

The invariant subgroup and quotient group structures become particularly important for applications to physics when they are considered in the context of the *direct product* of groups. Advanced quantum mechanics, particle physics and gauge theories generally depend heavily on these mathematical structures. Our purpose in this section is to describe the direct product structure, which follows from the Cartesian product of sets defined in Sect. 1.4.

There are two different types of direct products for groups. The first is essentially the Cartesian product of two groups A and B of order m and n, respectively, that is then endowed with a group structure. The Cartesian product is

$$G = A \times B = \{(a_1, b_1), (a_1, b_2), (a_1, b_3), ...(a_2, b_1), (a_2, b_2), ...(a_m, b_n)\}. \quad (2.6)$$

This set of mn ordered pairs (a_i, b_i) becomes a group when we define the group operation in G as

$$(a_i, b_i)(a_j, b_j) = (a_i a_j, b_i b_j). \quad (2.7)$$

We leave it as an end of chapter exercise (Problem 2.20) to show that the coordinate-wise rule of composition in Eq. 2.7 converts the set of ordered pairs in Eq. 2.6 into a group. As is true for any Cartesian product, the ordering of the groups is significant; in general we cannot say that $A \times B = B \times A$.

This version of the direct product is called the *external direct product*, a name that is motivated by the fact that the respective elements of A and B operate on parallel tracks; there is no intermingling of the elements of A with those of B.

Example 2.12 Let $A = \{e_A, a\}$ and $B = \{e_B, b\}$ be cyclic groups of order 2. Then their Cartesian product is

$$A \times B = \{(e_A, e_B), (e_A, b), (a, e_B), (a, b)\},$$

which becomes a group of order four when we apply the group operation in Eq. 2.7. We can express $G = A \times B$ in a 4×4 group multiplication table,

$$G = A \times B = \begin{array}{c|cccc} \square & (e_A, e_B) & (e_A, b) & (a, e_B) & (a, b) \\ \hline (e_A, e_B) & (e_A, e_B) & (e_A, b) & (a, e_B) & (a, b) \\ (e_A, b) & (e_A, b) & x & x & x \\ (a, e_B) & (a, e_B) & x & x & x \\ (a, b) & (a, b) & x & x & x \end{array} ,$$

whose completion we leave as an exercise (Problem 2.21). This group is called the *Klein four group*, and it is *not* isomorphic to $C_4 = \{p^0, p^1, p^2, p^3\}$ as one might at first expect. This becomes apparent once the table above is complete.   ▲

The second type of direct product for groups comes from stipulating the following conditions on the external direct product $G = A \times B$: (i) A and B are subgroups of G; (ii) $A \cap B = \{e\}$; (iii) $a_i b_j = b_j a_i$, for all $i = 1, 2, 3....m$ and $j = 1, 2, 3...n$.

With these conditions imposed, G is said to be the *internal direct product* of A and B and is designated as $G = A \otimes B$. The term "internal" is motivated by the fact that A and B are subgroups of G that are now intertwined. Once we have shown the external direct product is a group, there is essentially no additional work necessary to show that the internal direct product is a group as well.

Because of conditions (i)-(iii), the groups A and B in $G = A \otimes B$ must be invariant subgroups of G. We can show this by letting $a_i, \bar{a} \in A$, $b_j \in B$ and $g = a_i b_j \in G$. It follows that $g^{-1} = (a_i b_j)^{-1} = b_j^{-1} a_i^{-1}$. Forming the conjugation of $\bar{a}$ with g gives

$$g \bar{a} g^{-1} = a_i b_j \bar{a} b_j^{-1} a_i^{-1}.$$

Because all a_i and b_j commute, we can write this as

$$g \bar{a} g^{-1} = a_i \bar{a} b_j b_j^{-1} a_i^{-1} = a_i \bar{a} a_i^{-1},$$

which is an element of A. That is, the elements of A (of which $\bar{a}$ is an arbitrary choice) are self-conjugate with respect to all $g \in G$. This makes A a class, but A is also a group (by definition) and therefore A is an invariant subgroup. Consequently, $G/A \simeq B$, and by a similar procedure we find that $G/B \simeq A$ (see Sect. 2.7). In this regard, we consider the case of C_6 in Problem 2.22.

In physics, the most important applications of invariant subgroups, quotient groups and direct products involve continuous groups, usually represented as matrix groups which serve as operators. This, in turn, requires an understanding of the vector spaces upon which these matrix operators are defined, and a full treatment of this subject would take us into the depths of representation theory.

Those depths are beyond the scope of the present text, but we can get a good idea of how all this works by looking at the structures of vector spaces in Chap. 4 and algebras in Chap. 5. Once we do this, we will be in a better position to show how the concepts in this chapter connect to physics, and we will return to this in the context of Lie groups at the close of Chap. 8.

Problems

2.1 Show that the set of even integers is a group under addition. Is the set of odd integers a group under addition? Explain.

2.2 Let $a, b \in S = \{0, 1, 2, \ldots n - 1\}$ for integer $n > 0$. Define the $\square$ operation as

$$a \square b = a + b, \quad \text{if } a + b < n;$$
$$a \square b = r, \quad \text{if } a + b = n + r, \text{where} 0 \leq r < m.$$

(a) What type of arithmetic is represented by the $\square$ operator?

(b) Show $\Sigma = (S, \square)$ is a group;

(c) Write the group multiplication table for $m = 3$.

2.3 A thin square metal plate of uniform thickness lies in the x-y plane with its center of mass located at the origin. Its edges are parallel to the coordinate axes.

(a) What is the symmetry group of the plate *with respect to each of the three coordinate axes?*

(b) What is the symmetry group of the plate when all symmetries are taken into account?

(c) Does the initial orientation of the plate affect your answers in parts (a) and (b)? If so, explain.

2.4 A thin metal plate of uniform thickness and in the shape of an isosceles triangle lies in the x-y plane with its center of mass located on the y-axis and its short edge on the x-axis. What is the symmetry group of the plate *with respect to each of the three coordinate axes?*

2.5 We wish to consider a familiar problem from elementary mechanics in the context of a continuous transformation group. Let S be a "fixed" (inertial) laboratory reference frame and let S' be a frame moving with velocity v_0 as shown below,

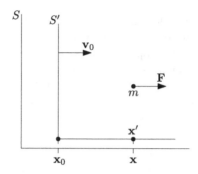

where the x axes of the two systems are aligned. The coordinates x and x_0 are measured in S, where x_0 marks the position of the origin of S'. The coordinate x' is measured in S'. A mass m located at x' is acted upon by a force $\mathbf{F}$ that is parallel to v_0.

(a) Write expressions for the position $\mathbf{x}$, velocity $\mathbf{v}$ and acceleration $\mathbf{a}$ of the mass m as measured in S in terms of the motion of S' (x_0, v_0, a_0) and the corresponding quantities ($\mathbf{x}'$, $\mathbf{v}'$, $\mathbf{a}'$) in S'.

(b) Under what conditions is S' an inertial frame of reference?

(c) The velocity expression you obtained as one of your answers in (a) is the *Galilean velocity transformation*. Show that these transformations form a continuous group. [*Hint:* Introduce another reference frame S'' that moves relative to S'.]

2.6 Refer to Fig. 2.5 and consider a vector $\mathbf{v} = \hat{\mathbf{i}}x + \hat{\mathbf{j}}y$ whose base point is at the origin. The rotation of this vector about the z-axis and through an angle θ may be written as a operation in matrix multiplication:

$$R_\theta \begin{pmatrix} x \\ y \end{pmatrix} = \begin{pmatrix} \cos\theta & -\sin\theta \\ \sin\theta & \cos\theta \end{pmatrix} \begin{pmatrix} x \\ y \end{pmatrix}.$$

A similar expression would describe a rotation around the same axis by an angle ϕ.

(a) Verify that the expression above is indeed correct for the rotation described.

(b) Find the inverse transformation R_θ^{-1} (use the easy method) and verify that

$$R_\theta^{-1} R_\theta = e = \begin{pmatrix} 1 & 0 \\ 0 & 1 \end{pmatrix}.$$

(c) Let $R_\phi R_\theta$ denote a succession of two rotations, first by an angle θ followed by an angle ϕ (following the right-to-left convention in compositions). Show that $R_\phi R_\theta = R_{\theta+\phi}$, thereby inferring the associative property holds for these rotations.

(d) Summarize these and any other necessary considerations so as to verify that these rotations constitute a one-parameter continuous group.

[*Note*: This group is the $SO(2)$ group, a subgroup of the general linear group $GL(n, \mathbb{R})$ with $n = 2$ and θ as the real-valued parameter. The "S" stands for "special," which means the matrix transformations have unit determinant. The "O" stands for "orthogonal." We'll discuss these topics more fully in Chaps. 5 and 8.]

2.7 A process known as *symmetry breaking* has consequences both modest and profound in Nature. Among the latter is the theory of *cosmological inflation*—a rapid expansion of the universe in the earliest stages of its evolution—that is thought to have been initiated though a break in the symmetry of a vacuum energy density, resulting in an energy release that then drove the inflation.

The concept behind this theory is analogous to that associated with everyday *phase transitions*, such as the freezing of a liquid. Before freezing, a liquid is essentially a random system with many symmetries. Upon freezing, these symmetries are broken (there is also a release of latent heat), and the resulting solid structure has a more limited set of symmetries.

Another example is found in *magnetism*. A *paramagnetic* material is one in which the atomic dipole moments are randomly oriented when no external magnetic field is present. When an external field is turned on, these dipoles align with the field. This is in contrast to *ferromagnetic* materials where there is a strong interaction (coupling) between neighboring dipoles. This coupling keeps the dipoles aligned even when no external field is present.

However, if a ferromagnet is heated above its *Curie temperature* it becomes paramagnetic. The reverse process—cooling a ferromagnet from above its Curie temperature to below—breaks the symmetries of that paramagnetic system and the ferromagnet is restored.

As an exercise (or extended project), examine more fully the processes and consequences of these or other forms of symmetry breaking in Nature.

2.8 Referring to Fig. 2.3 for the group S_3, (a) evaluate the parity of each element in S_3; (b) evaluate the products $\gamma\delta$, $\gamma\zeta$ and $\delta\zeta$; (c) find the elements of A_3.

2.9 Show that the order of A_n is $n!/2$.

2.10 By evaluating the number of inversions, find the parity of the following elements in S_6, S_4 and S_7. Write these elements in cycle notation.

(a) $\begin{pmatrix} 1\,2\,3\,4\,5\,6 \\ 5\,3\,1\,4\,2\,6 \end{pmatrix}$;   (b) $\begin{pmatrix} 1\,2\,3\,4 \\ 3\,4\,1\,2 \end{pmatrix}$;   (c) $\begin{pmatrix} 1\,2\,3\,4\,5\,6\,7 \\ 4\,1\,6\,7\,5\,3\,2 \end{pmatrix}$.

2.11 Construct the 6 x 6 group multiplication table for S_3.

2.12 Construct or verify the group multiplication table for D_2 given in Example 2.6.

2.13 In defining the conjugate classes of a group in Sect. 2.6, we asserted that conjugation is an equivalence relation. Show this to be true.

2.14 Work through the details in Example 2.9 and find the three conjugate classes K_1, K_2 and K_3 of S_3.

2.15 Work through the details in Example 2.10 and show that $H = \{\epsilon, \beta, \alpha\}$ is an invariant subgroup of S_3.

2.16 Verify the results given in Example 2.11 and the subsequent discussion by finding (a) the cosets of S_3 associated with the subsets B and C; (b) the coset of the invariant subgroup H.

2.17 It was stated in Definition 2.8 for a quotient group G/H that only an invariant subgroup H can partition a group G in such a way that G/H is a group and not just a partitioned set. The cosets formed with respect to the other subgroups A, B and C of S_3 in Example 2.11 are examples of partitions that are not groups. As an illustration of this last point, show that the set $\{A, M_{A1}, M_{A2}\}$ is not a group.

2.18 Using results from Problem 2.8, show that S_3/A_3 is isomorphic to ($\simeq$) C_2.

2.19 Let $C_4 = \{a^0, a^1, a^2, a^3\}$ be the cyclic group of order four. Show that $H = \{a^0, a^2\}$ is an invariant subgroup and that $C_4/H \simeq C_2$.

2.20 Show that the Cartesian product $G = A \times B$ in Eq. 2.6 is a group under the coordinate-wise rule of composition given in Eq. 2.7. [*Hint*: Let the identity element of G be (e_A, e_B).]

2.21 Complete the group multiplication table for $G = C_2 \times C_2$ (the *Klein four group*) in Example 2.12. [*Hint*: you will need to recall the structure of C_2 for both A and B when forming the coordinate-wise products, (e.g., $b^2 = e_B$).] Show G is *not* isomorphic to C_4 (a single example of non-isomorphism is sufficient), but to D_2 instead.

2.22 Show that the group $C_6 = \{a^0, a^1, a^2, a^3, a^4, a^5\}$ has two invariant subgroups $H_1 = \{a^0, a^3\} \simeq C_2$ and $H_2 = \{a^0, a^2, a^4\} \simeq C_3$. Unlike the case of the external direct product involving C_4 in Problem 2.21, the fact that H_1 and H_2 are invariant subgroups means $C_6 \simeq C_2 \otimes C_3$. Evaluate $C_2 \otimes C_3$ explicitly.

2.23 In Example 2.11 we found the subgroups of S_3 to be A, B and C and the invariant subgroup to be H. The quotient group S_3/H is isomorphic to each of the other subgroups (they are all isomorphic to C_2). However, $S_3 \neq H \otimes A$ (or B or C) because the elements of H do not commute with the elements of A (or B or C). Show this.

Guide to Further Study

The single most important criterion of an abstract algebra text for someone who may be engaged in self-study is to make sure it is user friendly and has lots of examples. Consequently, the choice is therefore very user-specific. Still, it is fair to say that most readers who are new to the subject would find the book by Pinter [10] to be very agreeable, but it is also fair to say that it stops short of many topics the reader may wish to explore further. For that deeper exploration at the undergraduate mathematics level, the works by Hungerford [5], Artin [1] and Birkhoff and Mac Lane [2] are standards. From there, one moves on to Hungerford [6] and the classic by Mac Lane and Birkhoff [9].

The classic works addressing "group theory in physics" are Hammermesh [4] for its emphasis on point groups, Weyl [13] for its emphasis on quantum mechanics, and Wigner [14] for its focus on atomic spectra. One bit of warning for the modern reader: the farther back you go in publication date (say, to mid-20th-century or earlier) for mathematics texts, the more likely you are to run into the Fraktur or Schwabacher calligraphic style for some of the notation. (Music students and students of German history should not have a problem!). The tradeoff is that the writing tends to be extremely clear and expansive in detail. The work by Tung [12], though not especially recent, is highly recommended for its focus on continuous groups and gauge theory. Its dense format is compensated by its numerous examples and very informative appendices.

As for category theory, the place for the physics student to begin is, without question, Geroch [3]. The broader mathematical context is addressed very well for non-mathematicians in the work by Spivak [11]. The classic work in category theory is the advanced-level text by Mac Lane [8].

One line of development that this text has not pursued more fully is with regard to group representation theory. This is because a full treatment is a text in itself, but we do somewhat mitigate this deficiency by giving some examples later in the text that are appropriate for the intended readership. The standard undergraduate mathematics texts cited above (and Tung's text for physics students) are the places to start if you wish to further explore representation theory in an organized way.

To repeat one other warning as you explore the seemingly infinite number of other references: not only does notation change among sources, but some definitions will shift slightly as well. It is always a good idea to check this carefully when jumping between sources.

References

1. Artin, M.: Algebra. Prentice-Hall, Upper Saddle River, NJ (1991)
2. Birkhoff, G., Mac, Lane S.: A Survey of Modern Algebra, 4th edn. A.K. Peters Ltd., Wellesley, MA (1997)
3. Geroch, R.: Mathematical Physics. Chicago Lectures in Physics, Univ. of Chicago Press, Chicago (1985)
4. Hammermesh, M.: Group Theory and Its Application to Physical Problems. Addison-Wesley, Reading, MA (1962); available in Dover reprint, New York (1989)
5. Hungerford, T.W.: Abstract Algebra—An Introduction. Saunders, Philadelphia (1990)
6. Hungerford, T.W.: Algebra. Springer, 12th printing, New York (2003)

7. Lanczos, C.: William Rowan Hamilton—An Appreciation. Am. Scientist **55**(2), 129–143 (1967)
8. Mac Lane, S.: Categories for the Working Mathematician, 2nd edn. Springer, New York (1998)
9. Mac Lane, S., Birkhoff, G.: Algebra, 3rd edn. Chelsea, New York (1993)
10. Pinter, C.C.: A Book of Abstract Algebra, 2nd edn. McGraw-Hill, New York (1990); available in Dover reprint, New York (2010)
11. Spivak, D.I.: Category Theory for the Sciences. MIT, Cambridge, MA (2014)
12. Tung, W.K.: Group Theory in Physics. World Scientific, Philadelphia and Singapore (1985)
13. Weyl, H.: The Theory of Groups and Quantum Mechanics, trans. from the revised 2nd German edition by H.P. Robertson. Dover, New York (1931)
14. Wigner, E.P.: Group Theory and Its Application to the Quantum Mechanics of Atomic Spectra, Expanded and Improved Edition, trans. from the German by J.J. Griffin. Academic Press, New York and London (1959)

Chapter 3
Rings and Fields

3.1 Rings

At the earliest stages of our education, when we are learning basic arithmetic, we are taught a set of rules about how to manipulate numbers in simple equations. Some students wonder about these rules, and are curious about such things as why numbers can be negative.[1] Perhaps they ask the same kinds of questions about the rules regarding the polynomials we later study in elementary algebra. If these students pursue these kinds of questions long enough they realize the answers are not so simple after all, and perhaps they become mathematics majors in college. The rest of us choose to accept the rules at face value, and then spend the rest our lives applying them to our professions and everyday experiences.

The same kind of thing happens early in our studies of physics and engineering with regard to complex numbers. Typically, the complex numbers are portrayed as a system that is "just like" the two-dimensional Euclidean plane. The only significant modification from the real plane appears to be that the y-coordinate axis is labeled with an "i" (or, in engineering, a "j"). Complex numbers are then identified as points in the plane, which we now call the "complex plane," and the y-component is the "imaginary" component of that complex number.

This geometric picture of complex numbers eventually leads to a very robust set of powerful mathematical methods, but when introduced in this way complex numbers can seem contrived. Nonetheless, their utility in articulating the laws of Nature and the elegant methods of analytic function theory are so profound that we learn the rules for how to work with them, and then proceed to use them robustly in our work in physics and engineering.

Still, some students—among them, perhaps, some future mathematicians— wonder about the implications that complex numbers have for functional analysis, the structure of complex product spaces, and what further applications there might be for higher-dimensional, hypercomplex number systems.

---

[1] A question of some historical significance; see the first few paragraphs of Sect. 3.4.2.

© Springer Nature Switzerland AG 2021
S. P. Starkovich, *The Structures of Mathematical Physics*,
https://doi.org/10.1007/978-3-030-73449-7_3

As it happens, the real, rational and complex numbers are special cases of a larger and more general algebraic structure called a *ring*. Another special case of a ring is the system of quaternions. Still-higher-dimensional, hyper-complex number systems also exist, although they eventually lose their ring character as their order increases. We know the role that some of these number systems play in modern physics; the others may await similar—but as yet undeveloped—applications.

The purpose of this chapter is to help the physics and engineering student gain an elementary understanding of rings. We develop the substructures, quotients and products corresponding to their group counterparts, and offer some historical background regarding complex numbers and quaternions. We close the chapter with an introduction to quaternion algebra. If nothing else, you should come away from this chapter with an appreciation of how truly specialized are the real and complex numbers we now take for granted.

3.1.1 Ring Axioms

The algebraic structure $\Sigma = (S, \oplus, \odot)$ is a *ring* if S is an abelian group with respect to $\oplus$, and if $\odot$ is associative and is distributive over $\oplus$. The $\oplus$ and $\odot$ operations are called "addition" and "multiplication," respectively, but these are generic labels and may not always correspond to our everyday understanding of those terms in ordinary arithmetic. Of course, closure of these operations within S is required as well.

Definition 3.1 Given $a, b, c \in S$ and internal binary operations $\oplus$ and $\odot$, the algebraic system $\Sigma = (S, \oplus, \odot)$ is a *ring* if:

1. Σ is closed under $\oplus$ and $\odot$;
2. Σ is associative under $\oplus$: $a \oplus (b \oplus c) = (a \oplus b) \oplus c$;
3. Σ is commutative under $\oplus$: $a \oplus b = b \oplus a$;
4. Σ contains an additive identity element, $e_\oplus$, such that $a \oplus e_\oplus = a$;
5. Σ contains an additive inverse element, $a_\oplus^{-1}$, such that $a \oplus a_\oplus^{-1} = e_\oplus$;
6. Σ is associative under $\odot$ so that $a \odot (b \odot c) = (a \odot b) \odot c$;
7. Multiplication is distributive over addition: $a \odot (b \oplus c) = (a \odot b) \oplus (a \odot c)$.

Note that Axioms (1)–(5) are those of an abelian group under $\oplus$. The additive identity element $e_\oplus$ is also called the *zero element* of the ring.   ∎

Additional conditions may be applied to the axioms in Definition 3.1 to obtain rings with different properties. For example, if S is commutative under $\odot$, then Σ is a *commutative ring*. Another possibility would be if a multiplicative identity $e_\odot$ exists such that $a \odot e_\odot = a$, in which case Σ is a *ring with unity (identity)*.[2] A third possibility would be if a multiplicative inverse $a_\odot^{-1}$ exists for all non-zero elements of S such that $a \odot a_\odot^{-1} = e_\odot$. In this case, Σ is a *ring with inverse*.

---

[2]Some authors (e.g., [9]) incorporate the multiplicative identity into the definition of a ring. Other authors (e.g., [6]) keep the identity separate. We follow the latter convention.

Fig. 3.1 The hierarchy of algebraic structures from rings to fields, where "commutative", "unity" and "inverse" refer to the *multiplicative* aspect of the ring. We adopt the convention of keeping the multiplicative identity (unity) separate from the definition of a ring, while some authors incorporate the identity into the definition

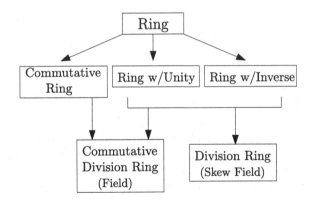

These conditions may be combined to yield even more specific structures. For example, if there is both a multiplicative identity *and* inverse, then the multiplicative part of a ring forms a group and Σ is called a *division ring*. In essence, a division ring is a set consisting of two groups, one of which is abelian under $\oplus$, the other non-abelian under $\odot$. These two groups are linked by the distributive property of $\odot$ over $\oplus$. Another name for a division ring is a *skew field*.

More specific still is the ring where all three additional conditions are applied, i.e., where there is a multiplicative identity and inverse, and where multiplication is commutative. In this case Σ is called a *commutative division ring*, for which another name is a *field*. We'll return to our discussion of fields and skew fields in Sect. 3.4. The hierarchy of algebraic structures from rings to fields is illustrated in Fig. 3.1.

Example 3.1 The natural numbers $\mathbb{N}$ do *not* form a ring; they don't even form an additive group. On the other hand, the integers $\mathbb{Z}$ form a commutative ring with unity although there is no multiplicative inverse. The rational, real and complex numbers each form a commutative division ring (a field). This example also is a reminder that the same set (e.g., $\mathbb{Q}$, $\mathbb{R}$ or $\mathbb{C}$) may satisfy the definitions of different algebraic structures depending on the associated binary operation(s). ▲

Example 3.2 The set of all 2×2 square matrices with complex entries, where $\oplus$ and $\odot$ are matrix addition and multiplication, respectively, form a ring. If A and B are two elements of the set of 2×2 matrices with

$$A = \begin{pmatrix} a_{11} & a_{12} \\ a_{21} & a_{22} \end{pmatrix} \quad \text{and} \quad B = \begin{pmatrix} b_{11} & b_{12} \\ b_{21} & b_{22} \end{pmatrix},$$

then the two binary operations are

$$A \oplus B = \begin{pmatrix} a_{11} + b_{11} & a_{12} + b_{12} \\ a_{21} + b_{21} & a_{22} + b_{22} \end{pmatrix} \quad A \odot B = \begin{pmatrix} a_{11}b_{11} + a_{12}b_{21} & a_{11}b_{12} + a_{12}b_{22} \\ a_{21}b_{11} + a_{22}b_{21} & a_{21}b_{12} + a_{22}b_{22} \end{pmatrix}.$$

The additive identity $e_\oplus$ is the zero matrix. A unit matrix would serve as an identity element. If all the matrices are non-singular, then each element may have a multiplicative inverse. Matrix multiplication is not necessarily commutative.   ▲

Example 3.3 Given some ring $\mathcal{R}$, we define a *polynomial ring* $\mathcal{R}[x]$ to be the set of *all* polynomials over $\mathcal{R}$ in some indeterminate variable $x \notin \mathcal{R}$. That is, the elements of $\mathcal{R}$ form the *coefficients* of the polynomials in $\mathcal{R}[x]$, and the indeterminate variable x is the "unknown" variable in the polynomial.   ▲

Example 3.3 greatly expands our usual sense of a polynomial and is worthy of a few comments. Typically $\mathcal{R} = \mathbb{R}$ or $\mathbb{C}$, and the binary operations $\oplus$ and $\odot$ are the corresponding rules of addition and multiplication, respectively. The indeterminate variable x is the unknown variable in the polynomial, and although x may be an element of $\mathbb{R}$ or $\mathbb{C}$, it is not taken to be an element of $\mathcal{R}$. For example, x might be complex, while $\mathcal{R} = \mathbb{R}$ (e.g., $x^2 + 1 = 0$). Many results (e.g., the division algorithm for polynomials) follow from the consideration of these familiar polynomials rings.

Now, however, we see that $\mathcal{R}$ can be *any* ring with well-defined binary operations, and x can be any indeterminate variable not in $\mathcal{R}$; we can still construct a polynomial $\mathcal{R}[x]$. We will not pursue this topic further in this text, but see Chap. 24 in [12] for an introduction, followed by other sources mentioned in the Guide to Further Study at the end of the chapter.

In our study of finite groups we found it helpful to construct a group multiplication table, and the same approach can be taken for finite rings. Here, however, two such tables are needed, one for $\oplus$ and the other for $\odot$. For example, consider the set $S = \{0, 1\}$ with $\oplus$ and $\odot$ defined by the tables

$$\begin{array}{c|cc} \oplus & 0 & 1 \\ \hline 0 & 0 & 1 \\ 1 & 1 & 0 \end{array} \quad \text{and} \quad \begin{array}{c|cc} \odot & 0 & 1 \\ \hline 0 & 0 & 0 \\ 1 & 0 & 1 \end{array}. \tag{3.1}$$

These tables incorporate modulo[3] 2 arithmetic. This structure $\mathcal{R} = (S, \oplus, \odot)$ is a ring (verification is left as an exercise) denoted as $\mathbb{Z}_2$.

The example of $\mathcal{R} = \mathbb{Z}_2$ illustrates the definition of the *characteristic* of a ring. Given $a \in \mathcal{R}$, if the n-fold sum $a + a + a + \ldots + a = na = e_\oplus$ (the zero element) for some integer n, then the ring is said to be of characteristic n. For $\mathbb{Z}_2$ with $S = \{0, 1\}$ we see that $0 + 0 = 0$ and $1 + 1 = 0$. Consequently, $\mathbb{Z}_2$ is a ring of characteristic 2. Compare that to $\mathbb{Z}$, where there is no n such that $na = e_\oplus$ for all $a \in \mathbb{Z}$. Consequently $\mathbb{Z}$ is said to be a ring of *characteristic zero*.

We leave as an exercise the construction of the tables for $\mathbb{Z}_3$ and $\mathbb{Z}_4$ and a verification of their characteristics. Generally, $\mathbb{Z}_n$ is a commutative ring with unity, though with no inverse. We'll see in Sect. 3.1.2 that the countably infinite set of integers $\mathbb{Z}$ forms a set of n equivalence classes that is isomorphic to $\mathbb{Z}_n$.

---

[3]Recall that *modulo* means "with respect to the modulus of," where "modulus" is taken to mean "standard of measurement." If p, q and n are integers with $n > 0$, we say that p and q are *congruent* (in the same equivalence class) if n (the modulus) divides $(p - q)$, or $p - q = kn$ for integer k. We write this as $p = q(\text{mod } n)$. Ex.: $15 = 3 \pmod{12}$; $16 = 4 \pmod{12}$.

3.1.2 Ring Morphisms

Morphisms between rings follow the same logic and terminology as that used for groups (see Sect. 2.4 and Table 2.2). For example, given two isomorphic division rings (skew fields) $\mathcal{R} \simeq \mathcal{R}'$, the bijections between them must include the unit elements $e_\oplus \leftrightarrow e'_\oplus$ and $e_\odot \leftrightarrow e'_\odot$, as well as $a \leftrightarrow a'$ for all $a \in \mathcal{R}$ and $a' \in \mathcal{R}'$ and their respective inverses. If $\mathcal{R}$ is commutative then so is $\mathcal{R}'$ and we have two isomorphic fields.

An example of a surjection (an onto, but not 1-1, map) between rings would be the mapping of all even integers $E \subset \mathbb{Z}$ to 0 and all odd integers $O \subset \mathbb{Z}$ to 1, or $\phi \colon \mathbb{Z} \to \mathbb{Z}_2$. Generally, $\phi \colon \mathbb{Z} \to \mathbb{Z}_n = \{[0], [1], [2], ...[n-1]\}$, a set containing n equivalence classes, where we have used the equivalence class notation [ ] from Sect. 1.2.2. An example of an injection (1-1, but not onto) between rings would be the insertion of the real numbers into the set of complex numbers, or $i \colon \mathbb{R} \to \mathbb{C}$.

As a structure-preserving map between rings, a ring homomorphism ϕ must preserve internal relationships involving both operations. Specifically, a ring homomorphism requires that $\phi(a \oplus b) = \phi(a) \oplus \phi(b)$ and $\phi(a \odot b) = \phi(a) \odot \phi(b)$.

3.2 Subrings, Ideals and Quotient Rings

Consider a ring $\Sigma = (S, \oplus, \odot)$, and let $S_0 \subset S$. Then $\Sigma_0 = (S_0, \oplus, \odot)$ is a *subring* if the defining axioms for a ring apply in Σ_0 with the same binary operations as in Σ. This line of reasoning parallels our consideration of subgroups, but now we have two binary operations to take into account.

In addition, as with the case of subgroups vis-a-vis groups, the precise nature of a subring may differ from that of the encompassing ring. We saw this in the example of the non-abelian group S_3 where one or more of its subgroups is abelian, and the same possibility applies here. However, as with groups, abelian rings have only abelian subrings.

Example 3.4 Several subrings are found among the familiar number systems. For example, the complex numbers $\mathbb{C}$ form a commutative division ring (a field) under ordinary addition and multiplication, and the real, rational and integer numbers form a succession of subrings. However, while $\mathbb{Q}$ and $\mathbb{R}$ are fields, $\mathbb{Z}$ is not.   ▲

Example 3.5 The even integers form a subring of $\mathbb{Z}$ under ordinary addition and multiplication, while the odd integers do not (Problem 3.5).   ▲

Recall that in Sect. 2.6 we defined an invariant subgroup $H \lhd G$ such that it is closed under conjugation with all elements $g \in G$. For two (generally distinct) elements $h_i, h_j \in H$ we wrote this as

$$H = \{h_i, h_j \colon h_i = g h_j g^{-1}, \text{ for all } g \in G\}, \tag{3.2}$$

or, equivalently, $H = \{h_i, h_j \colon g h_i = h_j g, \text{ for all } g \in G\}$.

Fig. 3.2 A universal
construction for quotient
rings

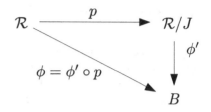

Table 3.1 Parallel constructions between groups and rings

Structure	Group G	Ring $\mathcal{R}$
Substructure	Subgroup	Subring
Kernel of ϕ^a	Invariant Subgroup H	Ideal[b] J
Quotient Structure	Quotient Group G/H	Quotient Ring $\mathcal{R}/J$

[a] See the definition of ϕ in the universal constructions in Figs. 2.7 and 3.2
[b] Occasionally referred to as the invariant subring

Given a ring $\mathcal{R}$, the goal here is to identify a subring J such that it is the ring-equivalent of an invariant subgroup. Although J is occasionally referred to as an "invariant sub*ring*," more often it is called an *ideal*. Once J is identified we may then use it to form a quotient structure. A complication is that for rings we have two binary operations ($\oplus$ and $\odot$) to consider, and these two operations may differ in some key respects, such as the presence of an identity or an inverse.

The first task is to define the criteria J must satisfy so that given a ring $\mathcal{R}$ we can identify J. Then with J in hand we'll be able to (i) define a *quotient ring* $\mathcal{R}/J$; (ii) identify a canonical projection p where $p\colon \mathcal{R} \to \mathcal{R}/J$; (iii) define an isomorphism ϕ' such that $\phi'\colon \mathcal{R}/J \to B$ where B is a ring chosen with a particular goal in mind; and (iv) identify the map $\phi = \phi' \circ p$ that maps J to the identity of the ring B. That is, $\phi\colon j \mapsto e_\oplus \in B$ for all $j \in J$. As was the case for invariant subgroups, J is called the kernel of ϕ, $J = \mathrm{Ker}(\phi)$. In short, the goal is to fill in the details of Fig. 3.2.

Table 3.1 shows the parallel constructions between groups and rings. We will expand this table as we develop additional structures in subsequent chapters.

We now proceed to identify the criteria to be satisfied by elements of J. Looking first at the $\oplus$ operation, we might be tempted to formulate a conjugation-type relationship among the elements of $\mathcal{R}$ as we did in Eq. 3.2 for invariant subgroups. However, because the group S in $\mathcal{R}$ is abelian under $\oplus$, the result of such a conjugation would simply be a restatement of the definition of $\mathcal{R}$. This is a consequence of the fact (Sect. 2.6) that an invariant subgroup of an abelian group is the group itself.

We also have to insure closure in J under the $\oplus$ operation, and this is done with modular arithmetic. Consequently, for two elements $j_1, j_2 \in J$ we tentatively define J to be such that $j_2 - j_1 \in J$. Again, this means $j_2 = j_1 + kn$ for integers k and n, or $j_2 = j_1 (\mathrm{mod}\, n)$.

For example, if $\mathcal{R} = \mathbb{Z}$ and $n = 2$, then $j_2 = j_1 (\text{mod } 2)$ means $J = E = \{0, \pm 2, \pm 4, \pm 6, ...\}$ so that $j_2 - j_1$ is some integral multiple of $n = 2$. Formulating J in this way insures closure in J under the $\oplus$ operation.

To complete the quotient structure we need to connect this tentative definition for J to those elements that belong to the rest of $\mathcal{R}$. We do this through the multiplicative operation by specifying that for any $j \in J$ and for all $r \in \mathcal{R}$ we require *both* $jr = j \odot r \in J$ and $rj = r \odot j \in J$. Only in the special case where $\odot$ is commutative will $jr = rj$, so we must keep both terms explicit to maintain generality.

We can now compile all these criteria into a formal definition for an ideal.

Definition 3.2 Given a ring $\mathcal{R}$, the subring J is an *ideal* if

$$J = \{j : jr \in J, \ rj \in J \text{ and } j_2 - j_1 \in J, \text{ for all } j \in J \text{ and } r \in \mathcal{R}\}.$$

∎

Continuing with our example of $\mathcal{R} = \mathbb{Z}$ and $J = E$, we conclude that J is an ideal and the quotient ring is

$$\mathcal{R}/J = \mathbb{Z}/E = \{[E], [O]\}.$$

Finally, if we let $B = \{0, 1\} = \mathbb{Z}_2$ (for example), then $\mathbb{Z}/E \simeq B$ and we have the tables

$\oplus$	E	O
E	E	O
O	O	E

$\oplus$	0	1
0	0	1
1	1	0

and

$\odot$	E	O
E	E	E
O	E	O

$\odot$	0	1
0	0	0
1	0	1

$E \simeq 0$, $O \simeq 1$ (3.3)

thereby meeting the goal of establishing the complete set of relationships shown in Fig. 3.2 among $\mathcal{R}$, $\mathcal{R}/J$ and B for $J = \text{Ker}(\phi)$.

3.3 Product Rings

Given two rings $\mathcal{R}$ and $\mathcal{R}'$ with elements $r \in \mathcal{R}$ and $r' \in \mathcal{R}'$, we define a *product ring* $\mathcal{S}$ as $\mathcal{S} = \mathcal{R} \times \mathcal{R}'$ with elements $s = (r, r') \in \mathcal{S}$. The task before us is to define addition and multiplication in such a way that the resulting set of ordered pairs is in fact a ring.

The rings $\mathcal{R}$ and $\mathcal{R}'$ need *not* have the same internal operations; a simple example would be if ordinary addition and multiplication applied to one of the rings while modular arithmetic applied to the other. However, in defining $\mathcal{S} = \mathcal{R} \times \mathcal{R}'$ as a ring we must insure that the elements $s \in \mathcal{S}$ not only adhere to the ring axioms via addition and multiplication operations in $\mathcal{S}$, but do so in a manner that respects the specific nature of those corresponding operations in $\mathcal{R}$ and $\mathcal{R}'$.

This means being careful to distinguish between the $\oplus$ and $\odot$ binary operations among the three rings — $\mathcal{R}$, $\mathcal{R}'$ and $\mathcal{S}$ — and being clear as to which operation in which ring is relevant at any given point in our development. This is facilitated with explicit notation, so we'll denote $\oplus_{\mathcal{R}}$ to be the addition operator on $\mathcal{R}$, with a corresponding notation for the other $\oplus$ and $\odot$ operations on the three rings.

First consider addition in $\mathcal{S}$. For elements $r_1, r_2, r_3 \in \mathcal{R}$ and $r_1', r_2', r_3' \in \mathcal{R}'$ we of course have

$$r_1 \oplus_{\mathcal{R}} r_2 = r_3 \in \mathcal{R}$$

and

$$r_1' \oplus_{\mathcal{R}'} r_2' = r_3' \in \mathcal{R}'.$$

Then on the ring $\mathcal{S}$ with elements $s_1 = (r_1, r_1')$, $s_2 = (r_2, r_2')$ and $s_3 = (r_3, r_3')$ we define the addition operation $\oplus_{\mathcal{S}}$ as

$$s_1 \oplus_{\mathcal{S}} s_2 = (r_1, r_1') \oplus_{\mathcal{S}} (r_2, r_2') = (r_1 \oplus_{\mathcal{R}} r_2, r_1' \oplus_{\mathcal{R}'} r_2') = s_3. \qquad (3.4)$$

Each element of the ordered pair s_3 is formed by an operation that is closed, and therefore addition in $\mathcal{S}$ is closed.

We distinguish among the additive identity elements of $\mathcal{R}, \mathcal{R}'$ and $\mathcal{S}$ by writing $e_{\oplus_{\mathcal{R}}}, e_{\oplus_{\mathcal{R}'}}$ and $e_{\oplus_{\mathcal{S}}}$, respectively. If we are dealing with ordinary numbers in all three rings, then zero is the additive identity (zero element) for $\mathcal{R}$ and $\mathcal{R}'$ and the ordered pair $(0, 0)$ is the zero element in $\mathcal{S}$. On the other hand, if the elements of all three rings are matrices, then the zero matrices for $\mathcal{R}$ and $\mathcal{R}'$ and an ordered pair of zero matrices for $\mathcal{S}$ would be the additive identity elements, and so forth. A similar degree of care must be taken in distinguishing the additive inverses of the rings involved in a product ring.

Next we define multiplication on $\mathcal{S}$ (see Definition 1.5 for the Cartesian product) as

$$s_1 \odot_{\mathcal{S}} s_2 = (r_1, r_1') \odot_{\mathcal{S}} (r_2, r_2') = (r_1 \odot_{\mathcal{R}} r_2, r_1' \odot_{\mathcal{R}'} r_2') = s_3. \qquad (3.5)$$

As with the addition operator, each element of the ordered pair s_3 is closed, and therefore multiplication in $\mathcal{S}$ is closed.

We distinguish among the multiplicative identity elements of $\mathcal{R}, \mathcal{R}'$ and $\mathcal{S}$ as $e_{\odot_{\mathcal{R}}}, e_{\odot_{\mathcal{R}'}}$ and $e_{\odot_{\mathcal{S}}}$, respectively. If we are dealing with ordinary numbers in all three rings, then the number 1 is the multiplicative identity (unity element) for $\mathcal{R}$ and $\mathcal{R}'$ and the ordered pair $(1, 1)$ is the multiplicative identity[4] in $\mathcal{S}$. On the other hand, if the elements of all three rings are matrices, then the unit matrices for $\mathcal{R}$ and $\mathcal{R}'$ and an ordered pair of unit matrices for $\mathcal{S}$ would be the multiplicative identity elements, and so forth. The notation for the multiplicative inverses within the rings are distinguished accordingly.

The following example illustrates the evaluation of a product ring when a different arithmetic applies to $\mathcal{R}$ and $\mathcal{R}'$.

Example 3.6 Let $\mathcal{R} = \mathbb{Z}$ and $\mathcal{R}' = \mathbb{Z}_3$. Ordinary arithmetic applies for the integers $\mathbb{Z}$ while mod(3) arithmetic applies to $\mathbb{Z}_3$. Elements of the product $\mathcal{S} = \mathbb{Z} \times \mathbb{Z}_3$ are

---

[4]Note that *both* $\mathcal{R}$ and $\mathcal{R}'$ must have a unity element if $\mathcal{S}$ is to have a unity element.

written as $s = (r, r')$ where $r \in \mathbb{Z}$ and $r' \in \mathbb{Z}_3$. For example, let $s_1 = (2, 2)$ and $s_2 = (5, 2)$. Then addition is evaluated as

$$s_3 = s_1 \oplus_S s_2 = (2, 2) \oplus_S (5, 2) = (2 \oplus_{\mathcal{R}} 5, 2 \oplus_{\mathcal{R'}} 2) = (7, 1)$$

and multiplication (using the same symbol s_3 for here) yields

$$s_3 = s_1 \odot_S s_2 = (2, 2) \odot_S (5, 2) = (2 \odot_{\mathcal{R}} 5, 2 \odot_{\mathcal{R'}} 2) = (10, 1).$$

▲

Example 3.7 Let $\mathcal{R} = \mathcal{R'} = \mathbb{Z}_2$ so that $S = \mathbb{Z}_2 \times \mathbb{Z}_2$. In this case, mod(2) arithmetic applies to both components of the element $s = (r, r')$. The elements of $\mathbb{Z}_2$ are $\{0, 1\}$ and we can write the four elements of S as $s_0 = (0, 0)$, $s_1 = (0, 1)$, $s_2 = (1, 0)$ and $s_3 = (1, 1)$. Applying mod(2) arithmetic we have, for example, $s_1 \oplus s_3 = (1, 0) = s_2$ and $s_1 \odot s_3 = (0, 1) = s_1$.

These results and the others may be entered into a table with the first term in the sum or product listed in the left column of the table and the second term listed in the top row.

$\oplus$	s_0	s_1	s_2	s_3
s_0	s_0	s_1	s_2	s_3
s_1	s_1	s_0	s_3	s_2
s_2	s_2	s_3	s_0	s_1
s_3	s_3	s_2	s_1	s_0

and

$\odot$	s_0	s_1	s_2	s_3
s_0	s_0	s_0	s_0	s_0
s_1	s_0	s_1	s_0	s_1
s_2	s_0	s_0	s_2	s_2
s_3	s_0	s_1	s_2	s_3

The table for $\oplus$ in shows S includes an additive abelian group that is not cyclic. Therefore, $\mathbb{Z}_2 \times \mathbb{Z}_2$ is not isomorphic to $\mathbb{Z}_4$. Given the result in Problem 2.21 on the Klein four group, this should not surprise us; $\mathbb{Z}_2 \simeq C_2$ and $\mathbb{Z}_4 \simeq C_4$. ▲

3.4 Fields

Because a skew field is a division ring and a field is a commutative division ring,[5] the case could be made that we need say nothing more about the algebra of fields beyond what we have said previously about rings. However, because the number systems most familiar to us are fields, we choose to list the field axioms explicitly and explore a few of the more important properties of these number fields.

Definition 3.3 For $a, b, c \in S$, the algebraic system $\Sigma = (S, \oplus, \odot)$ is a field if:

1. S is closed under $\oplus$ and $\odot$;
2. S is associative under $\oplus$: $a \oplus (b \oplus c) = (a \oplus b) \oplus c$;

---

[5]It is redundant to refer to "commutative fields" since a field is a commutative division ring. Still, the expression is often used to emphasize the distinction with skew fields.

Table 3.2 Order and completeness of $\mathbb{Q}$, $\mathbb{R}$ and $\mathbb{C}$

Field	Ordered	Complete
Rational numbers ($\mathbb{Q}$)	Yes	No
Real numbers ($\mathbb{R}$)	Yes	Yes
Complex numbers ($\mathbb{C}$)[(1)]	No	Yes

[(1)] See Sect. 3.4.2

3. S is commutative under $\oplus$: $a \oplus b = b \oplus a$;
4. S contains an additive identity element, $e_\oplus$, such that $a \oplus e_\oplus = a$;
5. S contains an additive inverse element, $a_\oplus^{-1}$, such that $a \oplus a_\oplus^{-1} = e_\oplus$;
6. S is associative under $\odot$: $a \odot (b \odot c) = (a \odot b) \odot c$;
7. S is commutative under $\odot$: $a \odot b = b \odot a$;
8. S contains a multiplicative identity element, $e_\odot$, such that $a \odot e_\odot = a$;
9. S contains a multiplicative inverse element, $a_\odot^{-1}$, such that $a \odot a_\odot^{-1} = e_\odot$;
10. Multiplication is distributive over addition: $a \odot (b \oplus c) = (a \odot b) \oplus (a \odot c)$ ∎.

Axioms (1)–(5) are those of an abelian group under $\oplus$, and Axioms (6)–(9) [(and, of course, (1)] are those of an abelian group under $\odot$. A field may be thought of as these two abelian groups linked by the distributive property in Axiom (10). The axioms of a skew field are the same as those for a field, but with the omission of Axiom (7).

3.4.1 Completeness and Order

We wish to consider two properties, *order* and *completeness*, of the rational, real and complex numbers. The results are previewed in Table 3.2.

Definition 3.4 Let $\Sigma = (S, \oplus, \odot)$ be a field F with the two binary operations $\oplus$ and $\odot$, and let $a, b \in F$. The field is *ordered*[6] if there exists a subset $P \subset S$ such that

1. If $a \in P$ and $b \in P$, then $a \oplus b \in P$ and $a \odot b \in P$;
2. For each $a \in F$, exactly one of the following holds (the *trichotomy law*)

    a. $a \in P$;
    b. $a = e_\oplus$ (the zero element)
    c. $(-a) \in P$.

∎

The notation in Defintion 3.4 is suggestive, and the reader should recognize that if $F = \mathbb{R}$ (or $\mathbb{Q}$) then P is the set of positive real (or rational) numbers. Complex

---

[6]Order relations were discussed generally in Sect. 1.2.1. The order relation defined here may be applied to any ring (e.g., the integers $\mathbb{Z}$) where P can be defined. See [9], pp. 261ff.

numbers will be discussed in Sect. 3.4.2, but the reader likely already knows them well enough to recognize that no such ordering relation exists for the complex field, i.e., it is not possible to define a subset $P \subset \mathbb{C}$ with the required properties.

Completeness for fields is defined in terms of whether the limit of every convergent sequence within the field is an element of that field.[7] A sequence $\{x_n\}$ in F is called a *Cauchy sequence* if for every $\epsilon > 0$ there is an integer N such that $d(x_n, x_m) < \epsilon$ for all $n, m \geq N$. That is, a Cauchy sequence converges, even if initially a finite set of N terms does not. The point to which it converges is called the *limit* of the sequence.

The question, though, is whether the limit x of the sequence $\{x_n\}$ lies inside or outside the field F.

Definition 3.5 If for *every* Cauchy sequence in the field F the limit x is in F, then F is said to be *complete*. Stated differently, if there is at least one Cauchy sequence in F such that the limit $x \notin F$, then F is not complete.   ■

Example 3.8 In the field $\mathbb{Q}$ of rational numbers, it is certainly possible to construct Cauchy sequences that converge to points in $\mathbb{Q}$. However, it is also possible to construct Cauchy sequences that converge to points that are not in $\mathbb{Q}$, i.e., points that represent irrational numbers, such as π or $\sqrt{2}$. Therefore, although we have shown $\mathbb{Q}$ to be an ordered field, it is not a complete field.   ▲

The real numbers *are* complete, and it is because they are both complete *and* ordered that makes them so amenable to describing continuous and deterministic processes in Nature. The construction of the reals is a standard exercise in advanced algebra, and we direct the interested reader to the algebra references cited in the Guide to Further Study at the end of this chapter. As we will see in the next section, the complex numbers are complete as well. Then, in Chap. 6, we will discuss the concept of completeness from a topological perspective.

3.4.2 The Complex Field $\mathbb{C}$ and Hamilton's Search for Number Triplets

When seen through their historical lenses, science and mathematics are replete with starts, stops, dead-ends and errant ways before a clean and logically organized framework eventually emerges. One example is the history of our familiar number systems, and it is a history recounted in many places. The comprehensive mathematical historical account given in [7] is highly recommended as a general reference for most areas of mathematics. Regarding complex numbers specifically the account given in [10] is similarly recommended. The historical accounts of complex numbers and quaternions presented (by a non-historian) in the remainder of this chapter rely on these two sources, as well as others which are cited as we proceed.

---

[7] We'll assume here that we have defined a distance $d(x_1, x_2)$ between two elements x_1 and x_2 of a field F. Compare this with the discussion on continuity in Sect. 1.3.2.

We trace the origin of our story of complex numbers to the work of Heron of Alexandria in the first century CE. It is reported that in his efforts to solve a geometry problem he encountered the square root of a negative number, and not knowing how to deal with this perplexing result he addressed it by the only means known at the time—he simply ignored the minus sign under the square root. He got the wrong answer to his problem.

When placed in historical context, Heron's difficulty in interpreting such things as $\sqrt{-1}$ becomes understandable when we consider that even as late as the 16th century mathematicians were still coming to grips with the concept of negative numbers as solutions to algebraic equations, let alone their square roots. Indeed, during this time the "-2" that we know to be one of the two roots of the equation $x^2 - 4 = 0$ would simply be ignored as a valid solution to the equation!

It was not until the year 1637 and the work of Descartes[8] that square roots of negative numbers were placed on stronger footing, and it is Descartes to whom we give credit for first using the word *imaginary* to describe them. Nonetheless, the preferred means for describing complex numbers and their associated methods remained a point of contention for about another 150 years.

In 1797 Caspar Wessel[9] published a paper with the Royal Academy of Denmark that essentially gives us the geometric picture of complex numbers that we have today. He identified complex numbers as points in a two-dimensional Cartesian plane, where $i = \sqrt{-1}$ served to label one of the two axes as the imaginary axis.

Unfortunately for Wessel's place in the history of mathematics, his paper (written in Danish) remained undiscovered by the wider mathematical community until it was translated into the French almost a century later, in 1895. Instead, it was a similar work by Jean-Robert Argand,[10] published in 1806, that got the attention of his contemporaries. Subsequent work by Gauss[11] advanced this geometrical interpretation (and the mathematical reputation) of complex numbers, and this picture of complex numbers spread quickly. By 1830 the *Argand diagram* (Fig. 3.3) that we know today became the generally accepted means for representing complex numbers.

Complex numbers in the Argand diagram may be expressed in two equivalent ways as $z = a + ib = re^{i\theta}$, where $a, b \in \mathbb{R}$, the latter expression being called the *polar form* of complex numbers.[12] When solving problems, the polar form is more convenient for multiplication, and the product of two complex numbers is interpreted as a rotation and scaling in the complex plane:

$$z_1 \cdot z_2 = r_1 e^{i\theta_1} \cdot r_2 e^{i\theta_2} = (r_1 r_2)e^{i(\theta_1 + \theta_2)} = r_3 e^{i\theta_3} = z_3.$$

---

[8] René Descartes (1596–1650), French mathematician and philosopher, and the work *La Geometrie*.

[9] Caspar Wessel (1745–1818), a self-taught Norwegian-born surveyor.

[10] Jean-Robert Argand (1768–1822), a self-taught Swiss bookkeeper.

[11] Carl Friedrich Gauss (1777–1855), one of the giants in the history of mathematics. Only a small fraction of his work was published during his lifetime, and many advancements in mathematics that are attributed to others were developed earlier by Gauss but remained unknown at the time.

[12] Recall Euler's equation: $e^{i\theta} = \cos\theta + i\sin\theta$, an expression that may be verified via a Taylor series expansion of $e^{i\theta}$. For $z = re^{i\theta}$, the quantity θ is called the *phase* of z and $r = |z|$ is its real-valued *modulus* (magnitude, or distance from the origin).

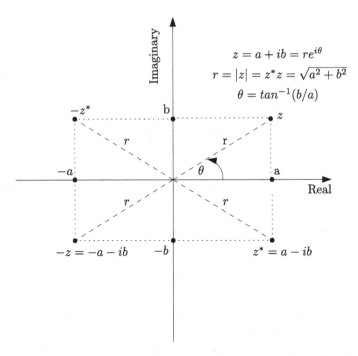

Fig. 3.3 The geometrical representation of complex numbers due originally to Wessel but named for Argand. Complex conjugation may be interpreted as reflection across the real line, but see the discussion on dual spaces in Sect. 4.3.2. Negation is an inversion through the origin

Addition, on the other hand, is most easily carried out using the *coordinate form* and is akin to component-wise vector[13] addition:

$$z_1 + z_2 = (a_1 + ib_1) + (a_2 + ib_2) = (a_1 + a_2) + i(b_1 + b_2) = a_3 + ib_3 = z_3.$$

The abelian group structure of complex numbers under addition and multiplication is readily apparent, and the distributive property $z_1 \cdot (z_2 + z_3) = z_1 \cdot z_2 + z_1 \cdot z_3$ follows directly, so $\mathbb{C}$ is indeed a field. The reader who seeks a review of elementary complex algebra will find some problems at the end of the chapter.

The Argand diagram helps us visualize why there is no total-ordering of the complex numbers as there is for the reals. Although there is an infinity of ordered *sub*sets, there is no ordering relation (Definition 3.4) that spans the full set of complex numbers.

The argument for concluding that the set of complex numbers is complete is the same as the earlier argument for the reals, except now the Cauchy sequences converge for both components of z (in geometric terms, from all possible directions

---

[13]The interpretation of vectors as "directed line segments" was already well established by 1830.

in the complex plane rather than simply along the real line). This completes the descriptions of the entries in Table 3.2.

In the 1830s, William Rowan Hamilton[14] began considering complex numbers from a different perspective. As Hamilton saw it, the depiction of complex numbers via the Argand diagram, while very useful, burdened these numbers with otherwise extraneous assumptions associated with the geometry of the Cartesian plane. Essentially he was asking whether it really was necessary to bring two-dimensional Euclidean geometry to bear on the definition and the algebra of these numbers, or whether there was a more purely algebraic formulation.

As an alternative to the Argand diagram, Hamilton offered the idea of a complex number as an "ordered number couplet" $z = (a, b)$—a number that *intrinsically*, by *definition*, has two parts to it, each part a real number, with no geometric meaning attached to either part. The question would then become how to define addition and multiplication of complex numbers using this couplet formulation.

Definition 3.6 For $a, b \in \mathbb{R}$, a *complex number* $z \in \mathbb{C}$ is defined as the couplet

$$z = (a, b) = a(1, 0) + b(0, 1).$$

The unit couplet $(1, 0)$ is the real number 1, and the unit couplet $(0, 1)$ is the imaginary unit i whose square (by the definition of couplet multiplication given below) is -1. The first entry in the couplet is the real part of z written as $a = \text{Re}(z)$, and the second entry is the imaginary part, or $b = \text{Im}(z)$. Addition is defined by

$$z_1 + z_2 = (a_1, b_1) + (a_2, b_2) = (a_1 + b_1, a_2 + b_2) = (a_3, b_3) = z_3.$$

Multiplication of complex numbers is defined as

$$z_1 \cdot z_2 = (a_1, b_1) \cdot (a_2, b_2) = (a_1 a_2 - b_1 b_2, a_1 b_2 + a_2 b_1) = (a_3, b_3) = z_3.$$

∎

As an exercise the reader should verify that $(0, 1) \cdot (0, 1) = -(1, 0) = -1$, and also that these number couplets, with the rules of addition and multiplication as defined, satisfy the field axioms (Problems 3.8 and 3.9). Further, these definitions yield results that are fully equivalent to those shown in Fig. 3.3, although expressed algebraically rather than geometrically. Today, it is second nature for us to think of $z = (a, b)$ as a point in a Cartesian complex plane with components a and b. Although the Argand diagram is both pedagogically pleasing and aids in the visualization of most operations in complex analysis, this geometric picture is not essential to the definition of complex numbers.

---

[14] See [8] for a biographical "mathematical appreciation" of the Irish mathematical physicist William Rowan Hamilton, (1805–1865). For a full biography of Hamilton see [4]. Known to physics students primarily for his eponymous equations in classical mechanics (equations he devised by the time of his 30th birthday), Hamilton actually spent most of his professional life on algebraic topics. He considered his development of quaternions to be his most significant achievement.

At this point, having recast the previously known complex numbers in algebraic terms, Hamilton set out to break new ground. If we think of a real number as a "number singlet" and a complex number as a "number couplet," then the logical next step would seem to be a "number triplet." Hamilton started his search for these "three-part numbers" in the early 1830s, and was driven in his quest by many considerations — some of them logical, mathematical and physical, others metaphysical.[15] Nonetheless, after years of on-and-off effort he was unable to define number triplets in such a way as to give closure under addition and multiplication.[16]

There are two ways of seeing the difficulty with number triplets; both difficulties are associated with the multiplication.[17] The first is a heuristic argument from a geometric perspective. If we recall that the product of two complex numbers (number couplets) may be interpreted as a rotation and scaling in the two-dimensional plane, then it might seem reasonable to expect that the product of two number triplets should be interpreted as a rotation and scaling in three-dimensions.

However, four numbers (not three) are needed to describe a rotation and scaling about an axis in three-dimensional space: we need two angles to orient the axis of rotation, one angle that describes the rotation around that axis, and another number that tells us something about the scaling. This is not a proof, but it strongly suggests (at least within this framework) that arbitrary rotations and scalings of three-dimensional vectors should be carried out using an algebraic object comprised of four independent components and not three.

Second, there is an algebraic perspective. Consider first the case of the multiplication of two complex numbers

$$z_1 z_2 = (a_1 + i b_1)(a_2 + i b_2) = (a_1 a_2 - b_1 b_2) + i(a_1 b_2 + b_1 a_2) = a_3 + i b_3 = z_3 \tag{3.6}$$

and the relationship between their corresponding squared moduli

$$(a_1^2 + b_1^2)(a_2^2 + b_2^2) = (a_1 a_2 - b_1 b_2)^2 + (a_1 b_2 + b_1 a_2)^2 = a_3^2 + b_3^2. \tag{3.7}$$

There are two relevant considerations here:

1. the product $z_1 z_2$ yields a result that is in the correct form for a complex number (i.e., we have closure): $z_3 = a_3 + i b_3$, and
2. the square modulus of z_3 is the sum of precisely two terms, not more and not fewer. In calculating the modulus of z_3 it would not occur to us to, say, simply ignore the $b_3^2 = (a_1 b_2 + b_1 a_2)^2$ term in Eq. 3.7.

---

[15] The full story is given in [4], particularly Part VII. See also [7], pp. 776–9, and [8]. Our purpose here is to illustrate the internal inconsistencies associated with number triplets.

[16] You might be tempted to ask: "Isn't a number triplet just a three-dimensional vector in component form?" The most direct answer is "A vector is not a number (scalar), and we are dealing here with a three-part *number*." More precisely, and in the present context, the ring axioms are not those of a vector space. We'll have more to say about vector spaces in Chap. 4.

[17] See [4], Chap. 22 and [7], Chap. 32.

Now consider the case of the multiplication of two triplets, each of the general form $p = a + ib + jc$, where j is meant to indicate a second complex component of the triplet and c is its coefficient. Carrying out the multiplication in the customary way, we find

$$
\begin{aligned}
p_1 p_2 &= (a_1 + ib_1 + jc_1)(a_2 + ib_2 + jc_2) \\
&= a_1 a_2 + i(b_1 a_2 + a_1 b_2) + j(c_1 a_2 + a_1 c_2) \\
&\quad + i^2(b_1 b_2) + j^2(c_1 c_2) + ij(b_1 c_2) + ji(c_1 b_2).
\end{aligned} \tag{3.8}
$$

The last four terms of Eq. 3.8 are a problem because, as written, the product $p_1 p_2$ is not in the correct form for a triplet. If we could limit our result to just the first three terms then we would have a solution.

One improvement we might make would be postulate that there is no difference whether we call i the first complex component and j the second component, or vice versa. Therefore, we assert a symmetry between i and j, and let $i^2 = j^2 = -1$. With this assertion in hand, the product $p_1 p_2$ becomes

$$
\begin{aligned}
p_1 p_2 &= (a_1 + ib_1 + jc_1)(a_2 + ib_2 + jc_2) \\
&= (a_1 a_2 - b_1 b_2 - c_1 c_2) + i(b_1 a_2 + a_1 b_2) + j(c_1 a_2 + a_1 c_2) \\
&\quad + ij(b_1 c_2) + ji(c_1 b_2).
\end{aligned} \tag{3.9}
$$

This is better, but it still leaves the ij and ji terms to deal with in Eq. 3.9.

Hamilton considered at least two alternatives:

1. let $ij = -ji$, but that still left four terms in the product $p_1 p_2$ rather than three, and
2. let $ij = ji = 0$, but that is analogous to ignoring the b_3^2 term in Eq. 3.7.

Neither alternative led to a solution, and he eventually gave up his search for a consistent and closed algebraic system of number triplets.

Consequently, after more than a decade of attempting to develop an internally consistent algebraic system for number triplets, Hamilton had little to show for it. However, we can start to see his train of thought evolve, and we see it most clearly in the first of the alternatives listed above, namely, that he was prepared to at least entertain the idea of surrendering the demand for commutativity of multiplication of numbers—something that was unheard of at the time.

3.4.3 The Quaternion Skew Field $\mathbb{H}$

One of the challenges a person faces when reading historical accounts of breakthroughs in science and mathematics is to distinguish the real from the apocryphal. However, when it comes to Hamilton's "quaternionic epiphany," we have the benefit of sufficient documentation[18] so as to be highly confident in the date and the circumstances under which he made the leap to four-component numbers.

While on their way to a council meeting of the Royal Irish Academy, Hamilton and his wife were walking along the Royal Canal in Dublin and were approaching Brougham Bridge. Suddenly (but, as we have seen, after considerable, yet unsuccessful, prior effort in a search for a consistent system of number triplets), the details of an entirely new algebraic structure suddenly became clear to him. It was the structure for the system of four-component numbers which came to be called *quaternions*.

On the spot, Hamilton wrote the algebraic relations into his notebook,[19] after which he and his wife continued on to the council meeting, where he announced his results. The date was October 16th, 1843.

Definition 3.7 Given $q_i \in \mathbb{R}, 0 \leq i \leq 3$, a *quaternion* $\mathbf{q}$ is defined as a four-component number

$$\mathbf{q} = \mathbf{1}q_0 + \mathbf{i}q_1 + \mathbf{j}q_2 + \mathbf{k}q_3,$$

with one real part and a three-component "pure quaternion" part. The unit quaternions are

$$\mathbf{1} = (1, 0, 0, 0)$$
$$\mathbf{i} = (0, 1, 0, 0)$$
$$\mathbf{j} = (0, 0, 1, 0)$$
$$\mathbf{k} = (0, 0, 0, 1).$$

Addition is carried out component-by-component. Multiplication of quaternions is defined such that the combinations $\mathbf{i}^2 = \mathbf{j}^2 = \mathbf{k}^2 = \mathbf{ijk} = -\mathbf{1}$, and where

$$\mathbf{ij} = \mathbf{k} = -\mathbf{ji}$$
$$\mathbf{jk} = \mathbf{i} = -\mathbf{kj}$$
$$\mathbf{ki} = \mathbf{j} = -\mathbf{ik}.$$

∎

---

[18]See [7], pp. 776-9, and [8] as secondary sources. We wonder: did Gauss discover them first?.

[19]The notebook survives, and an image of the relevant pages may be found at https://mathshistory. st-andrews.ac.uk/Bookpages/Hamilton9.gif. A portion of Hamilton's own account of this story may be found at https://www.irishphilosophy.com/2019/10/16/the-discover-of-quaternions/ (as of November, 2020). Whether on that same occasion, in 1843, Hamilton scratched key formulas into the stone of the bridge with his pocket knife (or did so later) is a subject of some debate among historians. Carvings in stone and a commemorative plaque exist at Brougham Bridge, which has been a pilgrimage destination for some of the world's greatest mathematicians.

Example 3.9 As an illustration of quaternion multiplication, consider the two quaternions $\mathbf{p} = \mathbf{1}p_0 + \mathbf{i}p_1$ and $\mathbf{q} = \mathbf{1}q_0 + \mathbf{k}q_3$. We find their product to be

$$\mathbf{pq} = \mathbf{1}(p_0 q_0) + \mathbf{k}(p_0 q_3) + \mathbf{i}(p_1 q_0) + (\mathbf{ik})(p_1 q_3)$$
$$= \mathbf{1}(p_0 q_0) + \mathbf{i}(p_1 q_0) - \mathbf{j}(p_1 q_3) + \mathbf{k}(p_0 q_3)$$
$$= p_0 q_0 \begin{pmatrix} 1 \\ 0 \\ 0 \\ 0 \end{pmatrix} + p_1 q_0 \begin{pmatrix} 0 \\ 1 \\ 0 \\ 0 \end{pmatrix} - p_1 q_3 \begin{pmatrix} 0 \\ 0 \\ 1 \\ 0 \end{pmatrix} + p_0 q_3 \begin{pmatrix} 0 \\ 0 \\ 0 \\ 1 \end{pmatrix}$$

$$\mathbf{pq} = \begin{pmatrix} p_0 q_0 \\ p_1 q_0 \\ -p_1 q_3 \\ p_0 q_3 \end{pmatrix}.$$

▲

As we can see in Definition 3.7, there is a symmetry among the unit quaternions $\mathbf{i}$, $\mathbf{j}$ and $\mathbf{k}$, and a non-commutativity of multiplication between any two of them. Therefore, unlike the real and complex number systems—each of which is both an abelian group and a field — quaternions form a non-abelian group (see Problem 3.17 and the comments there) in addition to forming a skew field (see Fig. 3.1).

Further, we define conjugation of quaternions by reversing the signs of the three unit quaternions in the pure quaternion part, that is,

$$\mathbf{q} = \mathbf{1}q_0 + \mathbf{i}q_1 + \mathbf{j}q_2 + \mathbf{k}q_3 \quad \Rightarrow \quad \mathbf{q}^* = \mathbf{1}q_0 - \mathbf{i}q_1 - \mathbf{j}q_2 - \mathbf{k}q_3. \tag{3.10}$$

From this we can define the norm of a quaternion as we would for a complex number. We will consider the matrix formulation of quaternions in Chap. 5. For a discussion of their role in Lie groups, please see [14]; their history is further recounted [7].

In the mid-19th century, the mathematical framework consisting of three spatial coordinates alongside a separate and independent absolute time coordinate was ingrained in how physicists thought about Nature; relativity theory was still a half-century in the future. Consequently, the impact of quaternions in physics came about slowly, but eventually with great successes. Among the applications of quaternions today are those we find in quantum field theory, special and general relativity and the kinematics of rigid body motion.

However, quaternions did play an immediate and very significant role in the late-19th-century development of vector analysis. We can begin to get a sense of this when we consider that the product of two pure quaternions $\mathbf{p} = \mathbf{i}p_1 + \mathbf{j}p_2 + \mathbf{k}p_3$ and $\mathbf{q} = \mathbf{i}q_1 + \mathbf{j}q_2 + \mathbf{k}q_3$ was defined early in the history of quaternions to be

$$\mathbf{pq} \equiv \mathbf{p} \times \mathbf{q} - \mathbf{p} \cdot \mathbf{q}. \tag{3.11}$$

The expressions on the righthand side of Eq. 3.11 correspond to the vector cross product and the scalar product of two ordinary, three-dimensional vectors.

In addition, the discovery of quaternions stimulated the axiomatic approach toward algebra, which previously had been focused primarily on finding the roots to algebraic equations. One interesting result has been the development of still higher-order, "hypercomplex" number systems. However, each new higher-order system comes at a cost — the surrender of some additional aspect of ordinary arithmetic.

For example, the next step beyond the quaternions (where we have already surrendered commutativity of multiplication) is the system of *octonions* — eight-part numbers where associativity is surrendered. At the next step, if we let a and b be 16-part numbers, we find that simple equations like $ax + b = 0$ no longer have unique solutions.[20] We leave this for you to explore further as you wish.

Problems

3.1 Referring to the ring axioms in Definition 3.1 and the subsequent additional conditions that may be applied, verify that $\mathbb{Q}$, $\mathbb{R}$ and $\mathbb{C}$ each satisfies the definition of a field.

3.2 Example 3.2 defined addition and multiplication for a ring of matrices. Verify the associative and distributive properties. [*Note*: This is an exercise that is heavy in algebraic manipulation, but otherwise is straightforward.]

3.3 Show that $\mathbb{Z}_2$ (the set of integers modulo 2 — see Eq. 3.1) is a ring.

3.4 Construct the tables for the rings $\mathbb{Z}_3$ and $\mathbb{Z}_4$, analogous to those in Eq. 3.1 for $\mathbb{Z}_2$. What are their respective characteristics?

3.5 Verify the statement in Example 3.5 that the even (but not the odd) integers form a subring of $\mathbb{Z}$.

3.6 Complete the tables for the product ring $\mathcal{S} = \mathbb{Z}_2 \times \mathbb{Z}_2$ in Example 3.7.

3.7 Construct the tables for the $\oplus$ and $\odot$ operations in $\mathcal{S} = \mathcal{R} \times \mathcal{R}'$, where
(a) $\mathcal{R} = \mathbb{Z}_2$ and $\mathcal{R}' = \mathbb{Z}_3$;
(b) $\mathcal{R} = \mathbb{Z}_3$ and $\mathcal{R}' = \mathbb{Z}_2$.

3.8 Use the couplet formulation of complex numbers (Definition 3.6) to verify that $i^2 = -1$.

3.9 Show that the couplet formulation of complex numbers (Definition 3.6) satisfies the field axioms.

---

[20] See, for example, the discussion and table in [3], pp. 17–18.

3.10 Express each of the following complex numbers in polar form, $z = re^{i\theta}$.

(a) $z_1 = (1 + 2i)^3$;
(b) $z_2 = i^{17}$;
(c) $z_3 = (1 + i)^n + (1 - i)^n$, for n a positive integer;
(d) $z_4 = 5/(-3 + 4i)$;
(e) $z_5 = [i/(i + 1)] + [(i + 1)/i]$.

3.11 Plot the locations of each of the five complex numbers in Problem 3.10 on an Argand diagram. [*Hint*: z_1, z_2, z_4 and z_5 fit nicely onto one diagram; z_3 has multiple points, so use a separate diagram and plot them for $1 \le n \le 8$.]

3.12 Referring to the complex numbers in Problem 3.10, express the following in Cartesian form, $z = x + iy$, where z^* is the complex conjugate of z.

(a) $z = z_1 + z_4^*$;
(b) $z = z_2 + z_5^*$.

3.13 Let $z_1 = x_1 + iy_1$ and $z_2 = x_2 + iy_2$. Explicitly show that $(z_1 z_2)^* = z_1^* z_2^*$. Rework this problem using $z_1 = r_1 e^{i\theta_1}$ and $z_2 = r_2 e^{i\theta_2}$.

3.14 Let $\mathbf{p} = 1 p_0 + \mathbf{i} p_1 + \mathbf{j} p_2 + \mathbf{k} p_3$ and $\mathbf{q} = 1 q_0 + \mathbf{i} q_1 + \mathbf{j} q_2 + \mathbf{k} q_3$ be two quaternions.

(a) Find the general expression for the product $\mathbf{pq}$;
(b) Show that the product of two quaternions is non-commutative, i.e., $\mathbf{pq} = -\mathbf{qp}$;
(c) Show that the conjugation (Eq. 3.10) of a product of two quaternions is such that $(\mathbf{pq})^* = \mathbf{q}^* \mathbf{p}^*$ (compare this result with that for complex numbers in Problem 3.13).

3.15 For a quaternion $\mathbf{p} = 1 p_0 + \mathbf{i} p_1 + \mathbf{j} p_2 + \mathbf{k} p_3$:

(a) Show that the quantity $\mathbf{pp}^*$ is a real number. [*Ans:* the quantity $\mathbf{pp}^* = p_0^2 + p_1^2 + p_2^2 + p_3^2 = |\mathbf{p}|^2$ is the *norm* of the quaternion $\mathbf{p}$];
(b) Find an expression for $\mathbf{p}^{-1}$. [*Hint*: Take the same approach you use when finding z^{-1}, for $z \in \mathbb{C}$.] This establishes the existence of a multiplicative inverse for quaternions, which is among the requirements for a skew field (Fig. 3.1).

3.16 Expand both sides of Eq. 3.11 and show that the equation is satisfied. [*Hint*: Treat $\mathbf{p}$ and $\mathbf{q}$ on the lefthand side as two *pure* quaternions, but as ordinary 3-vectors on the righthand side.]

3.17 The relations among the four unit quaternions may be summarized with a multiplication table. Complete the table shown below. [*Hint*: Follow the convention we have discussed previously, whereby the first term in the product is in the left-most column of the table, and the second term is in the top-most row. Ex: $\mathbf{i} \square \mathbf{j} = \mathbf{ij} = \mathbf{k}$.]

[*Note*: This is *not* the *quaternion group*, which has 8 elements, $\{\pm 1, \pm i, \pm j, \pm k\}$.]

Guide to Further Study

The stated purpose of this chapter was to introduce the physics and engineering student to an algebraic structure (the ring) in such a way as to help provide a contextual framework for those number systems that are widely used in those respective disciplines, namely, the real, complex and quaternion systems. Consequently, we did not explore many aspects of rings that would be central to an advanced course in abstract algebra aimed at budding mathematicians.

If you wish to further explore rings, the introductory works of Artin [1], Hungerford [5] and Pinter [12] are good places to proceed from here. More advanced treatments may be found in Hungerford [6] and Mac Lane and Birkhoff [9], with the cautionary note that these two sources define rings slightly differently.

For complex analysis, you may be interested in two works by Nahin, [10] and [11], an electrical engineer. The first starts as a history of complex numbers, but takes the reader through to an introduction to analytic function theory. The second will have great appeal for engineering and applied physics students with its discussion of Fourier transforms and electronic circuits. For a more formal mathematical account, the classic is Conway [2].

If you wish to see how quaternions play a role in group theory, Stillwell [14] makes use of them throughout his text on Lie groups. The biography of Hamilton by Hankins [4], a shorter piece by Lanczos [8] and the general mathematical historical reference by Kline [7] offer valuable insights to the history of one of the great leaps forward in mathematical physics.

References

1. Artin, M.: Algebra. Prentice-Hall, Upper Saddle River, NJ (1991)
2. Conway, J.B.: Functions of One Complex Variable I, 2nd edn. Springer, New York (1978)
3. Gilmore, R.: Lie Groups, Lie Algebras, and Some of Their Applications. Wiley, New York (1974), Dover reprint, Mineola, NY (2005)
4. Hankins, T.L.: Sir William Rowan Hamilton. Johns Hopkins University Press, Baltimore, MD (1980)
5. Hungerford, T.W.: Abstract Algebra—An Introduction. Saunders, Philadelphia (1990)
6. Hungerford, T.W.: Algebra. Springer, 12th printing, New York (2003)
7. Kline, M.: Mathematical Thought from Ancient to Modern Times, Published in 3 volumes. Oxford University Press, Oxford (1990)
8. Lanczos, C.: William Rowan Hamilton—An Appreciation. Am. Scientist **55**(2), 129–143 (1967)

9. Mac Lane, S., Birkhoff, G.: Algebra, 3rd edn. Chelsea, New York (1993)
10. Nahin, P.J.: An Imaginary Tale: The Story of $\sqrt{-1}$. Princeton University Press, Princeton, NJ (1998)
11. Nahin, P. J.: Dr. Euler's Fabulous Formula: Cures Many Mathematical Ills. Princeton University Press, Princeton, NJ (2006)
12. Pinter, C.C.: A Book of Abstract Algebra, 2nd edn. McGraw-Hill, New York (1990), available in Dover reprint, New York (2010)
13. Stillwell, J.: Numbers and Geometry. Springer-Verlag, New York (1998)
14. Stillwell, J.: Naive Lie Theory. Springer, New York (2008)

Chapter 4
Vector and Tensor Spaces

4.1 Modules and Vector Spaces

In an introductory physics course, vectors are framed as geometric objects—directed line segments representing physical quantities with both magnitude *and* direction, such as a velocity, a force or an electric field. This geometric perspective is encouraged by illustrations of vectors as "arrows" of a given length, originating at a particular point and pointing in a specified direction. A longer arrow corresponds to a greater magnitude of the quantity being described.

This picture is refined, but not made fundamentally different, by the representation of a vector as an *n*-tuple of numbers, which serve as its coordinates in an *n*-dimensional coordinate system. In this setting, vectors are written variously as row- or column-vectors, such as $\mathbf{B} = (B_x, B_y, B_z)$ for a magnetic field vector in Cartesian coordinates. The vector algebra we learn early in our studies is premised on these components, and in due course we come to think of vectors as objects which are *necessarily* defined in terms of a coordinate system.[1]

As our study of physics proceeds beyond the introductory course, the level of abstraction regarding vectors increases considerably. For example, in our study of quantum mechanics and special relativity terms like "bra vectors," "ket vectors" and "four-vectors" are used in contexts where there don't appear to be any "arrows." Later, when studying the special functions of mathematical physics that arise as solutions to ordinary and partial differential equations, we are told that certain operations may be performed on these scalar functions as though these functions are vectors.

This shows that while the geometric perspective on vectors has many applications in physics and engineering, by itself it is a very limited and specialized perspective. A general and more inclusive approach is necessary if we are to gain a full understanding of vectors in all their guises across the many branches of mathematical physics. This

---

[1] I have assumed you are familiar with vector algebra as described in these first two paragraphs. If a short practice session is needed, see Problem 4.1.

© Springer Nature Switzerland AG 2021
S. P. Starkovich, *The Structures of Mathematical Physics*,
https://doi.org/10.1007/978-3-030-73449-7_4

is the motivation for adopting an axiomatic approach to the study of vector spaces, and such is the approach taken in this chapter.

When we combine two algebraic structures to create a third, the composite structure must be internally consistent and be closed under its defined operations. In the particular case of a combination of a group with a ring we can imagine many possibilities, but the main focus among algebraists is to link an additive abelian group with a ring via a rule defined as multiplication. This new structure is called a *module*.

This still leaves several possibilities for the composite structure because of the different types of rings and various definitions for multiplication that might be invoked. In mathematical physics, the module of greatest importance is arguably the one that comes from specifying the ring to be a field (Chap. 3). In this case, the module is called a *linear vector space*, or just a *vector space*[2]—the linearity arising from the additive abelian nature of the associated group, and from the linear maps that are discussed later in this chapter.

Definition 4.1 Consider an additive abelian group G with addition of group elements symbolized as $\boxplus$. Also consider a field F whose elements act *internally* on each other via addition and multiplication operators denoted by $+$ and $\cdot$, respectively, and which act *externally* on the elements of G via multiplication denoted by $\odot$.

Let $|u\rangle, |v\rangle, |w\rangle \in G$, and $a, b, c \in F$. The algebraic system $\Sigma = (G, F, \boxplus, +, \cdot, \odot)$ is a *vector space* if:

1. Σ is closed under all operations on the elements of G;
2. Σ is associative under $\boxplus$: $|u\rangle \boxplus (|v\rangle \boxplus |w\rangle) = (|u\rangle \boxplus |v\rangle) \boxplus |w\rangle$;
3. Σ is commutative under $\boxplus$: $|u\rangle \boxplus |v\rangle = |v\rangle \boxplus |u\rangle$;
4. Σ contains an additive identity element, $|0\rangle$, such that $|u\rangle \boxplus |0\rangle = |u\rangle$;
5. Σ contains an additive inverse element, $|-u\rangle$, such that $|u\rangle \boxplus |-u\rangle = |0\rangle$;
6. $a \odot (|u\rangle \boxplus |v\rangle) = a \odot |u\rangle \boxplus a \odot |v\rangle$;
7. $(a + b) \odot |u\rangle = (a \odot |u\rangle) \boxplus (b \odot |u\rangle)$;
8. $(a \cdot b) \odot |u\rangle = (b \cdot a) \odot |u\rangle = a \odot (b \odot |u\rangle) = b \odot (a \odot |u\rangle)$.

∎

The elements of the group G are called *vectors*, the operation $\boxplus$ is called *vector addition*, and $|0\rangle$ is the *zero vector*. The elements of the field F are called *scalars*. Axioms (1)–(5) are those of an additive abelian group. Axioms (6) and (7) are distributive properties that define multiplication of vectors $|u\rangle, |v\rangle \in G$ by scalars $a, b \in F$. Axiom (8) shows the commutative and associative properties associated with the field F, where a or b may assume the zero or unit elements of F.

In addition, we have adopted the "bra" and "ket" vector notation due to Dirac.[3] We will write a *ket vector* as $|u\rangle$ and a *bra vector* as $\langle u|$. These two types of vectors

---

[2]We often shorten this even further to just "space." Once we get to our discussion of topology, "space" will have a different and broader meaning.

[3]Paul Adrien Maurice Dirac, (1902–1984), Swiss/British physicist. The origin of the terms "bra" and "ket" is a pun. Taken together in a scalar, or inner, product they form a "bracket" symbol such as $\langle u|v\rangle$. See Sect. 4.3

are "dual" to each other in a manner that will be defined below. On some occasions in this text we will use a bold-faced-letter (e.g., **u**) to denote a vector.

The notation used in Definition 4.1 is cumbersome. As an alternative, it is customary to use the "+" symbol to mean either scalar or vector addition depending on context, and to write $a|u\rangle$ for $a \odot |u\rangle$ and ab for $a \cdot b$. We will adopt this shorthand for the remainder of this text, but this can be risky if you are seeing this for the first time, and it is important to remember which operations apply to which elements.

Any combination of a group G and a field F that satisfies Definition 4.1 is a vector space. As such, we describe a vector space by saying "Σ is a vector space of the group G over the field F." If G is understood as a given, then we might say "Σ is a vector space over F." Most often, though, we say "G is a vector space over F."

In physics and engineering, the choices for G are wide and varied, but the choices for F are predominately either $F = \mathbb{R}$, in which case we say the vector space is *real*, or $F = \mathbb{C}$, in which case the vector space is said to be *complex*. The rationale for designating the space in terms of the properties of F rather than those of G will be become apparent as we proceed.

Example 4.1 Consider several familiar sets and operations in the context of the vector space axioms:

1. Let $G = \mathbb{R}$ and $F = \mathbb{R}$, and let the operations be those of ordinary arithmetic. Then G is a real vector space. We know this space as the real line.
2. Let G be the set of all vectors depicted as directed line segments or as ordered pairs (x, y) in the Cartesian plane, and let $F = \mathbb{R}$. Let the opertions be those of ordinary vector algebra in the plane. Then G is a real vector space. This is usually the context in which we first learn about vectors as "arrows."
3. Let $G = \mathbb{C}$ and $F = \mathbb{R}$, and let the operations be those of complex algebra. Then G is a real vector space that we know as the complex plane. Closure in this space is ensured because a complex number (an element of G) multiplied by a real number (an element F) yields an element of $\mathbb{C}$. Note, however, the converse is *not* true; $G = \mathbb{R}$ is *not* a vector space over $F = \mathbb{C}$ because we do not have closure in G when its elements are multiplied by a complex number in F.
4. Let $G = \mathbb{C}$ and $F = \mathbb{C}$, and let the operations be those of complex algebra. Then G is a complex vector space. How this space compares to the complex plane will be discussed in Sect. 4.2.2.
5. Let G be the set of all real 2×3 matrices (matrices with $m = 2$ rows, $n = 3$ columns and real-valued entries), and let $F = \mathbb{R}$. Define addition on G to be matrix addition (summation of corresponding entries) and adopt the usual method of multiplying a matrix by a scalar. Then G is a real vector space. This is true for any m and n.
6. Consider an interval on the real line, and let G be the set of all real-valued single-variable functions on that interval (e.g., all real-valued $f(x)$ on the interval (x_1, x_2)). Let the field $F = \mathbb{R}$. By the rules of ordinary arithmetic we have addition on G defined as $f_1(x) + f_2(x) = f_3(x)$ for $f_1, f_2, f_3 \in G$, and also $af(x) \in G$ for $a \in F$. G is a real vector space. ▲

It is important to note that the cross-product of two vectors is *not* accommodated by the vector space axioms, nor is there an axiom in Definition 4.1 that defines any type of multiplication of two elements of G.[4] For that, we need either an *inner product space* (Sect. 4.3) or an *algebra* (Chap. 5).

4.2   Linear Independence, Basis Vectors and Norms

The three properties of vector spaces that we describe in this section are among those which appear with greatest frequency in applications to physics and engineering. Though we tend to take them for granted in our everyday work, it is important to understand their meaning as we move to more abstract vector spaces.

4.2.1   Linear Independence

We first consider the definition of the linear independence of a set of vectors, from which many other vector space properties then follow.

Definition 4.2 Two vectors $|u\rangle$ and $|v\rangle$ are *linearly independent* if the equation $a|u\rangle + b|v\rangle = |0\rangle$ can be satisfied only if both scalars $a, b \in F$ are equal to zero. ∎

Example 4.2 The same test for linear independence applies across all types of vector spaces.

1. All elements in the vector space of the real numbers $\mathbb{R}$ over the field $\mathbb{R}$ are linearly *dependent*. Given vectors $|u\rangle$, $|v\rangle$ and $|w\rangle$ in $\mathbb{R}$ and scalars $a, b, c \in \mathbb{R}$, equations such as

$$a|u\rangle + b|v\rangle = |0\rangle,$$

$$a|u\rangle + b|v\rangle + c|w\rangle = |0\rangle$$

   may be satisfied by non-zero values of a, b and c.
2. When depicted as directed line segments, two collinear vectors in the two-dimensional Cartesian plane are linearly *dependent*. That is, nonzero values of the scalars $a, b \in \mathbb{R}$ can be selected such that $a|u\rangle + b|v\rangle = |0\rangle$.
3. For a vector space where G is a set of all real-valued single-variable functions $f(x)$ on the interval (x_1, x_2) and where $F = \mathbb{R}$, the functions

$$|u\rangle = 3 - 5x,$$

---

[4]Note that when we multiply two real numbers in the vector space $\mathbb{R}^1$, or two complex numbers in the vector space $\mathbb{C}^1$ (see Example 4.3 (1) and (4) below), one factor is assigned to G and the other is assigned to F.

$$|v\rangle = x(1+x),$$

$$|w\rangle = (x-3)(x-1)$$

are linearly *dependent* because

$$a|u\rangle + b|v\rangle + c|w\rangle = |0\rangle$$

for $a = 1$, $b = 1$ and $c = -1$.

4. As a counterexample to that just given, the set

$$|u\rangle = 1$$

$$|v\rangle = x,$$

$$|w\rangle = x^2$$

comprises a set of linearly *independent* vectors.   ▲

We return to a discussion of linear independence in our treatment of complete orthonormal sets of vectors in Sect. 4.4.

4.2.2 Basis and Dimension

For a vector space U, the maximum possible number of linearly independent vectors in the space defines its *dimension*, denoted as dim U. When normalized to unit magnitude (Sect. 4.2.3), these vectors form a set of *basis vectors* (often referred to simply as the *basis* for that space), and *any* vector in that space may be expressed as some linear combination of these basis vectors. Because of this latter property, the basis is said to form a *complete set of vectors* that *spans* the vector space.[5]

In Cartesian coordinates a vector $|u\rangle$ may be written variously as

$$|u\rangle = u^1\hat{\mathbf{i}} + u^2\hat{\mathbf{j}} + u^3\hat{\mathbf{k}} = u^1\hat{\mathbf{e}}_1 + u^2\hat{\mathbf{e}}_2 + u^3\hat{\mathbf{e}}_3 = u^1\begin{pmatrix}1\\0\\0\end{pmatrix} + u^2\begin{pmatrix}0\\1\\0\end{pmatrix} + u^3\begin{pmatrix}0\\0\\1\end{pmatrix}.$$

This practice of expressing a vector in terms of a *standard coordinate basis*, which in this example is

---

[5]See also Definition 4.7 in Sect. 4.4. A complete *set of vectors* in a space is not to be confused with a *complete vector space*. We discuss this latter concept in Sect. 4.4.4

$$\hat{\mathbf{e}}_1 = \begin{pmatrix} 1 \\ 0 \\ 0 \end{pmatrix} \quad \hat{\mathbf{e}}_2 = \begin{pmatrix} 0 \\ 1 \\ 0 \end{pmatrix} \quad \hat{\mathbf{e}}_3 = \begin{pmatrix} 0 \\ 0 \\ 1 \end{pmatrix} \; , \tag{4.1}$$

is one that we use often in mathematical physics. The concepts of basis and dimension are familiar from our dealings with Cartesian spaces, but it is important to realize that these concepts apply to all vector spaces regardless of whether their elements are directed line segments, functions, matrices, or something else entirely.

Example 4.3 Consider the dimensions of the vector spaces listed in Example 4.1.

1. If $G = \mathbb{R}$ and $F = \mathbb{R}$, then G is a *one*-dimensional real vector space. Its basis is simply the number 1, and we designate it as $\mathbb{R}^1$.
2. If G is the set of all vectors depicted as directed line segments, or as ordered pairs (x, y), in the Cartesian plane and if $F = \mathbb{R}$, then G is a *two*-dimensional real vector space. Its basis may be expressed in terms of a two-dimensional standard coordinate basis, or as the familiar $\hat{\mathbf{i}}$ and $\hat{\mathbf{j}}$. We designate this space as $\mathbb{R}^2$.
3. If $G = \mathbb{C}$ and $F = \mathbb{R}$, then G is a *two*-dimensional real vector space. This is the complex plane, and its standard coordinate basis is $1 = (1, 0)$ and $i = (0, 1)$ (see Definition 3.6 in Sect. 3.4.2).
4. If $G = \mathbb{C}$ and $F = \mathbb{C}$, then G is a *one*-dimensional complex vector space. Its basis is the number 1, and we designate it as $\mathbb{C}^1$.

The last two spaces in Example 4.1 are real, infinite-dimensional vector spaces.   ▲

When we discussed the historical origins of complex numbers in Chap. 3 we noted the close affinity between the complex plane and $\mathbb{R}^2$. Example 4.3(4) shows another way of representing a complex number, this time by framing it in the context of the space $\mathbb{C}^1$, a one-dimensional complex space, rather than a two-dimensional real space.

Therefore, in the same way that we could write a real number in the real vector space $\mathbb{R}^1$ as the vector $|r\rangle = r|1\rangle$, where r is a *real* scalar, a complex number in the complex space $\mathbb{C}^1$ may be written as $|z\rangle = z|1\rangle$, where z is a *complex* scalar. These types of constructions are why we discern real from complex spaces by the nature of the scalars in F rather than by the elements of G. For higher-dimensional spaces, the standard coordinate basis can be applied to all $\mathbb{C}^n$ as well as all $\mathbb{R}^n$.

In applications to physics and engineering, it is important to be clear as to the nature of the underlying space in which a problem is being stated. For example, quantum mechanics is built around complex spaces ($F = \mathbb{C}$) of complex functions, and we identify a set of basis functions in terms of which an arbitrary function may be expressed. This is what happens, for example, when we express a solution to Schrodinger's equation in terms of a complete set of orthonormal basis functions that are relevant to the system under consideration. As another example, when we study the special functions of mathematical physics, we frequently work with real ($F = \mathbb{R}$) infinite-dimensional spaces. Again, a set of basis functions—albeit an infinitely large set—is used to express an arbitrary function in the space.

Fig. 4.1 The distance function $d(u, v)$ in a two-dimensional space

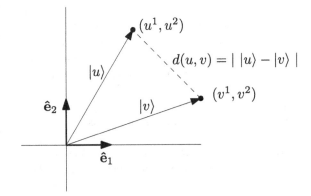

4.2.3 Norms and Distance Functions on Vector Spaces

The distance between two points in a two-dimensional space such as that shown in Fig. 4.1 is found from the *distance function* in Eq. (4.2), where $n = 2$.

$$d(u, v) = \sum_{i=1}^{n} \left(\epsilon_i |u^i - v^i|^2 \right)^{1/2} . \tag{4.2}$$

In Cartesian coordinates, $\epsilon_i = +1$ for each term in the sum and we get the familiar expression

$$d = \sqrt{(+1)(u^1 - v^1)^2 + (+1)(u^2 - v^2)^2} = \sqrt{(u_x - v_x)^2 + (u_y - v_y)^2} .$$

A topological space on which a distance function is defined is called a *metric space*.[6] If all the ϵ_i in Eq. (4.2) are ± 1, the space (or metric) is said to be *flat*. Flat spaces where all $\epsilon_i = +1$ (as in the Cartesian plane) are called *Eulidean*, and the geometry we use in such spaces is our familiar Euclidean geometry.

On the other hand, if there is a mix of $\epsilon_i = \pm 1$ in Eq. (4.2), then the metric is *pseudo-Euclidean*. For example, in special relativity (SR) the four-dimensional spacetime metric is pseudo-Euclidean and provides the means by which we calculate the spacetime distance (separation) between two points (events). The *Minkowski metric* of SR assigns the same sign to the three spatial coordinates and the opposite sign to the time coordinate, although whether we use the "−2 signature" (+ - - -) or the "+2 signature" (- + + +) is a matter of convention.

Spaces where the ϵ_i in Eq. (4.2) are replaced by functions are said to be *non-Euclidean*. Curved surfaces such as the surface of a sphere or the curved spacetime

---

[6]In Sect. 6.4.2 we show where metric spaces fit in the hierarchy of topological spaces. We define a metric tensor and offer several examples in Sect. 4.5.4

of general relativity (where the curvature is associated with gravity) employ non-Euclidean metrics.

A distance function is a special case of a more general concept called a *norm*. Although the two are often taken to be synonymous, they are distinguished from each other in that a norm does *not* depend on the existence of an inner product,[7] whereas an inner product is necessary to define a distance function.

For those of us in physics or engineering this may seem like a distinction without a difference since virtually all of the vector spaces we employ in our work have inner products defined on them. However, it often happens in the field of functional analysis that we encounter spaces which allow for definitions of norms (sometimes more than one) but where no inner product exists. Therefore, we need to distinguish the two definitions.

Definition 4.3 Consider a vector space G on a field F, and let $\mathbf{u}, \mathbf{v} \in G$ and $a \in F$. A *norm* in G is defined as a mapping, denoted by $\| \ \| : \mathbf{u} \mapsto \|\mathbf{u}\| \in \mathbb{R}$, such that

1. $\|\mathbf{u} + \mathbf{v}\| \le \|\mathbf{u}\| + \|\mathbf{v}\|$; (triangle inequality);
2. $\|a\mathbf{u}\| = |a| \|\mathbf{u}\|$; (homogeneous scaling);
3. $\|\mathbf{u}\| = 0$ if and only if $\mathbf{u} = 0$.

A vector space on which a norm is defined is called a *normed vector space*. If only the first two axioms apply, the space is said to have a *seminorm*. ∎

We contrast the definition of a norm with the definition of a distance function.

Definition 4.4 Consider a set X with $u, v, w \in X$. A *distance function* is defined as a mapping $d : X \times X \to \mathbb{R} \ge 0$ such that

1. $d(u, v) + d(v, w) \le d(u, w)$; (triangle inequality);
2. $d(u, v) = d(v, u)$; (symmetry);
3. $d(u, v) = 0$ if and only if $u = v$

Again, a set X on which a distance function is defined is called a metric space, and if only the first two axioms apply the set is said to have a pseudo-metric. ∎

Although we have said that a distance function and a norm are not synonymous, we can also say that in those spaces on which an inner product is defined they are related by $d(u, v) = \|u - v\|$. In this case, letting $v = 0$ in our discussion above leads to the definition of the norm of a vector as its magnitude, which we interpret as its extension from the origin of our coordinate system.

Generally, though, the class of normed spaces is larger than the class of inner product spaces, and even within the category of vector spaces different norms may be defined on any one particular space.[8] Another way of phrasing this is to say that if an inner product is defined on a space then it certainly induces a norm, but the converse is not true.

---

[7] Inner-product spaces are discussed in Sect. 4.3 and defined in Definition 4.6
[8] This will be important when we discuss Hilbert spaces in Sect. 4.4.4

4.3 Inner Product Spaces

A vector space on which an inner product is defined is called an *inner product space*, and inner product spaces that are defined over the fields $F = \mathbb{R}$ or $F = \mathbb{C}$ are centrally important to most of the things we do in physics and engineering,

We will consider the formulation of the *inner* (or *scalar*) *product* on a vector space from four different—but closely related—perspectives. First (and largely for context) we give a short summary of the approach taken in elementary vector algebra—an approach that relies heavily on coordinates and trigonometry. The second approach extends this perspective to coordinate spaces that may be either real or complex. It is here that we introduce the concept of a dual vector space.

In the third approach we introduce linear forms, linear functionals and sesquilinear (hermitian) maps. These are the building blocks of the inner product structure in complex function spaces. Finally, by narrowing this latter perspective to real function spaces we arrive at inner products by way of bilinear maps.

4.3.1 Inner Products in $\mathbb{R}^2$ Over $\mathbb{R}$

The inner product in the Cartesian plane should be very familiar to the reader, so this summary primarily serves to provide context for what follows.[9]

Example 4.4 Consider the space $\mathbb{R}^2$ over $\mathbb{R}$, and let $\mathbf{u}(x, y) = u_x(x, y)\,\hat{\mathbf{i}} + u_y(x, y)\,\hat{\mathbf{j}}$ and $\mathbf{v}(x, y) = v_x(x, y)\,\hat{\mathbf{i}} + v_y(x, y)\,\hat{\mathbf{j}}$ be two vectors *that are located at the same point*. The components (as elements of $F = \mathbb{R}^1$) are real-valued functions of x and y. The inner product of these two vectors is defined as

$$\mathbf{u} \cdot \mathbf{v} \equiv (u_x\hat{\mathbf{i}} + u_y\hat{\mathbf{j}}) \cdot (v_x\hat{\mathbf{i}} + v_y\hat{\mathbf{j}}) \equiv u_x v_x + u_y v_y \,, \tag{4.3}$$

and when combined with some plane trigonometry we get

$$\mathbf{u} \cdot \mathbf{v} = uv \cos\theta \,, \tag{4.4}$$

where u and v are the magnitudes of the respective vectors and θ is the angle between them. The magnitude of a single vector follows directly from taking the inner product of the vector with itself:

$$u = |\mathbf{u}| = \sqrt{|\mathbf{u}|^2} = \sqrt{\mathbf{u} \cdot \mathbf{u}} = \sqrt{u_x^2 + u_y^2} \,. \tag{4.5}$$

The requirement that two vectors must be located at the same point in order for their inner product to be defined is something we take for granted in these simple

---

[9]For notational clarity in this example, we use bold-faced letters rather than kets to represent vectors, and coordinate subscripts rather than numerical superscripts to distinguish components.

problems, but it becomes very important to remember this when we take up the more elaborate configurations in the following sections.   ▲

4.3.2   Inner Products in Coordinate Spaces

We can generalize the discussion in Sect. 4.3.1 by considering a two-dimensional vector space whose elements are written in a standard coordinate basis as

$$|v\rangle = v^1\hat{\mathbf{e}}_1 + v^2\hat{\mathbf{e}}_2 = v^1 \begin{pmatrix} 1 \\ 0 \end{pmatrix} + v^2 \begin{pmatrix} 0 \\ 1 \end{pmatrix} = \begin{pmatrix} v^1 \\ v^2 \end{pmatrix}. \tag{4.6}$$

This form of a two-dimensional vector in a coordinate space[10] can accommodate spaces such as

- $\mathbb{R}^2$ over $\mathbb{R}$ (Example 4.4) with basis vectors $\hat{\mathbf{i}}$ and $\hat{\mathbf{j}}$, and with $v^1, v^2 \in \mathbb{R}$, and
- $\mathbb{C}^2$ over $\mathbb{C}$, a complex space with two standard coordinate basis vectors, and with $v^1, v^2 \in \mathbb{C}$. [Recall that the space $\mathbb{C}$ over $\mathbb{C}$ is one-dimensional, as is $\mathbb{R}$ over $\mathbb{R}$.]

Note that we are now taking care to *write the ket basis vectors as column matrices* as we did in defining the standard coordinate basis in Sect. 4.2.2. As we proceed, you will see why we need to distinguish row vectors from column vectors.

We also need to reimagine the way we think about complex conjugation. Previously (see the caption to Fig. 3.3) we learned to think of complex conjugation as a reflection across the real axis of the complex plane. A more robust perspective is to think of *conjugation* as a bijection that maps the ket vector $|u\rangle$ to the bra vector $\langle u|$ in a different space. We write this map as $* : |u\rangle \in G \mapsto \langle u| \in G^*$, where G^* is a vector space that is separate from G, but which is "dual" or "conjugate" to G.

Importantly, what were *column* vectors in G become *row* vectors with conjugated components in G^*. This is so that we may calculate inner products using the established methods of matrix multiplication. Furthermore, complex conjugation, often called the *canonical isomorphism* between G and G^*, is its own inverse; the same conjugation operation maps G^* to G.[11]

Given the vectors $\mathbf{u} = |u\rangle \in G$ and $\mathbf{v} = |v\rangle \in G$, their inner product $\mathbf{u} \cdot \mathbf{v}$ is formed in two steps. First, we map one of the vectors (in this case, $|u\rangle$) to its conjugate $\langle u| \in G^*$. What were column vectors in G now become conjugated row vectors in G^*. Second, we perform the matrix multiplication between the row vector $\langle u|$ and

---

[10]We use the term "coordinate space" to mean that the basis vectors in the space may be written in a standard coordinate basis. We contrast this with function spaces, where the basis vectors are normalized functions.

[11]The concept of a dual space appears in many different guises across mathematics and physics. In the present context and because of the complex conjugation bijection, the space G^* is sometimes called a *conjugate space* to G. However, this term is sometimes used to denote dual spaces generally, regardless of whether complex conjugation is involved.

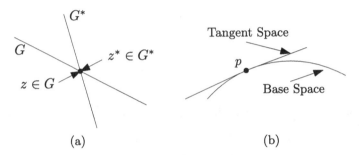

Fig. 4.2 Two conceptual sketches: in **a** the spaces G and G^* intersect at $z \in G$ and $z^* \in G^*$; in **b** we imagine G is a curve or surface in space, and the analog to G^* is called a tangent space

the column vector $|v\rangle$. The result will be a scalar quantity, as it must, which we write in bracket ("bra-ket") notation as $\langle u|v \rangle \equiv \mathbf{u}^*\mathbf{v}$.

Before working any examples, you should be at least somewhat bothered by this procedure because of the earlier admonition that two vectors must be at the same point in order for their inner product to be defined. How can that be the case here, if $\langle u|$ and $|v\rangle$ are in different spaces? The answer is that those two spaces intersect at the common base point of the two vectors.

In fact, every point $z \in G$ has attached to it a dual space G^* in which the point of intersection is $z^* \in G^*$ (Fig. 4.2a). This intersection is a consequence of the bijection between the two spaces.

In differential geometry, the dual space is called a *tangent space* and intersects a base space at a point p (Fig. 4.2b). Two vectors in the base space must be "transported" to the same point before their inner product can be evaluated. In the inner product $\langle u|v \rangle$, the base points of $\langle u|$ and $|v\rangle$ share that point of intersection.

The reason we have never previously thought about dual spaces when performing inner products in $\mathbb{R}^2$ over $\mathbb{R}$ (or $\mathbb{R}^n$ over $\mathbb{R}$) is that the dual space in that context is an identical copy of the base space. Still, in the inner-product calculations we have done in elementary physics (e.g., work integrals), we often move (parallel transport) vectors around so that they share a common base point, thereby allowing us to define the angle θ between them. Now we see that this "transporting" is permitted because of the canonical relationship between a space and its dual.

We can now summarize the procedure for taking the inner product $\langle u|v \rangle$ in a coordinate space, noting one additional consideration regarding the notation on components. Given two ket vectors $|u\rangle$ and $|v\rangle$ in the standard coordinate basis;

- Form the bra vector $\langle u|$ by

    1. changing the standard basis *column* vectors in $|u\rangle$ to *row* vectors;
    2. taking the complex conjugates of the components;[12] and

---

[12]In this text we use the "conjugation convention" as adopted by physicists. Mathematicians tend to reverse the conjugation between the bra and ket vectors.

3. switching the subscripts and superscripts from those in the ket vector $|u\rangle$. This step helps us keep track of which vectors are in which space, and is a convention we'll see again when we discuss tensors.

- Finally, the inner product $\langle u|v \rangle$ is found by multiplying the row matrix $\langle u|$ with the column matrix $|v\rangle$.

We'll apply this to inner products in the space $\mathbb{C}^2$ over $\mathbb{C}$, of which inner products in $\mathbb{R}^2$ over $\mathbb{R}$ would be a special case.

Example 4.5   Consider the vectors $\mathbf{u} = |u\rangle$ and $\mathbf{v} = |v\rangle$ in the *complex* vector space $\mathbb{C}^2$ over $\mathbb{C}$ written in the standard coordinate basis:

$$|u\rangle = u^1 \hat{\mathbf{e}}_1 + u^2 \hat{\mathbf{e}}_2 = u^1 \begin{pmatrix} 1 \\ 0 \end{pmatrix} + u^2 \begin{pmatrix} 0 \\ 1 \end{pmatrix},$$

$$|v\rangle = v^1 \hat{\mathbf{e}}_1 + v^2 \hat{\mathbf{e}}_2 = v^1 \begin{pmatrix} 1 \\ 0 \end{pmatrix} + v^2 \begin{pmatrix} 0 \\ 1 \end{pmatrix}.$$

Remembering that $u^i, v^i \in \mathbb{C}$, the bra vector $\langle u|$ becomes

$$\langle u| = u_1^* \hat{\mathbf{e}}^1 + u_2^* \hat{\mathbf{e}}^2 = u_1^*(1\ 0) + u_2^*(0\ 1).$$

The inner product is then found to be

$$\langle u|v \rangle = (u_1^*(1\ 0) + u_2^*(0\ 1)) \left(v^1 \begin{pmatrix} 1 \\ 0 \end{pmatrix} + v^2 \begin{pmatrix} 0 \\ 1 \end{pmatrix} \right)$$

$$= u_1^* v^1 (1\ 0) \begin{pmatrix} 1 \\ 0 \end{pmatrix} + u_2^* v^2 (0\ 1) \begin{pmatrix} 0 \\ 1 \end{pmatrix} + u_1^* v^2 (1\ 0) \begin{pmatrix} 0 \\ 1 \end{pmatrix} + u_2^* v^1 (0\ 1) \begin{pmatrix} 1 \\ 0 \end{pmatrix}$$

$$= u_1^* v^1 (1) + u_2^* v^2 (1) + u_1^* v^2 (0) + u_2^* v^1 (0)$$

$$\langle u|v \rangle = u_1^* v^1 + u_2^* v^2.$$

In general $\langle u|v \rangle$ is a complex number, and we see immediately that $\langle u|v \rangle = \langle v|u \rangle^*$.

The norm $\|\mathbf{u}\|$ of $|u\rangle$ is found by setting $|v\rangle = |u\rangle$ to yield a real number, as it must be:

$$\|\mathbf{u}\|^2 = \langle u|u \rangle = (u_1^* u^1 + u_2^* u^2) = |u_1|^2 + |u_2|^2.$$

This method of taking the inner product—by starting with vectors in the standard coordinate basis, forming a bra vector and performing the matrix multiplications—is directly applicable to inner products in $\mathbb{C}^n$ over $\mathbb{C}$ and $\mathbb{R}^n$ over $\mathbb{R}$ (as in Example 4.4, where $n = 2$ and all the components were real). For example, for the space $\mathbb{C}^3$ over $\mathbb{C}$ with

$$|u\rangle = \begin{pmatrix} u^1 \\ u^2 \\ u^3 \end{pmatrix} \quad \text{and} \quad |v\rangle = \begin{pmatrix} v^1 \\ v^2 \\ v^3 \end{pmatrix},$$

the inner product is

$$\langle u|v \rangle = (u_1^* \ u_2^* \ u_3^*) \begin{pmatrix} v^1 \\ v^2 \\ v^3 \end{pmatrix} = u_1^* v^1 + u_2^* v^2 + u_3^* v^3 \ .$$

▲

Unless otherwise specified, we will continue to assume a Euclidean metric when forming inner products as we have done here. Consequently these are referred to as *Euclidean inner products*.

4.3.3 Inner Products on Complex and Real Function Spaces—Sesquilinear and Bilinear Maps

In the previous section we applied the canonical isomorphism of complex conjugation to map the space G to its dual space G^*; the combination of these two spaces provided the structure for defining inner products in coordinate spaces. In this section we generalize those ideas and describe the structures that provide for a definition of the inner product in complex function spaces. Inner products in real function spaces then follow directly.

We start with the definition of a linear map, with further definitions being introduced by way of the examples.

Definition 4.5 Consider two spaces X and Y and a field F. The map $\phi : X \to Y$ is a *linear map* if for all $x_1, x_2 \in X$ and for all $a, b \in F$ the expression

$$\phi(ax_1 + bx_2) = a\phi(x_1) + b\phi(x_2).$$

is satisfied. The nature of the map ϕ will depend on the nature of X, Y and F. We denote the set of all linear maps from X to Y by the notation $L(X, Y)$. ■

Example 4.6 If $X = F$ and $Y = F$, then ϕ is called a *linear function* and the set $L(X, F)$ would be the set of all linear functions from X to F. Familiar examples include linear real-valued functions of a real variable x and complex-valued functions of a complex variable z. The function $\phi(z) = 3z$ is linear, whereas the functions x^2, $\sin z$ and e^x are examples of nonlinear functions. ▲

Example 4.7 If X is a coordinate vector space G and $Y = F$, then ϕ is said to be a *linear form*, or *one-form*. A one-form maps a vector to scalar in F, and this is the essence of an inner product.[13] In this case, $L(G, F)$ is the set of all one-forms relative to vectors in G, and $L(G, F)$ is called the *algebraic dual space* of G.

---

[13] Anti-symmetric combinations of one-forms are the objects of study in *exterior algebra* and *exterior calculus* We will return to this subject in Chap. 7

If $F = \mathbb{C}$, then $L(G, \mathbb{C})$ is the space G^* in Sect. 4.3.2, where the one-form ϕ was written as a bra vector $\langle u|$ that maps $|v\rangle$ to a complex number. If $F = \mathbb{R}$, then $L(G, \mathbb{R})$ is the set of all real-valued row vectors whose column vectors are in G.   ▲

Example 4.8 As an extension of Example 4.7, let G be a vector space where the elements of the space are *functions* and let $Y = F$. In this case, G is said to be a *function space* and ϕ is called a *linear functional*. Then $L(G, F)$ is the set of all linear functionals from G to F, and we call $L(G, F)$ the *functional dual space* of G.   ▲

When it comes to forming inner products on function spaces, linear functionals play the corresponding role that one-forms play in coordinate spaces. Among the many areas of mathematical physics in which we encounter linear functionals are quantum mechanics, probability theory, statistical mechanics, integral transforms and in applications of the special functions of mathematical physics.

Further, if a function space is an inner product space (as are virtually all of the function spaces we encounter in physics), the functions are said to be *square-integrable—* a terminology will become apparent as we proceed [(see Sect. 4.4.2 and Eq. (4.15)].

In practice, and when the context is clear, we tend to drop the adjectives "algebraic" or "functional" and refer instead to $L(X, Y)$ as the *dual space* of X whenever X is a vector space and $Y = F$ is a scalar. Later in this chapter we will expand this definition of a dual space so as to accommodate Cartesian products of X, thereby introducing multi-linear maps, which then lead to the study of tensors.

The linearity of a one-form or functional is not fully apparent until we consider its action on a vector sum. This consideration also reveals an important difference between linear maps in real and complex spaces.

Example 4.9 Let G be a complex vector space with the elements $|u\rangle$, $|v\rangle$ and $|w\rangle \in G$, and let $Y = \mathbb{C}$. Let $a, b \in \mathbb{C}$ be constants. The linearity of the inner product of $|u\rangle$ with the sum $(a|v\rangle + b|w\rangle)$ means that

$$\langle u|(av + bw)\rangle = a\langle u|v\rangle + b\langle u|w\rangle . \tag{4.7}$$

Definition 4.5 makes it clear that the linearity is in the "ket part" of the inner product in Eq. (4.7). However, reversing the order of the terms in the scalar product conjugates the result:

$$\langle (av + bw|u\rangle = a^*\langle v|u\rangle + b^*\langle w|u\rangle . \tag{4.8}$$

In this case the linearity is in the "bra part" of the inner product, but because of the conjugation it is not identical to the linearity in Eq. (4.7).

If Eqs. (4.7) and (4.8) were equal (as they would be in a real function space, where $F(= Y) = \mathbb{R}$), the map would be *bilinear*, i.e., the map would be linear (and in the same way) in *both* the "ket part" and the "bra part" of the inner product. However, in complex vector spaces the two results are conjugates and the map is called a *sesquilinear*, or *hermitian, mapping*.[14]

In physics, we usually refer to the inner product of complex vectors (whether in a coordinate or function space) as a *hermitian inner product*, and a sesquilinear map as a hermitian map. ▲

We summarize the results above in the following definition:

Definition 4.6 Consider a complex vector space G and vectors $|u\rangle$, $|v\rangle$, $|w\rangle \in G$. Let $a, b \in \mathbb{C}$ be constants. The *hermitian inner product* $\langle u|v\rangle$ of two vectors $|u\rangle, |v\rangle \in G$ is defined as a *sesquilinear map* $\phi : |v\rangle \to z \in \mathbb{C}$, where the map $\phi \in L(G, \mathbb{C})$ is a *one-form* or a *functional* depending on whether G is a coordinate space or function space, respectively. We write $L(G, \mathbb{C})$ to refer to the set of all such maps, and $L(G, \mathbb{C})$ forms a vector space G^* which is the dual space of G.

Writing $\phi = \langle u|$, a sesquilinear inner product $z = \langle u|v\rangle$ is one in which

$$\langle u|v\rangle = \langle v|u\rangle^*$$
$$\langle u|(av + bw)\rangle = a\langle u|v\rangle + b\langle u|w\rangle$$
$$\langle (av + bw|u\rangle = a^*\langle v|u\rangle + b^*\langle w|u\rangle .$$

In the special case of real vector spaces, the set of all bilinear maps on the real vector space G forms a dual vector space $L(G, \mathbb{R})$. In addition, we note that a sesquilinear or bilinear map is often described[15] according to whether the resulting inner product is positive or zero.

1. A sesquilinear map ϕ is said to be *positive* if $\langle u|u\rangle \geq 0$ for all $|u\rangle \in G$.
2. A sesquilinear map ϕ is said to be *strictly positive* if (a) it is positive, and (b) $\langle u|u\rangle = 0$ if and only if $|u\rangle = 0$.
3. If ϕ is strictly positive, then the inner product is said to be a *Euclidean* (see the comment following Example 4.5). Otherwise it is said to be *pseudo-Euclidean*.

Colloquially, it is convenient to think of ϕ as a "machine" that accepts a vector as "input" and generates a scalar as "output." ∎

Of course, nothing we have done in this section shows us how to actually *calculate* an inner product in a function space. We take up this topic in Sect. 4.4.

---

[14] The Latin prefix *sesqui*- means "one-half more, half again as much." A sesquilinear map, therefore, is "more than linear," but not so much as to be bilinear. Other terms used to describe this map are *conjugate bilinear* and *hermitian bilinear*.

[15] Terminology varies slightly among authors. We follow the convention in [6], pp. 10–11.

4.4 Orthogonality, Normalization and Complete Sets of Vectors

When we work with function spaces, one of the most important things we must do is identify a complete orthonormal set of vectors, in terms of which any vector in the space may be expressed. This concept is the same as what we considered in coordinate spaces; only the methodology is different, and we need to be clear on precisely what this requirement demands.

Definition 4.7 1. A vector is said to be *normalized* if its magnitude has been rescaled to unity. Typically, this is accomplished by dividing the vector by its norm.
2. Two vectors are said to be *orthogonal* if their inner product is zero.
3. A set of orthogonal vectors where each vector in the set has been normalized is said to form an *orthonormal set* of vectors.
4. A set of orthonormal vectors is said to be a *complete set*[16] if the number of vectors in the set equals the dimension of the space, in which case they form an orthonormal basis for the space (see Sects. 4.2.1 and 4.2.2). ∎

Orthogonality is a trivial concept in one-dimensional spaces, where any vector will be orthogonal only to the zero vector $|0\rangle$. This is apparent when we write $|r\rangle = r|1\rangle$ in $\mathbb{R}^1$, or $|z\rangle = z|1\rangle$ in $\mathbb{C}^1$, and form the inner product with any other vector in the space.

The familiar notion that two vectors in $\mathbb{R}^2$ are orthogonal if they are at 90 degrees to each other—or that a vector can be found that is perpendicular to a given plane in $\mathbb{R}^3$— needs to be generalized to coordinate and function spaces of arbitrary dimension. Before proceeding, however, it should be noted that there *are* instances where an orthonormal set is *not* preferred over some other *skew basis* that better aligns with the physical situation. One such place where this occurs is in crystallography, where directions are more usefully defined along crystal edges rather than in an orthonormal laboratory frame.[17]

4.4.1 Gram-Schmidt Orthogonalization—Coordinate Space

It often happens that we are given a set of non-orthogonal, non-normalized vectors in a space and need to find a set of orthonormal basis vectors; physics students frequently encounter this problem in quantum mechanics. We also need to do this in a way that avoids the use of "angles" in a final definition, as these may be difficult or impossible to imagine in higher-dimensional spaces.

---

[16] A *complete set* of vectors is *not* to be confused with a *complete vector space*. The latter relies on the Cauchy convergence of sequences in the space (see Definition 3.5). We take this up again in the context of Hilbert spaces (Sect. 4.4.4) and topology (Sect. 6.5)

[17] For more on this topic see, for example, [5], Chap. 3.

Fig. 4.3 A two-dimensional space X where we are given two non-orthonormal vectors $|\alpha_0\rangle$ and $|\alpha_1\rangle$, and need to find an orthonormal basis $\{|e_0\rangle, |e_1\rangle\}$ for the space, with no reference to angles

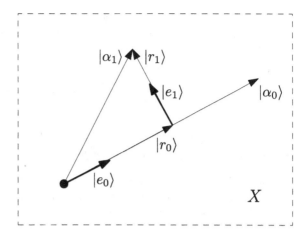

The process for doing this is called the *Gram-Schmidt orthogonalization process*,[18] which we describe in this section for coordinate spaces. In a parallel fashion, we will describe this process for function spaces in Sect. 4.4.3.

Given two non-orthogonal, non-normalized, linearly-independent vectors $|\alpha_0\rangle$ and $|\alpha_1\rangle$ in the two-dimensional space X, the goal is to find the complete orthonormal basis set $\{|e_0\rangle, |e_1\rangle\}$ for X as shown in Fig. 4.3.

First, the normalized basis vector $|e_0\rangle$ is just $|\alpha_0\rangle$ divided by its norm:

$$|e_0\rangle = \frac{|\alpha_0\rangle}{\||\alpha_0\rangle\|}. \tag{4.9}$$

Next, we project $|\alpha_1\rangle$ onto $|e_0\rangle$ to give $|r_0\rangle$, and then subtract $|r_0\rangle$ from $|\alpha_1\rangle$ to get

$$|r_1\rangle \equiv r_1|e_1\rangle = |\alpha_1\rangle - \langle\alpha_1|e_0\rangle|e_0\rangle. \tag{4.10}$$

Because everything on the righthand side of Eq. (4.10) is known, we can solve for the scalar r_1, from which we then have the normalized vector $|e_1\rangle$.

This process can be extrapolated iteratively for higher-dimensional spaces. For example, if the space in Fig. 4.3 were three-dimensional, with some vector $|\alpha_2\rangle$ (not shown in the figure) coming out of the plane, then we would project $|\alpha_2\rangle$ onto *both* $|e_0\rangle$ and $|e_1\rangle$, subtract the result from $|\alpha_2\rangle$ to get

$$|r_2\rangle \equiv r_2|e_2\rangle = |\alpha_2\rangle - \langle\alpha_2|e_0\rangle|e_0\rangle - \langle\alpha_2|e_1\rangle|e_1\rangle, \tag{4.11}$$

and then normalize (find the scalar r_2) to find $|e_2\rangle$.

---

[18] Jörgen Pedersen Gram (1850–1916), Danish number theorist and analyst; Erhard Schmidt (1876–1959), German analyst.

This process may be repeated in stepwise fashion for higher-dimensional spaces so long as we have n linearly-independent vectors with which to carry out the process. Generally, we can write

$$|r_n\rangle \equiv r_n|e_n\rangle = |\alpha_n\rangle - \sum_{i=0}^{n-1} \langle \alpha_n|e_i\rangle |e_i\rangle \tag{4.12}$$

for a specified n. As we find each new vector $|r_m\rangle$ for $m < n$, we normalize it to give $|e_m\rangle$ so that $|e_m\rangle$ may then be used in the next step.

Example 4.10 As an illustration of the Gram-Schmidt process, consider three vectors in Cartesian coordinates. Let $|\alpha_0\rangle = 4\hat{\mathbf{i}} + 3\hat{\mathbf{j}}$; $|\alpha_1\rangle = 3\hat{\mathbf{i}} + 4\hat{\mathbf{j}}$; and $|\alpha_2\rangle = \hat{\mathbf{i}} + \hat{\mathbf{j}} + 2\hat{\mathbf{k}}$. Then $|e_0\rangle$ is found directly from Eq. (4.9) to be

$$|e_0\rangle = \frac{|\alpha_0\rangle}{\||\alpha_0\rangle\|} = \frac{1}{5}(4\hat{\mathbf{i}} + 3\hat{\mathbf{j}}). \tag{4.13}$$

Next we find $|r_1\rangle$ from Eq. (4.10). The scalar product $\langle \alpha_1|e_0\rangle$ is found to be $(24/5)$, and we eventually find

$$|r_1\rangle \equiv r_1|e_1\rangle = -\frac{21}{25}\hat{\mathbf{i}} + \frac{28}{25}\hat{\mathbf{j}},$$

the norm of which is $r_1 = (7/5)$. From this, we obtain

$$|e_1\rangle = \frac{|r_1\rangle}{\||r_1\rangle\|} = \frac{1}{5}(-3\hat{\mathbf{i}} + 4\hat{\mathbf{j}}). \tag{4.14}$$

That the normalized vectors $|e_0\rangle$ and $|e_1\rangle$ are orthogonal is clear by inspection, so they form an orthonormal set in two dimensions.

Finally, we know $|e_2\rangle$ by inspection to be $\hat{\mathbf{k}}$ (why?). Still, following the process, we would use Eq. (4.11) to evaluate two inner products (one of which equals 2 and the other is equal to 1), find $|r_2\rangle$ and then its norm, and solve for $|e_2\rangle$. We leave this step (as well as filling in the detail above) as an exercise.   ▲

4.4.2   Orthonormalization in Function Spaces

As elements in a function space, functions may be normalized like any vector. Applying the framework in Definition 4.6, we can at least formally write an inner product of two functions $\psi(x)$ and $\xi(x)$, both of which are defined in a continuous domain D, as $\langle \psi(x)| : |\xi(x)\rangle \mapsto z \in \mathbb{C}$. The norm $\|\psi(x)\|$ follows from $\langle \psi(x)|\psi(x)\rangle = \psi^*(x)\psi(x) = \|\psi(x)\|^2$, with a similar expression for the norm of $\xi(x)$.

Therefore, we need to adopt a continuum version of the inner product rather than the component version. That said, we do this by first writing functions as n-component vectors and evaluating their inner product as

$$\langle \psi(x) | \xi(x) \rangle = \lim_{n \to \infty} \left(\sum_{i=1}^{n} \psi^*(x_i) \xi(x_i) \right) \Delta x_i,$$

where the functions are defined (albeit crudely) on intervals Δx_i rather than at points x_i. That changes, however, when we let $n \to \infty$ to yield an integral, and the inner product of these two functions, each defined on their common domain D, becomes

$$\langle \psi | \xi \rangle = \int_D w(x) \psi^*(x) \xi(x) \, dx = z \in \mathbb{C}. \tag{4.15}$$

We inserted the function $w(x)$, known as a *weight function*, because this general form for the inner product arises in consideration of the various special functions of mathematical physics (e.g., Legendre, Laguerre, Hermite) whose orthogonality properties are with respect to weight functions that generally are not unity.[19]

Equation 4.15 demonstrates why these functions are said to be *square-integrable*, and normalization of $\psi(x)$ follows directly by requiring

$$\langle \psi | \psi \rangle = \int_D w(x) \psi^*(x) \psi(x) \, dx = 1. \tag{4.16}$$

For well-behaved functions and domains with clear boundary conditions, the methods of Riemann integration[20] serve our needs when evaluating Eqs. (4.15) and (4.16).

Example 4.11 Consider the one-dimensional wave equation,

$$\frac{d^2 \psi}{dx^2} + k^2 x = 0, \tag{4.17}$$

defined for values of x on the closed interval $D = [-L/2, +L/2]$. The boundary conditions are $\psi(-L/2) = \psi(L/2) = 0$ and $\psi'(-L/2) = \psi'(L/2) = 0$, where $\psi'(x) = d\psi(x)/dx$. Physically, the solution is a standing wave, with $\psi(x)$ as the amplitude and with its endpoints fixed at $\pm L/2$. A general solution of Eq. (4.17) is

$$|\psi(x)\rangle = A \cos(kx) + B \sin(kx), \tag{4.18}$$

where A and B are constants of integration to be set by the boundary conditions.

---

[19] The special functions arise as solutions to differential equations that are solved using integrating factors, which then give the weight functions $w(x)$. See, for example, the treatment in [1]

[20] Alternative theories of integration are beyond the scope of this text, but see [20]) as a starting point if you are interested in exploring them.

Inasmuch as the domain of definition is symmetric about $x = 0$, we can identify the general solution as the sum of two parts: (a) an even part, where $|\psi(-x)\rangle = |\psi(x)\rangle$, and (b) an odd part, where $|\psi(-x)\rangle = -|\psi(x)\rangle$. These two parts correspond to the cosine and sine terms, respectively. Applying the boundary conditions, we find

$$|\psi_n(x)\rangle = A \cos(kx), \quad k = \frac{n\pi}{L} \quad \text{for} \quad n = 1, 2, 3... \tag{4.19}$$

as the even solution. The odd solution is found similarly:

$$|\psi_m(x)\rangle = B \sin(kx), \quad k = \frac{m\pi}{L} \quad \text{for} \quad m = 2, 4, 6.... \tag{4.20}$$

Here, k is the wavenumber, which is defined as $k = 2\pi/\lambda$ for wavelength λ.

Normalization follows from Eq. (4.16), with $w(x) = 1$ and $|\psi(x)\rangle = |\psi^*(x)\rangle$ a real-valued function. Upon normalizing we find the amplitudes A and B:

$$\langle \psi_n | \psi_n \rangle = A^2 \int_{-L/2}^{L/2} \cos^2(kx)\, dx = 1 \quad \Rightarrow \quad A = \sqrt{\frac{2}{L}}$$

$$\langle \psi_m | \psi_m \rangle = B^2 \int_{-L/2}^{L/2} \sin^2(kx)\, dx = 1 \quad \Rightarrow \quad B = \sqrt{\frac{2}{L}}. \tag{4.21}$$

Orthogonalization follows from Eq. (4.15), again with $w(x) = 1$. A straightforward calculation shows

$$\langle \psi_m | \psi_n \rangle = \int_{-L/2}^{L/2} \psi_m^*(x)\psi_n(x)\, dx = \delta_{mn}, \tag{4.22}$$

where δ_{mn} is the Kronecker delta, which equals 1 if $m = n$ but is otherwise zero. ▲

Therefore, $|\psi_m(x)\rangle$ and $|\psi_n(x)\rangle$ form a complete orthonormal set of functions on $D = [-L/2, +L/2]$, and as such they form the basis of the *Fourier series* expansion of an arbitrary function $f(x)$ on a closed interval—a discrete sum over wavenumbers, where each wavenumber contributes to the amplitude of $f(x)$. The continuum (integral) version of the Fourier series as $L \to \infty$ is the *Fourier transform*.

4.4.3 Gram-Schmidt Orthogonalization—Function Space

The logic of the Gram-Schmidt orthogonalization process for a coordinate space as described in Sect. 4.4.1 is the same for function spaces. The key results were those contained in Eqs. (4.9)–(4.12).

Therefore, articulating the Gram-Schmidt process in a function space becomes an exercise in symbol-switching, and Table 4.1 helps us make that notational transition from coordinate spaces to function spaces.

The results are:

$$|u_0\rangle = \frac{|\psi_0\rangle}{\||\psi_0\rangle\|} \,, \tag{4.23}$$

$$|\phi_1\rangle \equiv \phi_1|u_1\rangle = |\psi_1\rangle - \langle\psi_1|u_0\rangle|u_0\rangle \,, \tag{4.24}$$

$$|\phi_2\rangle \equiv \phi_2|u_2\rangle = |\psi_2\rangle - \langle\psi_2|u_0\rangle|u_0\rangle - \langle\psi_2|u_1\rangle|u_1\rangle \,, \tag{4.25}$$

$$|\phi_n\rangle \equiv \phi_n|u_n\rangle = |\psi_n\rangle - \sum_{i=0}^{n-1}\langle\psi_n|u_i\rangle|u_i\rangle \,. \tag{4.26}$$

Here the inner products are to be evaluated via integration over the domain of the functions that comprise the function space rather than by a simple component-by-component evaluation as before. Also, we make it a point to write the integrals as though the functions are complex, knowing that we can make the simplification later should they be real.

Example 4.12 Let a set of vectors be given as the set of linearly-independent functions $|\psi_n(x)\rangle = x^n$ for $n = 0, 1, 2, ...$, where the domain is the closed interval $D = [-1, 1]$. We wish to find a complete set of orthonormal basis functions for $w(x) = 1$.

For $n = 0$ and $|\psi_0\rangle = 1$, we can find $|u_0\rangle$ directly from Eq. (4.23) using:

$$\||\psi_0\rangle\|^2 = \langle\psi_0|\psi_0\rangle = \int_{-1}^{1} \psi_0^*\psi_0 \, dx = 2 \quad \Rightarrow \quad |u_0\rangle = \frac{1}{\sqrt{2}} \,.$$

Table 4.1 Comparison of Gram-Schmidt orthogonalization procedures[a]

Coordinate space (Fig. 4.3)	LI	LI + OR	LI + OR + N	Function space
$|\alpha\rangle$	X			$|\psi\rangle$
$|r\rangle$	X	X		$|\phi\rangle$
$|e\rangle$	X	X	X	$|u\rangle$

[a] Vectors (functions) whose labels denote whether they are *LI* linearly independent; *OR* orthogonal; or *N* normalized

For $n = 1$ and $|\psi_1\rangle = x$, we apply Eq. (4.24) and find

$$|\phi_1\rangle \equiv \phi_1|u_1\rangle = x - \left[\int_{-1}^{1} \psi_1^* u_0\ dx\right]|u_0\rangle \quad \Rightarrow \quad |\phi_1\rangle = x .$$

Then we normalize to find $|u_1\rangle$:

$$\||\phi_1\rangle\|^2 = \langle\phi_1|\phi_1\rangle = \int_{-1}^{1} \phi_1^* \phi_1\ dx = \frac{2}{3} \quad \Rightarrow \quad |u_1\rangle = \sqrt{\frac{3}{2}}\, x .$$

For $n = 2$ and $|\psi_2\rangle = x^2$, we apply Eq. (4.25) and evaluate

$$|\phi_2\rangle \equiv \phi_2|u_2\rangle = x^2 - \left[\int_{-1}^{1} \psi_2^* u_0\ dx\right]|u_0\rangle - \left[\int_{-1}^{1} \psi_2^* u_1\ dx\right]|u_1\rangle ,$$

from which we find

$$|\phi_2\rangle = x^2 - \frac{1}{3} .$$

Then we normalize to find $|u_2\rangle$:

$$\||\phi_2\rangle\|^2 = \langle\phi_2|\phi_2\rangle = \int_{-1}^{1} \phi_2^* \phi_2\ dx = \frac{8}{45} \quad \Rightarrow \quad |u_2\rangle = \sqrt{\frac{5}{2}} \cdot \frac{1}{2}\left(3x^2 - 1\right) .$$

This process continues indefinitely, and we can express the orthonormal functions $|u_n\rangle$ in terms of the *Legendre polynomials* $P_n(x)$ according to

$$|u_n\rangle = \left(\frac{2n+1}{2}\right)^{1/2} P_n(x) .$$

While the $|u_n\rangle$ are orthonormal, the $P_n(x)$ are orthogonal but not normalized because of the coefficient $\sqrt{(2n+1)/2}$. On occasion you will find definitions of the Legendre polynomials and other special functions where their corresponding coefficients are embedded in the function, but more often that is not the case.   ▲

Physics students are most likely to first encounter the Legendre polynomials $P_n(x)$ (albeit with an angular argument, e.g., $P_n(\cos\theta)$) in introductory quantum mechanics when solving Schrodinger's equation in spherical coordinates, such as for the hydrogen atom. Schrodinger's equation in spherical coordinates can be separated into a radial equation and an angular equation, and the associated Legendre polynomials (a close cousin of the $P_n(\cos\theta)$)) are one factor in the *spherical harmonics* that solve the angular equation.

Conceptually, these spherical harmonics represent standing waves on a spherical surface, and in this sense they bring us back full circle to the linear standing waves in Example 4.12. Because they form a complete orthonormal set, spherical harmonics can serve to represent arbitrary functions on the surface of a sphere. Any vibrating spherical surface, whether that of a bench-top-scale metal sphere or the surface of the Sun, lends itself to a description via spherical harmonics. They also find applications wherever it is necessary to model observational data on a spherical surface.

A detailed study of the special functions of mathematical physics and the closely related topic of integral transforms is not our purpose here. However, some additional examples of special functions are given in the problems at the end of the chapter, and we give a brief overview of *integral transforms* in Sect. 8.6.

4.4.4 Hilbert Spaces

Most of the vector spaces we encounter in physics are Hilbert spaces, and the goal of this short section is to summarize their defining characteristics and place them in the context of the inner product function spaces we have been discussing. Thorough treatments may be found in the cited references.

Just as the limit of a convergent sequence of rational numbers may lie outside of the set $\mathbb{Q}$, so in principle may a sequence of vectors in a vector space G converge to a vector not in G. In such a case, as with the rational numbers, we would say the vector space is not complete. If, however, every convergent sequence of vectors in G converges to a vector in G, then G is said to be a *complete vector space* in the same Cauchy-convergent sense as the real numbers are complete.[21] A complete normed vector space (Sect. 4.2.3 and Definition 4.3) is called a *Banach space*. If, in addition, an inner product is defined on the space, the space is called a *Hilbert space*.[22]

Definition 4.8 A *Hilbert space* $\mathcal{H}$ is an inner product space that is complete with respect to a specified norm.[23] ■

Physics students who have studied quantum mechanics often tend to think of Hilbert spaces as necessarily complex and infinite-dimensional, but this is not the case. A Hilbert space may be either a complex or real inner product space—the latter merely being a special case of the former. In addition, a Hilbert space may be of finite or infinite dimension. Indeed, a finite-dimensional inner product space with a

---

[21] See Sect. 3.4.1 and Definition 3.5

[22] If the inner product is strictly positive (Definition 4.6) the space is often called a *pre-Hilbert space*.

[23] A comprehensive introduction to Hilbert spaces is [7], and a rigorous (and very readable) account is given in the classic [10]. See also [19], Sect. II.7, and [6], Sect. 13.

norm defined by a metric is necessarily complete, and this describes essentially all of the vector spaces we encounter in elementary physics.

For function spaces the presence of an inner product means the functions are *square-integrable*. We encountered Hilbert function spaces in Examples 4.11 and 4.12 as infinite-dimensional vector spaces, each with a complete set of real, square-integrable basis functions over a closed interval. In physics, our first encounter with complex Hilbert spaces is typically in a first course in quantum mechanics, usually while solving Schrodinger's equation subject to specified boundary conditions.

The phrase "complete set of ...basis functions" used in the previous paragraph marks a different use of the word "complete" than in the Cauchy sense in Definition 4.8; the specification of a basis set is not the same as defining a convergent sequence. However, the two concepts merge in the context of finite-dimensional Hilbert spaces as we assess whether a particular sequence of vectors converges.

Two Hilbert spaces are isomorphic if their scalars are from the same field, their orthonormal bases have the same number of vectors and they share the same norm. This is straightforward for finite-dimensional spaces, but a full and careful treatment of the theory of infinite-dimensional Hilbert spaces is beyond the scope of this text; for this the Appendix in [11] is recommended.

4.5   Subspaces, Sums, and Products of Vector Spaces

Continuing with the pattern set in the previous chapters, this section introduces vector space substructures and products; of the latter, the tensor product is the one we focus on here. We also parse the distinctions between unions, sums and direct sums of vector spaces. Quotient structures are discussed in Sect. 4.6.

4.5.1   Vector Subspaces

Qualitatively, the principal consideration regarding subspaces is analogous to that for other algebraic structures, namely, that not every sub*set* of a space will be a sub*space*; the subset must satisfy the vector space axioms.

Definition 4.9 Given a vector space Σ over the field F, a subset Σ_0 is a *vector subspace* if for every $|u\rangle, |v\rangle \in \Sigma_0$ and $a, b \in F$ the linear combination $a|u\rangle + b|v\rangle$ is also in Σ_0.   ∎

The subspace Σ_0 must be closed with respect to the same operations—and be defined on the same field—as the space itself.

Among the other requirements we note particularly (for the benefit of our subsequent discussion of vector space sums) that the subspace must contain the zero vector. This requirement is implicit given the provision that any linear combination of vectors in the subspace must likewise be in the subspace; the linear combination could, in principle, yield the zero vector.

Examples of vector spaces were given in Sect. 4.1, from which a few examples of vector subspaces come to mind.

Example 4.13

1. Given that $\Sigma = \mathbb{C}$ over $\mathbb{C}$ (the *additive abelian group* $\mathbb{C}$ over the *field* $\mathbb{C}$) is a vector space, then $\Sigma_0 = \mathbb{C}$ over $\mathbb{R}$ is a subspace of Σ.
2. Following on (1), $\mathbb{R}$ over $\mathbb{R}$ (the real line) is a subspace of Σ and also of Σ_0.
3. A *finite* line segment in $\mathbb{R}^2$ (even one that contains the zero vector $|u\rangle = (0, 0)$) is *not* a vector space because it is not closed under addition.
4. Let the vector space Σ be the real Cartesian plane with vectors of the form $|u\rangle = (x, y)$. The x-axis and the y-axis are subspaces of Σ.
5. Continuing with (4), any infinite straight line that passes through the origin is a subspace of Σ because the zero vector $|0\rangle = (0, 0)$ is included. However, an arbitrary line that does not contain the origin is not a subspace of Σ. ▲

The two trivial subspaces of a vector space are the set containing only the zero vector $\{|0\rangle\}$, and the space itself. The concepts of linear independence, basis and dimension for subspaces carry over directly from what was described for spaces generally in Sect. 4.2.2. Clearly, if n is the dimension of a finite-dimensional space Σ, then a subspace Σ_0 will have dimension $m \leq n$.

4.5.2 Unions, Sums and Direct Sums of Vector Spaces

The union, sum and direct sum of two vector spaces may, in some very special circumstances, be the same thing. Generally, however, they are not, and it is important to know the differences among these three similarly-sounding composite structures.

Union ($\cup$) and Sum ($+$). The *union* of two vector spaces is, of course, the union of two *sets*. However, the union of two vector spaces is *not* generally a vector *space*; this is apparent from the fact that the sum of two vectors, with one vector from each space, may be (and very often is) outside the union. In this case, the union would not be closed under addition.

A simple example of this is shown in Fig. 4.4. Consider the two-dimensional vector space X with zero vector $|0\rangle_x$. Within X lie two one-dimensional subsets (the lines U and V), both of which are vector spaces in their own right with their own separate zero vectors. However, neither U nor V shares its zero vector with X, and therefore neither space taken alone is a subspace of X.

The union $W = U \cup V$ is the set of all vectors that lie in one line (vector space) or the other, or both. Nothing outside U and V is included in their union, and therefore W is not closed under vector addition. Therefore, even though U and V are vector

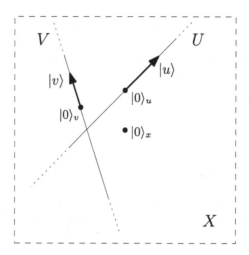

Fig. 4.4 Two vector spaces U and V with vectors $|u\rangle \in U$ and $|v\rangle \in V$. Both U and V are sub*sets*, but neither is a sub*space*, of X. The union $W = U \cup V$ is a set but not a vector space since the sum (the addition operation must be the same as that in X) of two vectors, one from each space, is in neither U nor V

spaces, their union W is not a vector space, and consequently the union of vector spaces is not a widely-used concept in mathematical physics.[24]

Referring to Fig. 4.4, the *sum* of two vector spaces U and V is equivalent to the Cartesian product of the two spaces. As such, the sum (usually written in this context as $W = U + V$ rather than $W = U \times V$) is the set of all ordered pairs of vectors in X, where one member of the pair is in U and the other is in V.

However, the sum is *not necessarily* a vector space. For example, in Fig. 4.4 the spaces U and V do not share a common zero vector, making the zero vector — an essential part of the definition of a vector space—ambiguous. Nonetheless, some references will refer to the sum as a vector space, and this happens when either the presumption is made (or the context is specified) that there is an unambiguous zero vector. These issues are made more clear with the direct sum.

Direct Sum ($\oplus$). Although the sum $W = U + V$ is not necessarily a vector space, it is a simple matter to define a slightly modified summation operation whereby vector spaces may be summed to give other vector spaces. This is accomplished with the *direct sum*.

The direct sum $W = U \oplus V$ of two vector spaces U and V is the sum $U + V$ with the additional constraint that $U \cap V = |0\rangle$; that is, the two spaces have only one vector in common, and that is the zero vector. Then (a) W is a vector space; (b) U and V are now subspaces of W (all three spaces share the same zero vector); and (c) U and V are said to be *complements* of each other with respect to W. Further, every

---

[24]This should sound familiar; the union of two groups is not necessarily a group (Example 2.8 in Sect. 2.5). However, unions of more general (i.e., not vector) spaces are important in topology (Chap. 6)

Fig. 4.5 Two vector spaces U and V with vectors $|u\rangle \in U$ and $|v\rangle \in V$. The two spaces intersect only at their corresponding zero vectors. Their direct sum $W = U \oplus V$ is a vector space, and as depicted here is a trivial subspace of X

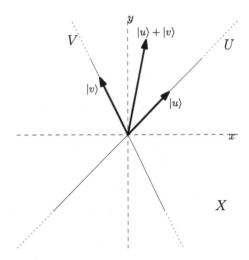

vector in W may be written in a unique way as $|w\rangle = |u\rangle + |v\rangle$, with $|u\rangle \in U$ and $|v\rangle \in V$. The dimension of W is given as dim $W = $ dim $U + $ dim V.

Definition 4.10 Consider two vector spaces U and V defined over a field F, with vectors $|u\rangle \in U$ and $|v\rangle \in V$. Further, let $a \in F$. The *direct sum* $W = U \oplus V$ is defined such that:

1. $W = U \times V = \{(|u\rangle, |v\rangle) : |u\rangle \in U, |v\rangle \in V\}$;
2. $U \cap V = |0\rangle$;
3. $a(|u\rangle, |v\rangle) = (a|u\rangle, a|v\rangle)$;
4. $(|u_1\rangle, |v_1\rangle) + (|u_2\rangle, |v_2\rangle) = (|u_1\rangle + |u_2\rangle, |v_1\rangle + |v_2\rangle)$.

The direct sum $W = U \oplus V$ is a vector space (Fig. 4.5). ∎

A finite-dimensional direct sum is often called a *direct product* and is denoted with the same $\oplus$ symbol. Consequently, the $\times$ symbol is sometimes used when the direct product terminology is employed.[25]

As we have seen, it is a separate question as to (a) whether a set is a vector space, and, if so, then (b) whether that space is then also a subspace of some other space. In our two-dimensional example (Fig. 4.5), $W = U \oplus V$ is shown to be a trivial subspace of X since $W = X$. Figure 4.5 also illustrates how the direct sum $W = U \oplus V$ does not depend on U and V being orthogonal; there is no orthogonality requirement in the definition of the direct sum.

Finally (and at the risk of stating the obvious), it is important to remember that the direct sum is a sum of spaces, not vectors. The ordered pair shown in Definition 4.10 is an ordered pair of vectors from *different* vector spaces of arbitrary dimension,

---

[25] See the comments above regarding the sum $W = U + V$. As with all terms and symbols, it is always a good idea to double check the definitions used (and the conditions assumed) in any given text or article.

and the dimensions of the two spaces may differ. For example, if $U = \mathbb{R}^2$ is a two-dimensional Euclidean plane and $V = \mathbb{R}^1$, then the direct sum $W = U \oplus V$ is the three-dimensional vector space $\mathbb{R}^3 = \mathbb{R}^1 \oplus \mathbb{R}^1 \oplus \mathbb{R}^1$.

4.5.3   Tensors and Tensor Spaces

Vector spaces may be multiplied as well as summed, and the most straightforward way of doing this is to take a vector from each space and combine them together in some consistent way. This operation, which we define below, is called the *tensor product*, and the new combination is called a *tensor* **t**. Repeating this process for all combinations of vectors between or among the vector spaces gives a set of tensors, which together describe a *tensor space T*. Tensors have both algebraic and differential properties, and we briefly examine the former in this and the following section. A more thorough treatment is given in Chap. 7.

For physicists, tensors are important mathematical structures because of how they transform under coordinate transformations. We know how a coordinate transformation can alter the outward appearance of an equation, even though its underlying meaning, perhaps a fundamental physical principle, remains unchanged.[26] Ideally, we would like a way of writing equations in a way that preserves the inherent properties of the system under study, irrespective of our *ad hoc* choice of coordinates.

Tensors are important to mathematical physics for precisely this reason; when differential equations are written in tensor format, they remain "form-invariant" under general coordinate transformations.[27]

We will focus our attention on tensors formed by drawing vectors from multiple copies of the same vector space. For example, if we have the Cartesian product $X \times X$ of the vector space X, then one vector is drawn from each factor in the product. The tensor is then said to be a tensor of *rank* two (or, a second-rank tensor) on X, because two vector spaces are used in its construction. If instead we used the Cartesian product $X^1 \times X^2 \times \cdots X^n$, the result would be an *nth*-rank tensor.

The same tensor product operation applies to one-forms, yielding *dual tensors* as elements of *dual tensor spaces*. We will have a bit more to say about these structures in this and the following section, but we will return to them more fully when we discuss antisymmetric tensors and differential forms in Sects. 7.3 and 7.4.

Up until now we have been speaking about tensors in broad generalities, and it is time to examine more precisely just *how* they are actually constructed from vector

---

[26] Just think of Newton's second law in Cartesian vs. spherical coordinates—the same physics, but two very different appearances.

[27] Relatedly, a particular choice of a coordinate system can cause one to conclude the presence or absence of a particular phenomenon, when in fact the "phenomenon" is nothing more than a coordinate effect. A famous example of this pertains to the event horizon of a black hole. For many years it was believed that a singularity occurred at the event horizon (within which nothing is visible to an outside observer). In fact, the event horizon "singularity" was just a coordinate effect, and the only "real" singularity is at the center of the black hole.

spaces. We start with a definition of the properties of the tensor product, where expressions like $\mathbf{u} \otimes \mathbf{v}$ are to be read as "$\mathbf{u}$ tensor $\mathbf{v}$."

Definition 4.11 Let $\mathbf{t}$, $\mathbf{u}$ and $\mathbf{v}$ be tensors and let c be a scalar. The *tensor product* is

1. Associative: $\mathbf{t} \otimes (\mathbf{u} \otimes \mathbf{v}) = (\mathbf{t} \otimes \mathbf{u}) \otimes \mathbf{v}$;
2. Homogeneous: $(c\mathbf{t}) \otimes \mathbf{u} = c(\mathbf{t} \otimes \mathbf{u}) = \mathbf{t} \otimes (c\mathbf{u})$;
3. Distributive (if $\mathbf{t}$ and $\mathbf{u}$ have the same order - see below): $(\mathbf{t} + \mathbf{u}) \otimes \mathbf{v} = \mathbf{t} \otimes \mathbf{v} + \mathbf{u} \otimes \mathbf{v}$.

It is very important to notice that commutativity is *not* a defining characteristic of the tensor product.[28]   ∎

The structure of a tensor space becomes apparent when we form the tensor product of two vectors, $\mathbf{u} \in X$ and $\mathbf{v} \in X$. Let

$$|u\rangle \equiv \mathbf{u} = u^1 \hat{\mathbf{e}}_1 + u^2 \hat{\mathbf{e}}_2 + u^3 \hat{\mathbf{e}}_3 = \sum_{i=1}^{3} u^i \hat{\mathbf{e}}_i \equiv u^i \hat{\mathbf{e}}_i \qquad (4.27)$$

$$|v\rangle \equiv \mathbf{v} = v^1 \hat{\mathbf{e}}_1 + v^2 \hat{\mathbf{e}}_2 + v^3 \hat{\mathbf{e}}_3 = \sum_{j=1}^{3} v^j \hat{\mathbf{e}}_j \equiv v^j \hat{\mathbf{e}}_j \,, \qquad (4.28)$$

where we have adopted the *Einstein summation convention*[29] as a shorthand notation. The fact that $\mathbf{u}$ and $\mathbf{v}$ are drawn from different factors in the Cartesian product $X \times X$ is indicated by their different summation indices.

Applying Definition 4.11 we expand the tensor product $\mathbf{t} = \mathbf{u} \otimes \mathbf{v}$ to yield

$$\begin{aligned}
\mathbf{t} = \mathbf{u} \otimes \mathbf{v} &= (u^1 \hat{\mathbf{e}}_1 + u^2 \hat{\mathbf{e}}_2 + u^3 \hat{\mathbf{e}}_3) \otimes (v^1 \hat{\mathbf{e}}_1 + v^2 \hat{\mathbf{e}}_2 + v^3 \hat{\mathbf{e}}_3) \\
&= u^1 v^1 (\hat{\mathbf{e}}_1 \otimes \hat{\mathbf{e}}_1) + u^1 v^2 (\hat{\mathbf{e}}_1 \otimes \hat{\mathbf{e}}_2) + u^1 v^3 (\hat{\mathbf{e}}_1 \otimes \hat{\mathbf{e}}_3) \\
&\quad + u^2 v^1 (\hat{\mathbf{e}}_2 \otimes \hat{\mathbf{e}}_1) + u^2 v^2 (\hat{\mathbf{e}}_2 \otimes \hat{\mathbf{e}}_2) + u^2 v^3 (\hat{\mathbf{e}}_2 \otimes \hat{\mathbf{e}}_3) \\
&\quad + u^3 v^1 (\hat{\mathbf{e}}_3 \otimes \hat{\mathbf{e}}_1) + u^3 v^2 (\hat{\mathbf{e}}_3 \otimes \hat{\mathbf{e}}_2) + u^3 v^3 (\hat{\mathbf{e}}_3 \otimes \hat{\mathbf{e}}_3)
\end{aligned}$$

---

[28]Further, the tensor product $\mathbf{u} \otimes \mathbf{v}$ should *not* be confused with the more familiar cross-product ($\times$) of two vectors. For one thing, the associative property does not hold for the cross-product; this is something you may already know from your earlier study of vectors, but we will see this when we discuss algebras in Chap. 5. More fundamentally, and in terms of structure, the cross-product $\mathbf{w} = \mathbf{u} \times \mathbf{v}$ yields a vector, with all three vectors being in the same vector space. This is to be contrasted with the tensor product, which yields tensors defined in a different (tensor) space from the vector spaces involved in the construction.

[29]For ket vectors, write the basis vector index as a subscript ($\hat{\mathbf{e}}_i$) and the component index as a superscript (u^i). Bra vectors use the opposite positional pattern for indices. The summation convention specifies that we sum over *repeated* indices, but only when one index is "up" and the other is "down." When applied "internally" to a single term (S_i^{ij}) it is called a *contraction* of tensor indices. We have used this convention previously in the text without calling it as such.

or

$$\mathbf{t} = \mathbf{u} \otimes \mathbf{v} = u^i v^j (\hat{\mathbf{e}}_i \otimes \hat{\mathbf{e}}_j) \equiv t^{ij} (\hat{\mathbf{e}}_i \otimes \hat{\mathbf{e}}_j) \in T \; , \qquad (4.29)$$

where the t^{ij} are the components of the tensor $\mathbf{t}$ in a tensor space T. The *basis tensors* in T are represented as $(\hat{\mathbf{e}}_i \otimes \hat{\mathbf{e}}_j)$. Conceptually, basis tensors are to a tensor space as basis vectors are to a vector space—they provide a means by which any tensor in that space may be expressed.

The tensor space T in Eq. (4.29) is nine-dimensional, and generally we can see that the tensor product of two vector spaces, each of dimension m, will give an m^2-dimensional tensor space whose elements are second-rank tensors (this contrasts with the direct sum ($\oplus$) of two spaces, where the *vector* space would have $m + m = 2m$ dimensions). More generally, if we considered n factors in the Cartesian product, with each vector space of dimension m, then the tensor space of nth-rank tensors would have m^n dimensions.

We now turn our attention to *dual tensors* and *dual tensor spaces*, which are multilinear versions of the maps and dual spaces discussed in Sect. 4.3.3. We can start by writing two one-forms as

$$\langle \alpha | \equiv \boldsymbol{\alpha} = u_1 \hat{\mathbf{e}}^1 + u_2 \hat{\mathbf{e}}^2 + u_3 \hat{\mathbf{e}}^3 = \sum_{i=1}^{3} u_i \hat{\mathbf{e}}^i \equiv u_i \hat{\mathbf{e}}^i$$

$$\langle \beta | \equiv \boldsymbol{\beta} = v_1 \hat{\mathbf{e}}^1 + v_2 \hat{\mathbf{e}}^2 + v_3 \hat{\mathbf{e}}^3 = \sum_{j=1}^{3} v_j \hat{\mathbf{e}}^j \equiv v_j \hat{\mathbf{e}}^j.$$

As we recall, a one-form maps a vector to a scalar and belongs to a vector space that is dual to the original space. If X is a vector space with $|v\rangle \in X$ and $\langle u |$ is an element of the dual space $X^* = L(X, \mathbb{R} \; or \; \mathbb{C})$, then their scalar product is $\langle u | v \rangle$; the asterisk (*) in X^* identifies the space as being dual to X even for real vector spaces. Analogously, an *nth*-rank dual tensor belongs to a dual tensor space T^* and maps an nth-rank tensor in T to a scalar.

We can form a second-rank dual tensor by applying the same method we used to derive Eq. (4.29), and we obtain

$$\mathbf{t} = \boldsymbol{\alpha} \otimes \boldsymbol{\beta} = \alpha_i \beta_j (\hat{\mathbf{e}}^i \otimes \hat{\mathbf{e}}^j) = t_{ij} (\hat{\mathbf{e}}^i \otimes \hat{\mathbf{e}}^j) \in T^* \; . \qquad (4.30)$$

Other combinations can lead to a variety of tensors and tensor spaces. Assuming the underlying vector spaces are three-dimensional, we may form a nine-dimensional space containing the tensor

$$\mathbf{t} = \boldsymbol{\alpha} \otimes \mathbf{u} = \alpha_j u^i (\hat{\mathbf{e}}^j \otimes \hat{\mathbf{e}}_i) = t^i_j (\hat{\mathbf{e}}^j \otimes \hat{\mathbf{e}}_i) \; , \qquad (4.31)$$

which differs from those in Eqs. (4.29) and (4.30) even though all three tensors are of the same rank. We can also form the 27-dimensional space containing the tensor

$$\mathbf{t} = \boldsymbol{\alpha} \otimes \mathbf{u} \otimes \mathbf{v} = \alpha_j u^i v^k (\hat{\mathbf{e}}^j \otimes \hat{\mathbf{e}}_i \otimes \hat{\mathbf{e}}_k) = t_j^{ik} (\hat{\mathbf{e}}^j \otimes \hat{\mathbf{e}}_i \otimes \hat{\mathbf{e}}_k) . \qquad (4.32)$$

The pattern of the indices in the tensor components invites a nomenclature that is related to the underlying spaces comprising the tensor space. The number of *non-repeating* indices in the expression for a tensor component is the tensor's *rank*. This is equivalent to the number of vector and dual spaces used in the construction of the tensor, after any contractions are carried out. A zeroth-rank tensor, i.e., a "tensor" that is "constructed" from zero vector spaces and zero dual spaces, is a scalar. When working with tensors, scalar quantities often arise when all tensor indices appear in pairs and we perform pairwise contractions (summations) on each such pair.

Each of the expressions t^{ij}, t_{ij} and t_j^i represents components of a second-rank tensor. However, these tensors differ from one another as reflected superficially in the location of their indices, and more substantively in the structure of the basis tensors in each space. We describe these differences by referring to the *order* of a tensor, which distinguishes between the number of vector and dual spaces involved in the construction of the tensor space.

The number of upper indices is called the *contravariant order* of the tensor and indicates the number of vector spaces involved in the construction, and the number of lower indices is its *covariant order* and tells us the number of dual spaces involved.[30] For example, t_j^{ik} would represent the components of a third-rank tensor, of contravariant order two and (by implication) covariant order one.

Another frequently used notation describes the second-rank tensors t^{ij} as a tensor of type $\begin{pmatrix} 2 \\ 0 \end{pmatrix}$, t_{ij} as a tensor of type $\begin{pmatrix} 0 \\ 2 \end{pmatrix}$ and t_j^i as a *mixed* tensor of type $\begin{pmatrix} 1 \\ 1 \end{pmatrix}$. The tensor t_k^{ij} in Eq. (4.32) is a third-rank mixed tensor of type $\begin{pmatrix} 2 \\ 1 \end{pmatrix}$. Similarly, a tensor of type $\begin{pmatrix} 1 \\ 0 \end{pmatrix}$ is just a vector, and a tensor of type $\begin{pmatrix} 0 \\ 1 \end{pmatrix}$ is a one-form. More generally, a tensor of type $\begin{pmatrix} m \\ n \end{pmatrix}$ is the tensor product of m vector spaces and n dual spaces.

We know that one-forms map vectors—and vectors map one-forms—to scalars. Therefore, a tensor of type $\begin{pmatrix} m \\ n \end{pmatrix}$ may be thought of as one which maps m one-forms and n vectors to a scalar. One of the more important examples is the metric tensor, a tensor of type $\begin{pmatrix} 0 \\ 2 \end{pmatrix}$, which maps two vectors to a scalar. We discuss the metric tensor and show a few examples of tensor algebra in the following section. Many more examples will be given in our discussion of differential forms in Chap. 7.

---

[30] The origin of these terms relates to how tensor components transform under coordinate transformations. We will return to this topic in Chap. 7

4.5.4   Metric and Associated Tensors

Perhaps more than any other *algebraic* structure, a *metric tensor* describes the *geometry* of a space.[31] It does this in several ways—most elaborately by being an important piece of the curvature tensor for that space (a topic we will not discuss in this text), but more simply by inducing an inner product on the space.

Formally, we may write the metric tensor as $\mathbf{g} = g_{ik}(\hat{\mathbf{e}}^i \otimes \hat{\mathbf{e}}^k)$, a second-rank covariant tensor that is constructed from two dual spaces and therefore has the capacity to map an ordered pair of vectors to a scalar:

$$\mathbf{g} : (\mathbf{u}, \mathbf{v}) \rightarrow \mathbb{R} .$$

If we write the two vectors as differential displacements, i.e., as $dx^i\hat{\mathbf{e}}_i$ and $dx^j\hat{\mathbf{e}}_j$, the resulting scalar quantity[32] is the squared length of a differential line element:

$$(ds)^2 = g_{ik}dx^i dx^k , \tag{4.33}$$

where the sum is taken over all values of the repeated indices i and k.

For a three-dimensional space X, Eq. (4.33) is a sum of nine terms:

$$\begin{aligned}(ds)^2 = {} & g_{11}dx^1 dx^1 + g_{12}dx^1 dx^2 + g_{13}dx^1 dx^3 \\ & + g_{21}dx^2 dx^1 + g_{22}dx^2 dx^2 + g_{23}dx^2 dx^3 \\ & + g_{31}dx^3 dx^1 + g_{32}dx^3 dx^2 + g_{33}dx^3 dx^3.\end{aligned} \tag{4.34}$$

However, in all of the most familiar physical applications the metric is diagonal—not just symmetric, with $g_{ik} = g_{ki}$, but diagonal—so that the only non-zero terms in Eq. (4.34) are those where $i = k$.

Consequently, Eq. (4.34) reduces to

$$(ds)^2 = g_{11}dx^1 dx^1 + g_{22}dx^2 dx^2 + g_{33}dx^3 dx^3 . \tag{4.35}$$

If we are given a metric on a space, we can find the expression for the differential displacement on that space. Before examining examples of metrics themselves and seeing how the differential displacements are found, let us first review a few examples of displacements that you will recognize.

Perhaps the simplest is that for a three-dimensional Euclidean space expressed in terms of Cartesian coordinates ($x^1 = x$, $x^2 = y$ and $x^3 = z$), where the differential displacement is

$$(ds)^2 = (dx)^2 + (dy)^2 + (dz)^2 . \tag{4.36}$$

---

[31] Metric spaces are considered in a topological context in Sect. 6.4.2

[32] How the basis tensors combine in this and other situations will be examined more closely in Chap. 7. Here, we lose nothing by focusing solely on the components.

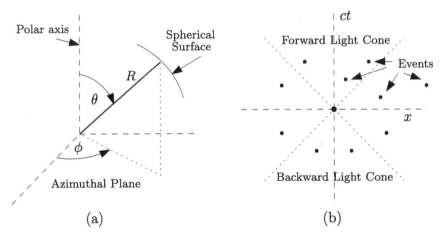

Fig. 4.6 **a** Spherical coordinates, and **b** spacetime coordinates in one spatial dimension

Among other familiar examples (FIg. 4.6) is the differential displacement on the surface of a sphere ($x^1 = r$, $x^2 = \theta$ and $x^3 = \phi$; on the surface, omit x^1),

$$(ds)^2 = (R\,d\theta)^2 + (R\sin\theta\,d\phi)^2 \ , \tag{4.37}$$

where θ is the polar (or zenith) angle and ϕ is the azimuthal angle as measured around the polar axis. If you have studied special relativity ($x^0 = t$, $x^1 = x$, $x^2 = y$, $x^3 = z$), you will recognize the differential spacetime displacement to be

$$(ds)^2 = (cdt)^2 - (dx)^2 - (dy)^2 - (dz)^2 \ , \tag{4.38}$$

where $(ds)^2$ represents the square of the spacetime interval between two events.

These expressions for differential displacements come about through the inner product that is defined on each space. If we are given a vector, we need to first find its corresponding, or "associated," one-form. The inner product of a vector with its associated one-form yields the square of the vector's magnitude.

Consider the example of a vector $|u\rangle$ in a three-dimensional space. We obtain its associated one-form by acting on the vector with the metric tensor,

$$u_i = g_{ik}u^k = (g_{11}u^1, g_{22}u^2, g_{33}u^3) \equiv (u_1, u_2, u_3) \ . \tag{4.39}$$

Generally, any two tensors—not just vectors and one-forms—are called *associated tensors* if they are related to one another through the action of a metric tensor.

In our example, the inner product $\langle u|u \rangle$ may now then be written variously as

$$\langle u|u \rangle = u_i u^i = (u_1 \; u_2 \; u_3) \begin{pmatrix} u^1 \\ u^2 \\ u^3 \end{pmatrix} = g_{11} u^1 u^1 + g_{22} u^2 u^2 + g_{33} u^3 u^3 = |u|^2 \,. \quad (4.40)$$

We can now examine the metrics that pertain to the examples of displacements in Eqs. (4.36)–(4.38). First, we have the Euclidean metric associated with three-dimensional Cartesian coordinates,

$$g_{ik} = \begin{pmatrix} 1 & 0 & 0 \\ 0 & 1 & 0 \\ 0 & 0 & 1 \end{pmatrix} \Rightarrow dx_i = (dx, dy, dz) \,. \quad (4.41)$$

Next is the non-Euclidean metric associated with the geometry on the surface of a sphere,

$$g_{ik} = \begin{pmatrix} R^2 & 0 \\ 0 & R^2 \sin^2 \theta \end{pmatrix} \Rightarrow dx_i = (R^2 d\theta, R^2 \sin^2 \theta \; d\phi) \,. \quad (4.42)$$

Finally, we have the four-dimensional, pseudo-Euclidean *Minkowski metric* of special relativity,

$$g_{ik} = \begin{pmatrix} 1 & 0 & 0 & 0 \\ 0 & -1 & 0 & 0 \\ 0 & 0 & -1 & 0 \\ 0 & 0 & 0 & -1 \end{pmatrix} \Rightarrow dx_i = (cdt, -dx, -dy, -dz) \,. \quad (4.43)$$

The differential displacements follow directly from the matrix multiplication shown in Eq. (4.40), and we leave this as an exercise.

The inverse operation (changing a bra to a ket) is carried out by the inverse metric tensor g^{ik}, which in matrix algebra is the inverse of the g_{ik} matrix. The method for finding matrix inverses is discussed in Chap. 5 in the event the reader is unfamiliar with them.

The applications of tensors to physics are vast. Newtonian mechanics employs Euclidean metric tensors, while special relativity, electromagnetism, quantum mechanics and quantum field theory are built upon the gravity-free Minkowski metric. However, in general relativity (GR—*Einstein's equations*), the components of the metric tensor are something to be *found*, not stipulated at the outset.

In particular, Einstein's equations connect the geometry of spacetime (specifically, its curvature) with the distribution and motion of matter. Spacetime curvature (in the guise of what we call "gravity") affects the distribution and motion of matter, which in turn affects the spacetime curvature. The result is a highly nonlinear set of coupled partial differential equations.

Remarkably, a few simplifying assumptions about the spacetime metric in general relativity have yielded a plethora of astrophysical results. These include the standard cosmological model, the theories of compact massive objects (black holes, neutron stars and white dwarfs), a precise description of the motions in our solar system, accurate models of the gravitational radiation associated with massive rotating binary objects and much more.

More locally, global positioning satellites would be useless in a matter of hours if the GR-induced effects of time dilation were not taken to account. Tensors are used in the study of more terrestrial applications as well, such as the study of turbulence in fluid dynamics, and the stresses and strains in materials.

We will expand our discussion of tensors in Chap. 7 in the context of differentiable manifolds. In particular, we will explore antisymmetric tensors, p-forms and differential forms, and in so doing we will introduce the exterior calculus, one of the more essential tools in modern mathematical physics.

4.6 Cosets and Quotient Spaces

A quotient structure may be defined for vector spaces by applying the same principle used earlier for sets, groups and rings, namely, the application of an equivalence relation so as to partition the space into equivalence classes. Those equivalence classes then form the elements of the quotient space, onto which the original space may be surjectively mapped.[33]

The process for creating quotient structures in those earlier cases began by identifying a substructure[34] and then forming cosets relative to that substructure.[35] At least for groups and rings we were somewhat constrained by the substructure around which the cosets could be formed. Things are somewhat less constrained for vector spaces, where any subspace will serve the purpose of a "base space" around which cosets may be formed to create the quotient space.

A two-dimensional example of this is shown in Fig. 4.7, where we imagine the space $X = \mathbb{R}^2$ filled with (and thereby partitioned by) an infinitely large set of infinite parallel lines. We might identify the equivalence relation as "points passing through a specified point on the y-axis," in which case each line would be its own equivalence class. Each line is a vector space, but only lines that pass through the origin are subspaces of X (Sect. 4.5.1). In Fig. 4.7, that subspace is the line M.

Having identified a *set*—the set of lines—that partitions X, the question remains as to whether this set of lines somehow satisfies the axioms of a *vector space*. What would "addition of lines" or "multiplication of lines by a scalar" mean? Which line would serve as the "zero vector?"

---

[33] As stated in Sect. 1.2.2 "...wherever there is an equivalence relation, there is a quotient structure, and vice versa."

[34] For sets, an identifying characteristic; for groups, an invariant subgroup; for rings, an invariant subring (ideal).

[35] For a review, see the discussion on quotient sets in Sect. 1.5, quotient groups in Sect. 2.7 and quotient rings in Sect. 3.2

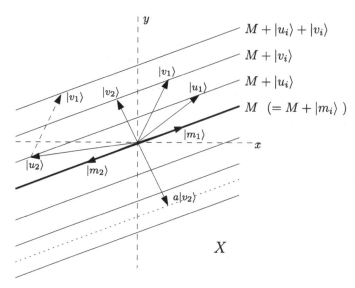

Fig. 4.7 The space $X = \mathbb{R}^2$ is partitioned by an infinite set of infinite parallel lines. Note, for example, that $|m_1\rangle + |u_1\rangle = |m_2\rangle + |u_2\rangle$; both operations yield the same line. This figure and the surrounding discussion in this section are variations of those in Chap. 4 of [17]. See also the discussions in Chap. 10 of [9] and Appendix A.4 of [14]

We now proceed to answer these questions. Geometrically, the "addition of two parallel lines" may be thought of as a combination of positional shifts in a direction perpendicular to the lines. Algebraically, the addition is between two vectors, and by this construction the space M plays the role of a zero vector.

We can see this by considering any of the vectors $|u_i\rangle$ which have the same component perpendicular to M. When added to a vector $|m_i\rangle \in M$ the result is a member of the equivalence class $M + |u_i\rangle$; the equivalence relation being that all vectors in the line $M + |u_i\rangle$ are at the same perpendicular distance from M. This defines addition. In the same way $M + (|u_i\rangle + |v_i\rangle) = (M + |u_i\rangle) + (M + |v_i\rangle)$, which shows this definition of addition is associative, and it is also commutative.

If the sum is between two vectors in M (e.g., $|m_1\rangle + |m_2\rangle$), the result is another vector in M, which we write as $M + |m_i\rangle = M + M = M$, showing that M is the zero vector. These results establish the set of parallel lines as an additive abelian group. Further, multiplication of a line by a scalar is a multiplicative shift in a direction perpendicular to the line. Clearly, all $|x_i\rangle \in X$ are contained within the set of parallel lines, so closure is apparent as well.

Taken together these properties show that the set of parallel lines (with base space M as the zero vector) constitutes a vector space under addition and scalar multiplication of the elements (the lines) in the space. Consequently, the set of lines is a vector space that not only geometrically, but now also algebraically, partitions X. This makes the set of parallel lines a quotient space, denoted as X/M.

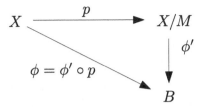

Fig. 4.8 A universal construction for vector spaces

Table 4.2 Parallel quotient constructions between groups, rings and vector spaces[a]

Structure	Group G	Ring $\mathcal{R}$	Vector space X
Substructure	Subgroup	Subring	Subspace
Kernel of ϕ	Invariant subgroup H	Ideal J	Base subspace M
Quotient structure	Quotient group G/H	Quotient ting $\mathcal{R}/J$	Quotient space X/M

[a] See Figs. 2.7, 3.2 and 4.8

The quotient structure shown in Fig. 4.8 for vector spaces corresponds to the quotient structures for sets (Fig. 1.9), groups (Fig. 2.7) and rings (Fig. 3.2). Recall from our earlier discussions that the map p is the canonical map from the original structure to the quotient structure, and the map ϕ' is an isomorphism from the quotient structure X/M to B whose particular definition depends on the application. Then the map ϕ is a composition of these two maps, $\phi = \phi' \circ p : X \to B$, with M as its kernel. Table 4.2, an expansion of Table 3.1, summarizes these results.

One additional consideration concerns the dimension of the base space M and quotient space X/M under different circumstances. In the example of Fig. 4.7 it is clear that dim $X = 2$ and dim $M = 1$. The quotient space X/M (again, the set of lines) is likewise one-dimensional; it has only one basis vector, where the single parameter that distinguishes one line from the other is, for example, the y-intercept of each line. The result

$$\dim X/M = \dim X - \dim M \tag{4.44}$$

in this example applies to finite-dimensional vector spaces generally. Further, the value of the y-intercept is one example of the vector space B to which X/M is mapped; another example of B would be the value of the x-intercept. With the space B thusly defined, the kernel M of ϕ is mapped to the identity element (in this case, zero) as was also the case for groups and rings.

Problems

4.1 Here are a few problems if you need to brush up on some elementary vector algebra. The space is the real plane, we use Cartesian coordinates and each of the following vectors should be expressed in the form $\mathbf{v} = \hat{\mathbf{i}}x + \hat{\mathbf{j}}y$.

(a) The vector $\mathbf{v}_1$ from the point $A = (1, 3)$ to the point $B = (3, -5)$;

(b) The vector $\mathbf{v}_2$ from the origin to the midpoint C of the vector $\mathbf{v}_1$ in part (a);

(c) The vector(s) whose base point is at the point C in part (b) and that is (are) perpendicular to the vector $\mathbf{v}_1$ in part (a);

(d) The unit vector that makes an angle of 60 degrees below the positive x-axis;

(e) The unit vector that is tangent to the curve of $y(x) = 6x^2 + 18x - 108$ when x equals the roots of $y(x)$.

4.2 Referring to Example 4.4, show that the geometric form of the scalar product as given in Eq. (4.4) follows from the coordinate form in Eq. (4.3).

4.3 How would the procedure described in Sect. 4.3.2 and demonstrated in Example 4.5 be different for evaluating an inner product in the space $\mathbb{C}$ over $\mathbb{R}$? [*Hint*: Consider the basis vectors. Are they real or complex? How do they map under conjugation?]

4.4 In Example 4.5 we showed that if $|u\rangle$ and $|v\rangle$ are complex, then the inner product $\langle u|v\rangle$ is in general a complex number. The purpose of this problem is to show that we may express the inner product solely in terms of its real part.

(a) First, show that if z is a complex number, then $\mathrm{Im}(z) = \mathrm{Re}(-iz)$;

(b) Next, let $|u\rangle$ and $|v\rangle$ be two complex vectors as in Example 4.5. Applying the result in part (a), show that

$$\mathrm{Im}\,(\langle u|v\rangle) = \mathrm{Re}(-i\langle u|v\rangle) = \mathrm{Re}(\langle u|iv\rangle) ;$$

(c) Applying your result part (b), show that an inner product may be expressed solely in terms of its real part, as

$$\langle u|v\rangle = \mathrm{Re}(\langle u|v\rangle) + i\,\mathrm{Re}(\langle u|iv\rangle) .$$

4.5 Another approach is to describe inner products solely in terms of norms.

(a) Consider the two real vectors $\mathbf{u} = \hat{\mathbf{i}}u_x + \hat{\mathbf{j}}u_y$ and $\mathbf{v} = \hat{\mathbf{i}}v_x + \hat{\mathbf{j}}v_y$ in the Cartesian plane as in Example 4.4. Show that

$$\langle u|v\rangle = \frac{1}{4}\|\mathbf{u} + \mathbf{v}\|^2 - \frac{1}{4}\|\mathbf{u} - \mathbf{v}\|^2 .$$

(b) Next consider two complex vectors as in Example 4.5. Applying the result in Problem 4.4(c), show that

$$\langle u|v\rangle = \frac{1}{4}\|u + v\|^2 - \frac{1}{4}\|u - v\|^2 + \frac{i}{4}\|u + iv\|^2 - \frac{i}{4}\|u - iv\|^2 .$$

The results in (a) and (b) are called the *polarization identities*. See [14], Sect. 8.1 for a longer discussion. See also [11], Sect. 71 for a related discussion.

4.6 Apply Definition 4.5 and show that the function $f(x) = e^x$ is nonlinear.

4.7 Let $\psi(x)$ be a normalized complex function, and let z_0 be a complex number of unit magnitude. Evaluate the following:
  (a) $\langle \psi + z_0 | \psi + z_0 \rangle$;
  (b) $\langle z_0 \psi | z_0 \psi \rangle$.

4.8 Fill in the missing steps in the Gram-Schmidt orthogonalization process shown in Example 4.10, and formally complete the process to find $|e_2\rangle$.

4.9 Continuing with Gram-Schmidt orthogonalization process in Example 4.12 [note the domain of integration],
  (a) Find the normalized vector $|u_3\rangle$;
  (b) From your answer in part (a), find the Legendre polynomial $P_3(x)$;
  (c) Show that $|u_3\rangle$ and $|u_2\rangle$ are orthogonal;
  (d) If we are given P_0 and P_1, successive Legendre polynomials may be found through the *recursion relation*

$$(2n + 1)xP_n(x) = (n + 1)P_{n+1}(x) + nP_{n-1}(x) , \quad \text{for} \quad n = 1, 2, 3, \ldots$$

Using this recursion relation and the values of $P_1(x)$ and $P_2(x)$ in Example 4.12, show that you get the same result for $P_3(x)$ as you obtained in part (b), and the same $|u_3\rangle$ as you found in part (a);
  (e) The *Rodrigues formulas* are differential relations for finding many of the orthogonal polynomials. For the Legendre polynomials, the Rodrigues formula is

$$P_n(x) = \frac{1}{2^n n!} \left(\frac{d}{dx} \right)^n (x^2 - 1)^n .$$

Use this expression to find $P_3(x)$.
  (f) Show that $P_2(x)$ and $P_3(x)$ are orthogonal;
  (g) The Legendre polynomials $P_n(x)$ solve the Legendre differential equation

$$(1 - x^2)P_n''(x) - 2xP_n'(x) + n(n + 1)P_n(x) = 0, \quad \text{for} \quad n = 0, 1, 2, \ldots .$$

Show that $P_3(x)$ solves Legendre's equation.

[*Note*: In addition to solving their own unique differential equations, the polynomials that are important to mathematical physics are found via a Gram-Schmidt orthogonalization process using the same underlying vector space $|\psi_n(x)\rangle = x^n$ for $n = 0, 1, 2, \ldots$ as we used to find the Legendre polynomials in Example 4.12. The differences among these special functions are their weight functions and their domains of integration in Eq. (4.15). Further, these special functions all have recursion relations that allow for straightforward calculation of higher-order terms in their polynomials once the lowest-order terms are known. Many special functions are dealt with to varying degrees of completeness in specialized areas, such as quantum mechanics. However, for a complete treatment of the special functions see [1] or an equivalently comprehensive text in mathematical methods.]

4.10 Among the special functions you are likely to encounter in elementary quantum mechanics are the *Hermite polynomials*, which arise in the course of solving the one-dimensional quantum mechanical harmonic oscillator problem. The Hermite equation is

$$H_n''(x) - 2xH_n'(x) + 2nH_n = 0 ,$$

and the Rodrigues formula for Hermite polynomials is

$$H_n(x) = (-1)^n e^{x^2} \left(\frac{d}{dx}\right)^n e^{-x^2} .$$

(a) Using the Rodrigues formula, find H_0, H_1 and H_2;

(b) Show that each of your three answers in part (a) satisfies the Hermite equation;

(c) The domain over which the Hermite polynomials are defined is $(-\infty, +\infty)$. Show that H_0 and H_1 are orthogonal. [*Hint:* The integral is not trivial, and you may wish to use an integral table or your favorite computer algebra software.]

4.11 Compare $R^1 \oplus R^2$ and $R^1 \otimes R^2$.

4.12 Fill in the details of the derivations for the differential displacements given in Eqs. (4.36)–(4.38). [*Hint:* The metrics are given later in Sect. 4.5.4.]

4.13 The energy-momentum four-vector for a particle of mass m and total energy E in special relativity is

$$p^\mu = \begin{pmatrix} p^0 \\ p^1 \\ p^2 \\ p^3 \end{pmatrix} ,$$

where $p^0 = E/c$, and the p^i for $i = 1, 2, 3$ are the components of the ordinary spatial three-momentum (e.g., the x, y and z components with magnitude p.

(a) Find p_μ using the Minkowski metric Eq. (4.38), and then evaluate the inner product $p_\mu p^\mu$ to show that

$$E^2 = p^2 c^2 + m^2 c^4 .$$

(b) What is the expression for the rest energy of a particle of mass m (you know this, of course, but use the result above)? What is the expression for the momentum of a photon?

[*Note:* Here we are using the Greek alphabet for the 0, 1, 2, 3 components, and the Latin alphabet solely for the 1, 2, 3 components—this is one of the conventions you will see used often. Therefore, you will need to change the notation given for the metric in Eq. (4.38) to Greek letters.]

4.14 Qualitatively compare the concept of a quotient space with the concept of a projection map. How would you define a quotient space for a three-dimensional Euclidean space? How would you describe an elevation contour map in terms of quotient spaces?

Guide to Further Study

Much of this chapter and the next falls within the purview of linear algebra, for which there are essentially two paths for further study. First, among the several options of linear algebra texts designed primarily for mathematics students, the text by Hoffman and Kunze [14] would be a reasonable choice for a next step after our text.

The second path leads to applications of matrix methods for solving differential equations. In this regard, an especially interesting approach focuses on dynamical systems theory, for which the two-volume work by Hubbard and West [15], and the text by Braun [3], are recommended for their hands-on approaches. Although it is a bit more advanced than the others, Hirsch and Smale [13] remains the standard.

If you wish to explore vector spaces further, then you have many choices. From among them, I'll mention just two: the short classic by Halmos [11] is highly recommended for its succinct clarity, but so too are the early chapters of Hassani [12] in which you will find many examples and a deeper development beyond what we have included in this text.

The above-mentioned text by Hassani lies within the genre of "comprehensive approaches" to mathematical physics. However, "comprehensive" typically means "massive," so be prepared! Two others (among many) in this same category are Arfken, et al. [1] and Riley, et al. [16]. Placing these three texts on a spectrum of abstractness, my view is that [12] is the most abstract but well within reach after our text, [16] is the least abstract and [1] is somewhere between them. A discussion of tensors is found in these and most other comprehensive texts, but Bishop and Goldberg [2] deserves your attention for its geometric flavor, particularly once we complete Chap. 7.

There is a large pool of candidates among the early works on the methods of mathematical physics that serious students should explore over time. For now, and again choosing just two, Courant and Hilbert [8] begins where this chapter has left off and is particularly noteworthy for its focus on special functions and partial differential equations, while Byron and Fuller [4] is slightly more user-friendly and is more directly relevant to to the kinds of methods encountered at the advanced undergraduate level in physics.

Finally, for their breath of scope I mention again the works of Geroch [9] and Roman [17], but now I add to them the excellent work by Simmons [18]. These are highly recommended if you wish to extend your reading along many of the lines of development we have started in this text.

References

1. Arfken, G.B., Weber, H.J., Harris, F.E.: Mathematical Methods for Physicists—A Comprehensive Guide, 7th edn. Academic Press, Waltham, MA (2013)
2. Bishop, R.L., Goldberg, S.I.: Tensor Analysis on Manifolds. Macmillan, New York (1968)
3. Braun, M.: Differential Equations and Their Applications, 4th edn. Springer, New York (1993)
4. Byron, Jr., F.W., Fuller, R.W.: Mathematics of Classical and Quantum Mechanics, (Two volumes bound as one). Dover, New York (1992); an unabridged, corrected republication of the work first published in two volumes by Addison-Wesley, Reading, MA, (Vol. 1 (1969) and Vol. 2 (1970))
5. Callister, W.D.: Materials Science and Engineering—An Introduction, 7th edn. Wiley, New York (2007)
6. Choquet-Bruhat, Y., DeWitt-Morette, C., Dillard-Bleick, M.: Analysis, Manifolds and Physics, Part I: Basics, 1996 Printing. Elsevier, Amsterdam (1982)
7. Debnath, L., Mikunsiński, P.: Introduction to Hilbert Spaces with Applications, 3rd edn. Elsevier Academic Press, Burlington, MA (2005)
8. Courant, R., Hilbert, D.: Methods of Mathematical Physics, First, English edn. Second Printing. Interscience Publishers, New York (1955)
9. Geroch, R.: Mathematical Physics. Chicago Lectures in Physics. University of Chicago Press, Chicago (1985)
10. Halmos, P.R.: Introduction to Hilbert Space and the Theory of Spectral Multiplicity, 2nd edn. AMS Chelsea, Providence, RI (1957)
11. Halmos, P.R.: Finite-Dimensional Vector Spaces, 2nd edn. D.Van Nostrand, Princeton, NJ (1958)
12. Hassani, S.: Mathematical Physics—A Modern Introduction to its Foundations, 2nd edn. Springer, Switzerland (2013)
13. Hirsch, M.W., Smale, S.: Differential Equations, Dynamical Systems, and Linear Algebra. Academic Press, San Diego, CA (1974)
14. Hoffman, K., Kunze, R.: Linear Algebra, 2nd edn. Prentice-Hall, Englewood Cliffs, NJ (1971)
15. Hubbard, J.H., West, B.H.: Differential Equations: A Dynamical Systems Approach, Part I. Springer-Verlag, New York (1991); Differential Equations: A Dynamical Systems Approach—Higher-Dimensional Systems. Springer-Verlag, New York (1995)
16. Riley, K.F., Hobson, M.P., Bence, S.J.: Mathematical Methods for Physics and Engineering, Third Edition, Eighth printing with Corrections. Cambridge University Press, Cambridge (2012)
17. Roman, P.: Some Modern Mathematics for Physicists and Other Outsiders, 2 Volumes. Pergamon Press, Elmsford, NY (1975)
18. Simmons, G. F.: Introduction to Topology and Modern Analysis. McGraw-Hill, New York (1963), now in reprint by McGraw-Hill, India (2003)
19. Taylor, A.E., Lay, D.C.: Introduction to Functional Analysis, 2nd edn. Wiley, New York (1980)
20. Wilcox, H.J., Myers, D.L.: An Introduction to Lebesgue Integration and Fourier Series. Robert E. Krieger Publishing Co., Huntington, N.Y (1978); unabridged corrected edition, Dover, New York (1994)

Chapter 5
Algebras and Operators

5.1 Algebras

An algebra combines the features of a ring with those of a vector space, and consequently there are two approaches to studying their properties. One approach—by far the most common in mathematics texts—emphasizes the ring aspect. Our approach—less common, but likely more amenable to physics students—will take a vector space perspective. Both approaches get us to the same destination.

We start by considering what is *not* included in the definition of a vector space. Specifically, although the vectors can be summed, there is no provision for multiplying them in a manner that insures closure in the vector space.[1] An *algebra* is the structure that addresses this issue.

Recall that a vector space (or, *linear* vector space) is a combination of an additive abelian group with a field (Definition 4.1). The field ($\mathbb{R}$ or $\mathbb{C}$) brought with it a multiplicative operation, and now we incorporate a new multiplicative operation among the elements of the additive abelian group as well (hence, the ring aspect). The result is an algebra (or, *linear* algebra). We require this new multiplicative operation (we'll label it as $\triangle$) to be distributive over addition, but *not* necessarily associative. If $\triangle$ *is* associative, then we have an associative algebra; otherwise the algebra is non-associative. All of this is summarized in Definition 5.1.

Definition 5.1 Consider a vector space $\mathcal{A}$ with vector addition denoted as $\boxplus$, but now also with a multiplicative operation[2] denoted by $\triangle$. Let F be a field whose elements act *internally* on each other via addition and multiplication denoted by $+$ and $\cdot$, respectively, and which act *externally* on the elements of $\mathcal{A}$ via multiplication denoted by $\odot$.

---

[1] Although we *did* define the tensor product between two vectors (Sect. 4.5.3), the result was a different kind of object (a tensor) in another space entirely (the tensor space); the multiplication was not closed with respect to a given vector space.

[2] The precise nature of the multiplication will depend on the type of vector space.

© Springer Nature Switzerland AG 2021
S. P. Starkovich, *The Structures of Mathematical Physics*,
https://doi.org/10.1007/978-3-030-73449-7_5

Let $X, Y, Z \in \mathcal{A}$ and $a, b \in F$. The algebraic system $\Sigma = (\mathcal{A}, F, \boxplus, \triangle, \cdot, \odot)$ is an *associative algebra* if:

1. Σ is closed under all operations;
2. Σ is associative under $\boxplus$: $X \boxplus (Y \boxplus Z) = (X \boxplus Y) \boxplus Z$;
3. Σ is commutative under $\boxplus$: $X \boxplus Y = Y \boxplus X$;
4. Σ contains an additive identity element, X_0, such that $X \boxplus X_0 = X$;
5. Σ contains an additive inverse element, $-X$, such that $X \boxplus -X = X_0$;
6. $a \odot (X \boxplus Y) = (a \odot X) \boxplus (a \odot Y)$;
7. $(a + b) \odot X = (a \odot X) \boxplus (b \odot X)$;
8. $(a \cdot b) \odot X = (b \cdot a) \odot X = a \odot (b \odot X) = b \odot (a \odot X)$;
9. Σ is associative over $\triangle$: $X\triangle(Y\triangle Z) = (X\triangle Y)\triangle Z$;
10. The operation $\triangle$ is distributive over $\boxplus$: $X\triangle(Y \boxplus Z) = (X\triangle Y) \boxplus (X\triangle Z)$;
11. $a \odot (X\triangle Y) = (a \odot X)\triangle Y = X\triangle(a \odot Y)$.

Notational differences notwithstanding, these axioms are the same as those for a vector space, with the addition of the "triangle" operator in Axioms (9)–(11). ∎

The element X_0 is the *zero element*. Axioms (6) and (7) are distributive properties that define multiplication of an element $X \in \mathcal{A}$ by a scalar $a \in F$. Axioms (8) and (11) illustrate the commutative and associative properties associated with the field F, where a and b may assume the zero or unit elements of F. As defined above, Σ is *associative* because of Axiom (9). If this axiom is omitted, then the algebra is said to be *non-associative*. Some of the most important algebras in physics are non-associative, among them being the *Lie* and *Poisson algebras*.

The multiplicative operation $\triangle$ in Definition 5.1 is not necessarily commutative, but if it happens that $X\triangle Y = Y\triangle X$ then we have a *commutative algebra*. Neither does the definition require a unit multiplicative element (such that $1_X\triangle X = X$), nor a multiplicative inverse (such that $X\triangle X^{-1} = 1_X$), but if 1_X and X^{-1} are present then we have an *algebra with unity* and an *algebra with (multiplicative) inverse*, respectively.

In order to simplify the notation it is customary to use the "+" symbol to mean either $+$ or $\boxplus$ depending on context, and to write aX for $a \odot X$ and ab for $a \cdot b$. We also will drop the $\triangle$ symbol in most cases and write $X\triangle Y$ as XY. It is important to keep track of which operations apply to which elements, and to also pay attention to the order in which multiplication is applied. A convenient shorthand description for algebras is to say, for example, "$\mathcal{A}$ is an algebra over F" (or an "F-algebra"). In this text, the field F will be either the real or complex numbers, and accordingly the algebra will be said to be either real or complex.

Among the more common applications of algebras in physics are quantum mechanical operator algebras that are used to represent physical quantities; matrix operators that effect linear transformations on vector spaces[3]; functional operators

---

[3]From the linear transformations we can often identify characteristic vectors (eigenvectors) and their corresponding characteristic scalar quantities (eigenvalues) that may have significant physical interpretations.

as they appear in integral transforms; and differential operators in differential equations. Certain functions of operators are important as well, with one example being $\exp A$ of an operator A, which plays a defining role in Lie groups.

Because an algebra is built on a vector space, these two structures share a number of properties as well as the concepts of *basis* and dimension; the dimension of an algebra is more often called its *order*. Examples of algebras may be drawn from the examples of vector spaces in Examples 4.1 and 4.3.

Example 5.1

1. The one-dimensional vector spaces $\mathbb{R}^1$ and $\mathbb{C}^1$ are real and complex algebras, respectively, of order one. They are associative and commutative algebras.
2. The two-dimensional vector space $\mathbb{C}$ over $\mathbb{R}$ is a real associative algebra of order two. This is an associative and commutative algebra. Both the zero and unit vectors—the numbers 0 and 1, respectively—are present in this and the previous example.
3. The set of "vectors as arrows" (directed line segments) in $\mathbb{R}^3$ is a real, non-commutative and non-associative algebra of order three, where multiplication is defined as the vector cross-product (the Lie product, see Sect. 5.3.1). Given the vectors $\mathbf{A}$, $\mathbf{B}$ and $\mathbf{C}$ in $\mathbb{R}^3$, recall how non-commutativity follows from

$$\mathbf{A} \times \mathbf{B} = -\mathbf{B} \times \mathbf{A}, \tag{5.1}$$

which is an *antisymmetry*. Non-associativity is reflected in the relations

$$\mathbf{A} \times (\mathbf{B} \times \mathbf{C}) = \mathbf{B}(\mathbf{A} \cdot \mathbf{C}) - \mathbf{C}(\mathbf{A} \cdot \mathbf{B})$$
$$(\mathbf{A} \times \mathbf{B}) \times \mathbf{C} = \mathbf{B}(\mathbf{A} \cdot \mathbf{C}) - \mathbf{A}(\mathbf{B} \cdot \mathbf{C}) \tag{5.2}$$

The zero and multiplicative unit vectors are both present in this algebra.
4. The set of all n × n matrices with multiplication defined as matrix multiplication such that $A(BC) = (AB)C$ is an associative algebra but generally is not commutative. The zero matrix is the matrix where all entries are zero. The multiplicative unit matrix (if present) is the identity matrix (with a 1 for each diagonal entry and zeros elsewhere). The order of the algebra is n^2; the complete set of linearly-independent basis matrices would consist of those n^2 distinct matrices with a 1 in a single location and zeros elsewhere. The matrix entries may be either real or complex, with the algebra described accordingly.
5. The set of all real functions on the closed interval $[-1, 1]$ constitute a real algebra under the operations of ordinary arithmetic. It is an algebra of infinite order. ▲

5.2 Structure Constants

The different kinds of multiplicative operations on different algebras may be described in terms of scalar quantities known as *structure constants*. The number of structure constants within an algebra will be dictated by the order of the algebra, as shown in several examples below. We will describe structure constants by using the index notation and summation convention introduced in Sect. 4.5.3 for tensors, and we will proceed (for now) with a generic type of multiplication.

Consider two algebraic operators A, $B \in \mathcal{A}$ written as $A = \alpha^i e_i$ and $B = \beta^j e_j$, where α^i and β^j are scalar components with respect to the *basis operators* e_i and e_j. This is a general notation that allows for the possibility that A and B would be ordinary vectors in some circumstances, but they are not restricted as such; A and B might just as easily be matrices. Therefore, we refrain from using vector notation (e.g., the bra and ket notation) for the elements of an algebra.

With these preliminary comments, we may now write

$$AB = \alpha^i e_i \beta^j e_j = \alpha^i \beta^j e_i e_j \in \mathcal{A}. \tag{5.3}$$

The sum over repeated indices spans the order of the algebra. Further, the term $e_i e_j$ (which is *not* a tensor product—*that* operation would take us into a tensor space) is a combination of two elements of $\mathcal{A}$ and (by the closure property) must be in $\mathcal{A}$ as well. This requires that we write

$$e_i e_j = s_{ij}^k e_k \tag{5.4}$$

so that Eq. 5.3 becomes

$$AB = \alpha^i \beta^j s_{ij}^k e_k = (AB)^k e_k. \tag{5.5}$$

The expression $(AB)^k$ in Eq. 5.5 is the k-component of $AB \in \mathcal{A}$, and the scalar parameter s_{ij}^k is the set of structure constants[4] for the algebra when evaluated over all values of i, j and k. Specifically,

$$s_{ij}^k = \left\{ \begin{array}{l} \text{The scalar coefficient of} \\ \text{the } \alpha^i \beta^j \text{ term as that} \\ \text{term appears in the k-} \\ \text{component of } AB. \end{array} \right\}. \tag{5.6}$$

Because there are three indices in the structure coefficient, and because each index takes on as many distinct values as there are dimensions in the vector space (the order of the algebra), an algebra of order n will have n^3 structure coefficients. However, typically the structure constants are not independent, and for lower-order algebras we can usually find them by performing relatively few calculations.

Example 5.2 Consider the second-order algebra $\mathbb{C}$ over $\mathbb{R}$, i.e., the complex plane, where the two basis vectors are 1 and i. Let two complex numbers be given as

---

[4]In the literature and most texts, the more common notation for our s_{ij}^k is c_{ijk}. We have selected a notation that is consistent with the Einstein summation convention. The cyclic permutation on the indices is the same, namely, $i \to j \to k \to i$ for a set of three indices.

$$A = \alpha^1 e_1 + \alpha^2 e_2 = \alpha^1 + i\alpha^2 = x_A + iy_A$$
$$B = \beta^1 e_1 + \beta^2 e_2 = \beta^1 + i\beta^2 = x_B + iy_B. \tag{5.7}$$

There are two ways of evaluating the product AB. Straightforward algebra yields

$$AB = 1[\alpha^1 \beta^1 - \alpha^2 \beta^2] + i[\alpha^1 \beta^2 + \alpha^2 \beta^1]. \tag{5.8}$$

Evaluation of Eq. 5.5 by summing over all values of the indices gives

$$
\begin{aligned}
AB = &\underbrace{e_1[\alpha^1 \beta^1 s_{11}^1 + \alpha^1 \beta^2 s_{12}^1 + \alpha^2 \beta^1 s_{21}^1 + \alpha^2 \beta^2 s_{22}^1]}_{k=1} \\
&+ \underbrace{e_2[\alpha^1 \beta^1 s_{11}^2 + \alpha^1 \beta^2 s_{12}^2 + \alpha^2 \beta^1 s_{21}^2 + \alpha^2 \beta^2 s_{22}^2]}_{k=2}.
\end{aligned} \tag{5.9}
$$

Comparison of Eqs. 5.8 and 5.9 yields the $n^3 = 2^3 = 8$ structure constants that define multiplication of two complex numbers in the algebra $\mathbb{C}$ over $\mathbb{R}$:

$$s_{11}^1 = s_{12}^2 = s_{21}^2 = +1$$

$$s_{12}^1 = s_{21}^1 = s_{11}^2 = s_{22}^2 = 0 \tag{5.10}$$

$$s_{22}^1 = -1.$$

▲

Example 5.3 Next consider the non-associative, non-commutative third-order algebra of ordinary vectors in $\mathbb{R}^3$. The multiplication operation is the vector cross product, which is familiar from elementary vector algebra, and as such we'll use the usual vector notation. Let $\mathbf{A} = \hat{\mathbf{i}} A_x + \hat{\mathbf{j}} A_y + \hat{\mathbf{k}} A_z$ and $\mathbf{B} = \hat{\mathbf{i}} B_x + \hat{\mathbf{j}} B_y + \hat{\mathbf{k}} B_z$. Then

$$
\begin{aligned}
\mathbf{A} \times \mathbf{B} &= \hat{\mathbf{i}}(A_y B_z - A_z B_y) + \hat{\mathbf{j}}(A_z B_x - A_x B_z) + \hat{\mathbf{k}}(A_x B_y - A_y B_x) \\
&= \underbrace{e_1(A^2 B^3 - A^3 B^2)}_{k=1} + \underbrace{e_2(A^3 B^1 - A^1 B^3)}_{k=2} + \underbrace{e_3(A^1 B^2 - A^2 B^1)}_{k=3}
\end{aligned}
$$

$$
= \underbrace{e_1(A^2 B^3 s_{23}^1 + A^3 B^2 s_{32}^1)}_{k=1} + \underbrace{e_2(A^3 B^1 s_{31}^2 + A^1 B^3 s_{13}^2)}_{k=2} + \underbrace{e_3(A^1 B^2 s_{12}^3 + A^2 B^1 s_{21}^3)}_{k=3}
$$

$$\implies \mathbf{A} \times \mathbf{B} \equiv AB = A^i B^j s_{ij}^k e_k = (AB)^k e_k. \tag{5.11}$$

The $n^3 = 3^3 = 27$ structure constants are the components of the *permutation tensor*,[5] $s_{ij}^k = \epsilon_{ij}^k$, which are equal to $+1$ for a cyclic permutation of the indices $(ijk) = (123)$; equal to -1 for a non-cyclic permutation; and equal to 0 for the case where any two indices are equal. An equivalent way of framing this problem is in terms of the Lie product (Sect. 5.3.1). ▲

We can see from these examples that although the number of structure constants in an algebra might at first appear daunting, it is often the case that the number of unique values turns out to be relatively small. Further, it is frequently possible to find relationships among the structure constants within a given algebra, and one of the more common of these relationships is found in associative algebras generally.

Example 5.4 Consider any associative algebra such as $\mathbb{C}$ over $\mathbb{R}$ (which also happens to be commutative) or the set of n × n matrices in Example 5.1(4) (which generally is not a commutative algebra). We can find the relationship among the structure constants of an associative algebra by expanding the expression $(AB)C = A(BC)$ and applying the condition that the combinations of structure constants on both sides of the equation must be equal.

Using Eq. 5.5 we can write $(AB)C$ as

$$(AB)C = [\alpha^i \beta^j s_{ij}^k e_k]\gamma^l e_l = \alpha^i \beta^j \gamma^l s_{ij}^k e_k e_l, \tag{5.12}$$

where we let $C = \gamma^l e_l$. Following the pattern in Eq. 5.4, we can write $e_k e_l = s_{kl}^m e_m$ so that Eq. 5.12 becomes

$$(AB)C = \alpha^i \beta^j \gamma^l s_{ij}^k s_{kl}^m e_m. \tag{5.13}$$

Repeating this methodology, we write (BC) as

$$(BC) = \beta^j \gamma^l e_j e_l = \beta^j \gamma^l s_{jl}^k e_k. \tag{5.14}$$

With $A = A^i e_i$, we can write $A(BC)$ as

$$A(BC) = A^i e_i [\beta^j \gamma^l s_{jl}^k e_k] = \alpha^i \beta^j \gamma^l s_{jl}^k s_{ik}^m e_m. \tag{5.15}$$

Comparing Eqs. 5.13 and 5.15 gives the relationship among structure constants for an associative algebra:

$$s_{ij}^k s_{kl}^m = s_{jl}^k s_{ik}^m \tag{5.16}$$

The meaning of these expressions is similar to that described in Eq. 5.6. ▲

The structure constants are just scalars, and although their number grows as n^3 for algebras of order n, it often happens that most of the structure constants are zero and the others are ± 1 (at least for real algebras).

---

[5] Another name for the permutation tensor is the *epsilon tensor*.

5.3 Lie and Poisson Algebras

In forming a non-associative algebra, we replace Axiom 9 in Definition 5.1—the associative property of the multiplication operation $\triangle$—with some new structure or condition; merely omitting the axiom would offer little or no guidance on how to proceed. Among the most important of these algebras as measured by their contributions to physics are the Lie and Poisson algebras,[6] and our first encounter with them is typically in our study of quantum and classical mechanics. We will get a taste of that encounter in this section.

5.3.1 Lie Algebras

The defining characteristic of a Lie algebra is the replacement of the associative multiplicative operation with that of the *Lie bracket*.

Definition 5.2 The Lie bracket (*commutator bracket*, or simply the *commutator*) of two operators A and B, denoted as $[A, B]$, defines multiplication within a Lie algebra to be such that

1. $[A, B] \equiv (AB - BA) = -[B, A]$, and
2. $[A, [B, C]] + [B, [C, A]] + [C, [A, B]] = 0$,

The first condition describes the antisymmetry of the Lie bracket and defines the operator $(AB - BA)$. The second condition is the *Jacobi identity*.[7] An algebra that satisfies Definition 5.2(1) is called a *commutator algebra*, and an algebra that satisfies Definition 5.2(2) is called a *Lie algebra*. The two elements within the Lie bracket are said to *commute* if their Lie bracket equals zero. ∎

All commutator algebras are Lie algebras. However, the converse is not true; there are algebras with bracket relations that are not strictly commutators, but which nonetheless satisfy the Jacobi relation and are thereby Lie algebras. One such example is the Poisson algebra of Hamiltonian mechanics, which we will discuss in Sect. 5.3.2.

Example 5.5 Consider two operators

$$A = \frac{\mathrm{d}}{\mathrm{d}x}() \quad \text{and} \quad B = ()^3,$$

where A differentiates a function $f(x)$ and B cubes it. We want to ask whether A and B commute. As a particular example, let $f(x) = x$. Then, remembering to operate

---

[6]Marius Sophus Lie (pron. "Lee") (1842–1899), Norwegian analyst, geometer and group theorist; Siméon Denis Poisson (1781–1840) French analyst. Each particular example of a Lie or Poisson algebra—and there are many—may be identified by its structure constants.

[7]Karl Gustav Jacob Jacobi (1804–1851), German algebraist and analyst. Note the cyclic permutation $A \to B \to C \to A$ among the three terms in the sum; if you can remember the pattern in any one of the three terms, then you instantly have all three.

right-to-left we have

$$[A, B]f(x) = (AB - BA)f(x), \text{ so that}$$

$$AB(f(x)) = A(x^3) = 3x^2;\ BA(f(x)) = B(1) = 1 \Rightarrow [A, B]x = 3x^2 - 1.$$

Therefore, A and B do not commute generally, but do commute if $x = 1/\sqrt{3}$. ▲

The Jacobi identity involves terms like $[A, [B, C]]$, which expand fully[8] into four terms, each comprised of three operators. For example,

$$[A, [B, C]] = [A, (BC - CB)] = (ABC - ACB - BCA + CBA).$$

There are two closely-related (but subtly different) interpretations of the Lie bracket, and it is important to draw the distinction between them. In the first interpretation—of which Example 5.5 is an example—A, B and C are elements of an *associative* algebra $\mathcal{A}$, and the Lie bracket maps $A, B, C \in \mathcal{A}$ to elements in a non-associative algebra $\mathcal{A}_L$ (the subscript "L" for "Lie"). Consequently, $\mathcal{A}_L$ is often called the *commutator algebra of* $\mathcal{A}$. For example, if $A, B \in \mathcal{A}$, then

$$[,] : A, B \in \mathcal{A} \rightarrow [A, B] \equiv (AB - BA) \in \mathcal{A}_L. \tag{5.17}$$

In the second interpretation, the Lie bracket $[A, B]$ is a direct replacement for the associative $\triangle$ operation in $A \triangle B$. In this instance, A, B and C are elements of a *non-associative* algebra $\mathcal{A}_L$. Perhaps the most familiar example of this is the vector cross product in $\mathbb{R}^3$, in which case the two provisions in Definition 5.1 become what we recognize as the vector identities

$$\mathbf{A} \times \mathbf{B} = -\mathbf{B} \times \mathbf{A}$$

$$\mathbf{A} \times (\mathbf{B} \times \mathbf{C}) + \mathbf{B} \times (\mathbf{C} \times \mathbf{A}) + \mathbf{C} \times (\mathbf{A} \times \mathbf{B}) = 0.$$

Under this interpretation the Lie bracket $[,]$ is usually referred to as the *Lie product* within a non-associative algebra (see Example 5.3).

As noted, the difference between these two interpretations is a subtle one, but it is important because it speaks to the difference in the fundamental nature of the elements A and B. In the first interpretation, A and B are fundamentally elements of an associative algebra (like the operators in Example 5.5) with which a non-associative algebra can then be constructed via the commutator. In the second interpretation, the elements A and B (shown above as vectors) fundamentally, first and foremost, belong to a non-associative algebra.

The first interpretation is the manner in which Lie algebras appear in quantum mechanics; the matrix and differential operators in quantum mechanics correspond to

---

[8]The distributive property of multiplication over addition still holds in a Lie algebra.

physical quantities, and are fundamentally elements of an associative algebra $\mathcal{A}$. The commutator then acts on these operators and establishes a non-associative algebra $\mathcal{A}_L$ on the same underlying space of operators as $\mathcal{A}$.

If two operators commute, then the quantum mechanical interpretation is that each corresponding physical quantity may be measured or specified with a precision that is independent of the uncertainty in the other. Indeed, the fact that the momentum and position operators *do not* commute leads directly to the Heisenberg Uncertainty Principle. We will explore this application in the end-of-chapter problems.

5.3.2 Poisson Algebras

Given an associative algebra $\mathcal{A}$, a Poisson algebra $\mathcal{A}_P$ ("P" for "Poisson") may be constructed in a manner analogous to that used for a Lie algebra in Eq. 5.17. The structural difference lies in the nature of the bracket operator.

Definition 5.3 The *Poisson bracket* of two operators A and B, denoted as $\{A, B\}$, defines multiplication within a Poisson algebra to be such that

$$\{A, BC\} = B\{A, C\} + \{A, B\}C. \tag{5.18}$$

The Poisson bracket $\{A, B\}$ maps the elements A, B of the associative algebra $\mathcal{A}$ to elements in $\mathcal{A}_P$. ∎

If it happens that $\{A, B\} = [A, B] \equiv (AB - BA)$, i.e., if the Poisson bracket is a commutator, then the Eq. 5.18 reduces to the Jacobi identity. Therefore, a Poisson algebra contains a Lie algebra as a special case. We leave the demonstration of this as an end-of-chapter problem.

The example of a Poisson algebra that physics students are most likely to encounter is a part of the Hamiltonian formulation of classical mechanics. Consider a single particle with momentum p and coordinate position q, and let $f(p, q)$, $g(p, q)$ and $h(p, q)$ be real-valued functions as defined over some domain of p and q. These functions are elements of an associative algebra (see Example 5.1(5)).

Next consider a Poisson bracket[9] $\{f, g\}$ defined as

$$\{f, g\} \equiv \frac{\partial f}{\partial p}\frac{\partial g}{\partial q} - \frac{\partial g}{\partial p}\frac{\partial f}{\partial q} \equiv f_p g_q - g_p f_q. \tag{5.19}$$

Note that the Poisson bracket in Eq. 5.19 is *not* a commutator. That is, this bracket does not act as an operator on a function in the same manner as the Lie bracket, which defines "commutator."

Still, we wish to test whether the bracket defined in Eq. 5.19 is consistent with Definition 5.3 for a Poisson algebra. Applying Eq. 5.19 to Eq. 5.18, we ask whether

---

[9]Sign conventions vary among authors; we are following the convention in [8].

$$\{f, gh\} \overset{?}{=} g\{f, h\} + \{f, g\}h. \tag{5.20}$$

We leave it as an exercise to show that the answer to the question in Eq. 5.20 is in fact "yes." We leave it as a further exercise to show that this same $\{f, g\}$ also satisfies the Jacobi identity. Therefore this Poisson algebra (it is an algebra once derivatives of all orders are included) is also a Lie algebra, but it is a Lie algebra that is not a commutator algebra.

In classical mechanics we usually focus our attention on the time-evolution of the primary elements of the system (e.g., how particle positions vary over time). However, we also frequently ask how some other function (e.g., momentum or energy) varies with time; we are especially interested in identifying any conserved quantities that may exist in a dynamical system. These questions are addressed in way that makes full use of Poisson algebras as they apply to Hamiltonian phase space.[10]

If we take the total time-derivative of the function $f(p, q, t)$ and apply Hamilton's equations with Hamiltonian $H(p, q)$, the result describes the time-variation of that function as the system (taken here as a single particle) evolves in Hamiltonian phase space:

$$\begin{aligned}
\frac{df}{dt} &= \frac{\partial f}{\partial t} + \frac{\partial f}{\partial p}\dot{p} + \frac{\partial f}{\partial q}\dot{q} \\
&= \frac{\partial f}{\partial t} + \frac{\partial f}{\partial p}\left(-\frac{\partial H}{\partial q}\right) + \frac{\partial f}{\partial q}\frac{\partial H}{\partial p} \\
&= \frac{\partial f}{\partial t} - \{f, H\} = \frac{\partial f}{\partial t} + \{H, f\}.
\end{aligned} \tag{5.21}$$

We see that if $f \neq f(t)$ explicitly and commutes with H, then f is a conserved quantity. A similar result holds in the Heisenberg representation of quantum mechanics for physical operators that commute with the Hamiltonian operator of the system.

5.4 Subalgebras, Quotients and Sums of Algebras

Subalgebras and their closely-related algebra ideals lead to quotient algebras. The mathematics literature tends to present this material in the most general terms of rings and modules, but because we have been limiting our discussion to specific rings

---

[10] We can think of the Hamiltonian as the total energy of the system. For a single particle, Hamiltonian phase space is one in which the coordinates are the momentum p and position q rather than position and time. That is, we describe the time-evolution of the particle in terms of how its position and momentum change, rather than just its change in position. Also, Hamilton's equations are $\dot{p} = -\partial H/\partial q$ and $\dot{q} = \partial H/\partial p$, where the "dot" denotes the time-derivative. If you have not seen this before, these are all the details you need for now to walk through Eq. 5.21. We will explore Hamilton's equations and the p–q phase space of a few simple mechanical systems more fully in Sect. 7.6 in the context of differential forms and symplectic manifolds.

($\mathbb{R}$ and $\mathbb{C}$) and modules (vector spaces) we will focus our attention there. The goal of this section is to give you a basic sense of how quotient algebras are constructed, as these have the most direct applications to physics. We close the section with a brief summary of the properties of the direct sum of algebras.

5.4.1 Subalgebras, Algebra Ideals and Quotients

Given an associative F-algebra $\mathcal{A}$ (i.e., an associative algebra $\mathcal{A}$ over the field F), a *subalgebra* $\mathcal{U}$ is a subset of $\mathcal{A}$ that satisfies all of the axioms for an algebra, including closure within $\mathcal{U}$.

In our discussion of groups in Chap. 2, we saw that what distinguishes an invariant subgroup H from an otherwise arbitrary subgroup is how composition between $h \in H$ and *any* g in the group G yields an element of H. An analogous idea distinguishes the *ring ideal* J of a ring $\mathcal{R}$ from other subrings. The same concept once again describes the distinction between an *algebra ideal* $\mathcal{U}_I$ and a subalgebra $\mathcal{U}$ (Fig. 5.1).

We can see how this works for an associative algebra by comparing Fig. 5.1a, b. Figure 5.1a shows the subset $\mathcal{U}$ as a subalgebra of $\mathcal{A}$. In Fig. 5.1b, we must also have $u \triangle v, v \triangle u \in \mathcal{U}_I$ for *any* $v \in \mathcal{A}$.

Fig. 5.1 The difference between **a** a subalgebra, and **b** an algebra ideal, where $a \in F$ and $\triangle$ is an associative multiplication operation

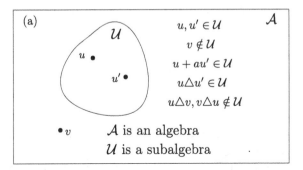

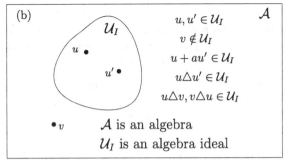

Example 5.6 Referring to Example 5.1(5),[11] let $\mathcal{A}$ be the set of all real-valued functions (both bounded and unbounded) on some interval $X \subset \mathbb{R}$, and let $\mathcal{U}$ be the set of all bounded functions on X. Then $\mathcal{U}$ is a subalgebra because for $u, u' \in \mathcal{U}$ we have

$$u + u' = \text{(bounded)} \in \mathcal{U} \qquad \text{and} \qquad u \triangle u' = \text{(bounded)} \in \mathcal{U}.$$

However, $\mathcal{U}$ is not an algebra ideal because an element of $\mathcal{U}$ multiplied by an unbounded function $v \in \mathcal{A} - \mathcal{U}$ is unbounded, that is,

$$u \triangle v = \text{(bounded)(unbounded)} = \text{(unbounded)} \notin \mathcal{U},$$

and we don't have closure in $\mathcal{U}$.

As an alternative, if $\mathcal{U}_I$ were the set of all functions that are identically zero on the interval X, then $\mathcal{U}_I$ is an algebra ideal because

$$u + u' = 0 \in \mathcal{U}_I, \qquad u \triangle u' = 0 \in \mathcal{U}_I \qquad \text{and} \qquad u \triangle v = 0 \in \mathcal{U}_I.$$

▲

We now turn our attention to quotient algebras.

Definition 5.4 Given (a) an algebra $\mathcal{A}$ over the field F; (b) an algebra ideal $\mathcal{U}_I$; (c) $v \in \mathcal{A}$; and (d) $a \in F$, we define a *quotient algebra* $\mathcal{A}/\mathcal{U}_I$ of $\mathcal{A}$ by $\mathcal{U}_I$ as consisting of those subsets of $\mathcal{A}$ that satisfy the following criteria:

$$(\mathcal{U}_I + v_1) + (\mathcal{U}_I + v_2) = \mathcal{U}_I + (v_1 + v_2)$$
$$a(\mathcal{U}_I + v) = \mathcal{U}_I + av$$
$$(\mathcal{U}_I + v_1)(\mathcal{U}_I + v_2) = \mathcal{U}_I + v_1 \triangle v_2.$$

■

The subsets satisfying Definition 5.4 are the cosets with respect to the algebra ideal $\mathcal{U}_I$. The first criterion identifies an "additive coset" with respect to $\mathcal{U}_I$. The first two criteria are carried over from the definition of a quotient vector space in Sect. 4.6 and Fig. 4.7; the subspace M in that context is analogous to the algebra ideal $\mathcal{U}_I$ in this present context. The third criterion identifies a "multiplicative coset" with respect to $\mathcal{U}_I$ and incorporates the multiplication operation that is now part of the algebra that (by definition) was not part of the vector space in Sect. 4.6.

Example 5.7 Continuing with Example 5.6 (where the set of all functions that are identically zero on the interval X forms the ideal $\mathcal{U}_I$), the criteria in Definition 5.4 are trivially satisfied. In this instance, there is only one coset of $\mathcal{U}_I$, namely, $\mathcal{A} - \mathcal{U}_I$. ▲

Table 5.1 summarizes the parallel constructions among the quotient structures we have considered in this text. We previously have seen how sets, groups, rings and

---

[11] See also the example in [4], p. 101, and the discussion there.

Table 5.1 Parallel quotient constructions between groups, rings, vector spaces and algebras[a]

Structure	Group G	Ring $\mathcal{R}$	Vector space X	Algebra $\mathcal{A}$
Substructure	Subgroup	Subring	Subspace	Subalgebra
Kernel of ϕ	Inv. subgrp. H	Ring ideal J	Subspace M	Algebra ideal $\mathcal{U}_I$
Quotient (Q)	Q. group G/H	Q. ring $\mathcal{R}/J$	Q. space X/M	Q. algebra $\mathcal{A}/\mathcal{U}_I$

[a]See Figs. 2.7, 3.2, 4.8 and 5.2

Fig. 5.2 A universal construction for quotient algebras

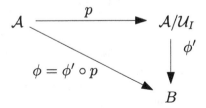

fields can be partitioned into cosets via an equivalence relation (the kernel). In this section, we have shown that the same now holds true for algebras. Consequently, we are able to sketch a quotient diagram (Fig. 5.2) for algebras as we have for the other structures, where here $\mathcal{A}/\mathcal{U}_I$ is a *quotient algebra*.

Conceptually, what we have said about subalgebras and algebra ideals for associative algebras applies equally well to non-associative algebras, but with a different definition of multiplication. For example, given a Lie commutator algebra $\mathcal{A}_L$, the subset $\mathcal{U} \subset \mathcal{A}_L$ is a subalgebra if $u + u'$ and $[u, u']$ belong to $\mathcal{U}$. If in addition we have $[u, v] \in \mathcal{U}$ for all $v \in \mathcal{A}_L$, then $\mathcal{U}$ is an algebra ideal, $\mathcal{U}_I$. These topics lie beyond the scope of the present text, but we give some direction for further study in the Guide at the end of the chapter.

5.4.2 Direct Sums of Algebras

The discussion of the direct sum of vector spaces in Sect. 4.5.2 is largely replicated with regard to the direct sum of algebras, so we will limit ourselves here to a definition. The only amendment is the inclusion of a provision (Definition 5.5(5)) for the internal multiplication that distinguishes an algebra from a vector space.

Definition 5.5 Consider two associative algebras $\mathcal{U}$ and $\mathcal{V}$ defined over a field F, with vectors $u \in \mathcal{U}$ and $v \in \mathcal{V}$. Further, let $a \in F$. The *direct sum* $\mathcal{W} = \mathcal{U} \oplus \mathcal{V}$ is defined such that:

1. $W = \mathcal{U} \times \mathcal{V} = \{(u, v) : u \in \mathcal{U}, v \in \mathcal{V}\}$;
2. $\mathcal{U} \cap \mathcal{V} = 0$;
3. $a(u, v) = (au, av)$;
4. $(u_1, v_1) + (u_2, v_2) = (u_1 + u_2, v_1 + v_2)$;
5. $(u_1, v_1)(u_2, v_2) = (u_1 \triangle u_2, v_1 \triangle v_2)$. ∎

The most common applications of direct sums to physics involve the direct sums of matrix algebras. We offer a basic example in the end-of-chapter problems (Problem 5.16). More advanced applications appear in the context of representation theory for Lie algebras.

5.5 Associative Operator Algebras on Inner Product Spaces

From among the algebras with particular applications to physics that we mentioned in Sect. 5.1, we wish to consider associative matrix operator algebras that operate on inner product spaces. These structures play central roles in virtually all branches of physics, and they are the primary means for representing continuous groups and for describing linear transformations in coordinate spaces.

As appropriate, this section may serve as a short review or introduction to elementary matrix algebra, linear transformations and the kinds of matrix operations that you are most likely to need in your studies in physics and engineering. If you are already very familiar with this material, you should at least do a quick survey of the notation, as these ideas will be needed later in the text.

5.5.1 Definitions, Notations and Basic Operations with Matrices

We will use a 2×2 matrix (a matrix with 2 rows, and 2 columns) to illustrate our notation, and unless otherwise specified we will assume the individual matrix entries are complex numbers. We write a matrix A as

$$A = A^i_j = \begin{pmatrix} A^1_1 & A^1_2 \\ A^2_1 & A^2_2 \end{pmatrix} = \begin{pmatrix} a & b \\ c & d \end{pmatrix},$$

(5.22)

where the index i represents the row and the index j represents the column, whether they are "up" or "down." We adopt this notation for this section only so as to be able to apply the Einstein summation convention.

Example 5.8 Our first example concerns the formation of the *transpose* A^T of A, where we switch rows and columns:

$$A^T = \begin{pmatrix} a & c \\ b & d \end{pmatrix}.$$

(5.23)

▲

Example 5.9 Next, if we take the complex conjugate (*) of A^T, the result is (unsurprisingly) called the *transpose conjugate* (or *conjugate transpose*) of A. Denoted as $A^\dagger$ and read as "A-dagger," we have

$$(A^T)^* \equiv A^\dagger = \begin{pmatrix} a^* & c^* \\ b^* & d^* \end{pmatrix}. \tag{5.24}$$

▲

The *trace*[12] of an $n \times n$ matrix A is the sum of its diagonal elements from $A_1^1 \to A_n^n$. Denoted as Tr A, the trace can be defined only for square matrices.

The *determinant* of a matrix A, denoted as det A, is a scalar quantity that is defined for any $n \times n$ matrix. We will explore its geometric meaning in Sect. 5.5.2, but first we describe the *general* method for its evaluation, using a 2×2 matrix for illustrative purposes.[13] This method may then be applied to any square matrix.

First, associated with each element A_j^i of an $n \times n$ matrix is its *minor*, which is the $(n-1) \times (n-1)$ matrix that remains after "crossing out" the ith row and the jth column in A. We denote this minor as M_i^j—note that we are switching the up-down index locations.

Next, if we multiply the minor by $(-1)^{i+j}$ we obtain the *cofactor*, or *signed minor*, of A_j^i which we write as $D_i^j = M_i^j(-1)^{i+j}$. Table 5.2 tabulates these quantities for the 2×2 matrix A in Eq. 5.22.

The determinant det A is then found by selecting *any* row of A (i.e., keeping i fixed) and performing the sum $A_j^i D_i^j$ over all values of j (i.e., over all columns).

Example 5.10 Using Table 5.2, we find the determinant det A as follows:

For $i = 1$: $\det A = A_j^1 D_1^j = A_1^1 D_1^1 + A_2^1 D_1^2 = (a)(d) + (b)(-c) = ad - bc;$

For $i = 2$: $\det A = A_j^2 D_2^j = A_1^2 D_2^1 + A_2^2 D_2^2 = (c)(-b) + (d)(a) = ad - bc.$

▲

Table 5.2 Matrix elements, minors and cofactors[a]

A_j^i	M_i^j	D_i^j
$A_1^1 = a$	$M_1^1 = d$	$D_1^1 = +d$
$A_2^1 = b$	$M_1^2 = c$	$D_1^2 = -c$
$A_1^2 = c$	$M_2^1 = b$	$D_2^1 = -b$
$A_2^2 = d$	$M_2^2 = a$	$D_2^2 = +a$

[a]For the 2×2 matrix A in Eq. 5.22

---

[12]In the older literature, the trace is also called the *spur*.

[13]We note here that the "determinant" of a 1×1 "matrix" (i.e., a number) is just the number itself. If det $A = 0$, then A is said to be a *singular* matrix.

If A is a 3×3 matrix, the minors M and cofactors D are now 2×2 matrices, so we use their respective determinants in the sum for det A. The process expands to higher-order $n \times n$ matrices, where the first step yields a set of $(n-1) \times (n-1)$ matrices for the minors and cofactors, and their determinants are used in the sum for the evaluation of det A. Matrices with unit determinant are said to be *special*.

Next we consider the *inverse* A^{-1} of the matrix A, defined to be that matrix such that $AA^{-1} = I$, where I is the *identity matrix*—an $n \times n$ matrix where each diagonal element is 1 and all other elements are zero. The evaluation of the inverse involves forming the *adjoint* of A, denoted as Adj A, which is defined as the transpose of the matrix of cofactors.

For example, given the matrix A in Eq. 5.22, and using Table 5.2, we have

$$D = D_i^j = \begin{pmatrix} D_1^1 & D_1^2 \\ D_2^1 & D_2^2 \end{pmatrix} = \begin{pmatrix} d & -c \\ -b & a \end{pmatrix}, \tag{5.25}$$

from which we find Adj $A = D^T$:

$$\text{Adj } A \equiv D^T = \begin{pmatrix} d & -b \\ -c & a \end{pmatrix}. \tag{5.26}$$

If we then form the product $A(\text{Adj } A)$, we find in our example

$$A(\text{Adj } A) = \begin{pmatrix} a & b \\ c & d \end{pmatrix} \begin{pmatrix} d & -b \\ -c & a \end{pmatrix} = \begin{pmatrix} \det A & 0 \\ 0 & \det A \end{pmatrix} = (\det A)I. \tag{5.27}$$

Again, because $AA^{-1} = I$, we have the result for the inverse of A:

$$A^{-1} = \frac{\text{Adj } A}{\det A}. \tag{5.28}$$

Finally, Definition 5.6 provides a nomenclature that is widely used and helps to clarify whether we are speaking of real or complex matrices. Note especially the corresponding roles played by the transpose A^T (for real matrices) and the conjugate transpose $A^\dagger$ (for complex matrices). Matrices where $A = (A^T)^*$ are said to be *self-adjoint*, of which orthogonal and unitary matrices are two examples. We discuss these two types of matrices, as well as Hermitian matrices, in Sect. 5.7.

Definition 5.6 A real matrix A is said to be:

Symmetric, if $A^T = A$;
Skew-symmetric, if $A^T = -A$;
Orthogonal, if $A^T = A^{-1}$.

A complex matrix A is said to be:

Hermitian, if $A^\dagger = A$;
Skew-hermitian, if $A^\dagger = -A$;
Unitary, if $A^\dagger = A^{-1}$.

∎

Special orthogonal and unitary $n \times n$ matrices are elements in the Lie groups $SO(n)$ and $SU(n)$, respectively. We discuss these further in Chap. 8.

5.5.2 Linear Transformations, Images and Null Spaces

A *linear transformation* on a vector space is a linear map[14] from one vector space to another. For vector spaces U and V (as in Fig. 5.3), we write the linear transformation from U to V as $T : U \to V$, and the mapping of $|u\rangle \in U$ to $|v\rangle \in V$ as

$$T : |u\rangle \to |v\rangle \quad \text{or} \quad T(\mathbf{u}) = \mathbf{v}. \tag{5.29}$$

Those vectors in U that are mapped to the zero vector in V form the *kernel*, or *null space*, of the transformation T. The *rank* of T is the dimension of the space constituting the range, or *image*, of T. Together, these concepts combine to form one of the more important results in linear algebra:

$$\dim(\text{Im } T) + \dim(\text{Ker } T) = \dim U. \tag{5.30}$$

Occasionally this is written with "rank T" in place of "dim (Im T)." A formal proof of this theorem is given in most linear algebra texts.[15] We will demonstrate it shortly with some examples, after which it may seem more intuitive.

If U is a space of m dimensions and V is a space of n dimensions, then the vectors $|u\rangle$ and $|v\rangle$ may be treated as column vectors with m and n components, respectively.

Fig. 5.3 The linear transformation T from vector space U to vector space V

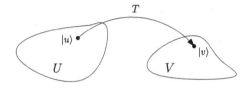

---

[14] See Sect. 4.3.3 and Definition 4.5.
[15] See, for example, [7], pp. 71–72.

The transformation T then becomes an exercise in matrix multiplication, with T represented as an $n \times m$ matrix A:

$$
\begin{pmatrix} A_1^1 & \cdots & A_m^1 \\ \vdots & & \vdots \\ A_1^n & \cdots & A_m^n \end{pmatrix} \begin{pmatrix} u^1 \\ \vdots \\ u^m \end{pmatrix} = \begin{pmatrix} v^1 \\ \vdots \\ v^n \end{pmatrix}.
\tag{5.31}
$$

The *rank of a matrix* is its number of linearly independent columns,[16] but the rank of an $n \times m$ matrix is not always obvious by inspection. The most reliable way of determining the rank is through a series of *elementary row operations*—a process akin to the method of Gaussian elimination—so as to place the matrix in *echelon form*. In this form the matrix's rank is obvious. We leave this as a topic for you to explore further on your own.[17]

Let us consider the options in Eq. 5.30:

- If A is nonsingular (det $A \neq 0$), then dim(Ker T) = 0, which is to say that the only vector in U that is mapped to the zero vector in V is the zero vector, which is of zero dimension. Equation 5.30 becomes dim(Im T) + 0 = dim U;
- If the matrix A is singular (det $A = 0$), there are two possibilities:

  1. First, if A is the zero matrix ($A = 0$ means all entries are zero), then all vectors in U are mapped to the zero vector in V. The full space U is the kernel of T, and Eq. 5.30 becomes 0 + dim(Ker T) = dim U;
  2. Second, if det $A = 0$ but $A \neq 0$, then both terms on the left-hand side of Eq. 5.30 are non-zero. The effect of the transformation T is to collapse the space U onto a space of a smaller dimension.

We can illustrate these ideas by considering the real, two-dimensional case where $\mathbf{u} \in U = \mathbb{R}^2$ and $\mathbf{v} \in V = \mathbb{R}^2$. Limiting ourselves to real spaces presents limitations, and we will loosen this constraint in the next section. For now, however, it serves our purposes. Using Cartesian coordinates, and with A real, we write

$$
A\mathbf{u} = \begin{pmatrix} a & b \\ c & d \end{pmatrix} \begin{pmatrix} x \\ y \end{pmatrix} = \begin{pmatrix} ax + by \\ cx + dy \end{pmatrix} \in V.
\tag{5.32}
$$

If A is a nonsingular matrix (det $\neq 0$, rank = 2), then we have situations like those shown in Fig. 5.4 where the columns in A tell us how the corresponding unit vectors in U are mapped to V (Fig. 5.5). The value of det A is the multiplicative factor by which areas in U are mapped to V.

Generally, if T is nonsingular, then dim(Im T) = dim U. Further, the transformations in Fig. 5.4 are isomorphisms (i.e., bijections), so T^{-1} exists as an inverse map. Algebraically, this is manifested by the inverse matrix A^{-1}.

---

[16]This is the column rank. A corresponding row rank can be shown to equal the column rank.

[17]See, especially, [7], Sects. 1.3 and 1.4.

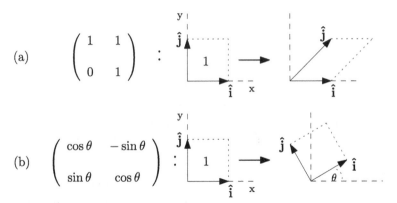

Fig. 5.4 The **a** shearing, and **b** rotation of the unit square in $\mathbb{R}^2$, with det = 1

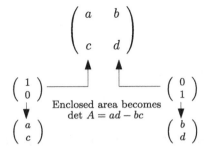

Fig. 5.5 The effect of a non-singular transformation on a unit square in $\mathbb{R}^2$. The area enclosed by the unit vectors in U is scaled by a factor of det A in V

As an aside, we note that nonsingular transformations that preserve distances between points (and hence angles between lines) as in Fig. 5.4b are called *Euclidean* transformations. Rotations, translations and reflections are examples of Euclidean transformations. So, for example, line segments, circles and triangles are mapped to congruent segments, circles and triangles, respectively.

Nonsingular transformations that preserve straight lines but not necessarily distances between points, as in Fig. 5.4a, are called *affine* transformations. For example, under affine transformations triangles may be mapped to other triangles that are neither congruent nor even similar. Regardless, in all these cases, the dimension of the image is equal to the dimension of the domain.

Returning now to Eq. 5.32, the situation for singular matrices (det = 0) in $\mathbb{R}^2$ is much different than for nonsingular matrices. As noted previously, the simplest singular matrix is the zero matrix $A = 0$. It is of zero rank, and the image of this "map" is the zero vector — dim(Im T) = 0 so that dim(Ker T) = 2; the entire plane (i.e., every vector) in U is mapped to the zero vector in V.

The rank of a *non-zero* singular matrix in $U = \mathbb{R}^2$ must equal one (why?), which means dim(Im T) = 1 (a line). An example is the *projection transformation* in Fig. 5.6.

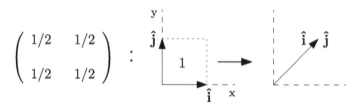

Fig. 5.6 A projection transformation that maps the plane of $\mathbb{R}^2$ to a line. The line $y = -x$ (not shown in the figure) is the kernel of the transformation and is mapped to the zero vector

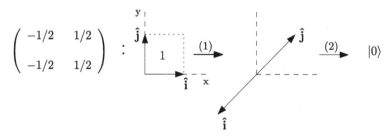

Fig. 5.7 A nilpotent transformation—one whose repeated application (here, twice) eventually yields the zero vector. The kernel of this transformation is the line $y = x$

You should verify that repeated application of the map in Fig. 5.6 (now with $U = \mathbb{R}^1$) yields the same line. Equation 5.30 also tells us that $\dim(\text{Ker } T) = 1$, and indeed the line $y = -x$ in U (not shown) is mapped to the zero vector in V. Geometrically, projecting the line $y = -x$ onto the line $y = x$ yields zero. Algebraically,

$$\begin{pmatrix} 1/2 & 1/2 \\ 1/2 & 1/2 \end{pmatrix} \begin{pmatrix} -1 \\ 1 \end{pmatrix} = \begin{pmatrix} 0 \\ 0 \end{pmatrix} \in V. \tag{5.33}$$

The singular map shown in Fig. 5.7 is also of rank = 1. Here, T maps $U = \mathbb{R}^2$ to the line $y = x$, but it also maps the line $y = x$ to the zero vector:

$$\begin{pmatrix} -1/2 & 1/2 \\ -1/2 & 1/2 \end{pmatrix} \begin{pmatrix} 1 \\ 1 \end{pmatrix} = \begin{pmatrix} 0 \\ 0 \end{pmatrix} \in V. \tag{5.34}$$

Geometrically, we can see that the first transformation in Fig. 5.7 maps the kernel, $y = x$, to the zero vector while also mapping the rest of the plane to the kernel! A second application of the transformation then maps the remaining line (the kernel) to zero. A transformation whose repeated application on a vector space eventually yields $|0\rangle$ is called a *nilpotent transformation*.

5.5.3 Eigenvectors, Similarity Transformations and Diagonalization of Matrices in Real Spaces

Given two real vector spaces,[18] U and V, and a linear transformation $T: U \rightarrow V$, there exists a set of vectors that maintain their orientations while being mapped from U to V, although their magnitudes may change. These vectors can also be made to serve as a basis for V, and they play a central role in a broad range of physical applications of operator algebras. Much of the reason for their efficacy in physical applications is that these vectors carry with them certain scalar quantities that we often associate with observable physical parameters.

These vectors are called *eigenvectors*, and their corresponding scalar quantities are called *eigenvalues*.[19] For any given space, they are a characteristic of the transformation and not solely of the space, *per se*. When we express linear transformations in terms of matrices, the practical goal of finding the eigenvectors and eigenvalues amounts to an exercise in matrix algebra.

Equation 5.35 codifies the features described above. For $|u\rangle = \mathbf{u} \in U$, $|v\rangle = \mathbf{v} \in V$ and the transformation T represented by the matrix A, we have

$$A\mathbf{u} = \mathbf{v} = \lambda\mathbf{u} \in V. \tag{5.35}$$

The scalar parameter λ is the eigenvalue corresponding to the eigenvector $\mathbf{u}$ and serves to change the vector's magnitude without changing its orientation. Equation 5.35 is called an *eigenvalue equation*, and for a given A its solution is found by first solving for λ, and then solving for the corresponding eigenvector $\mathbf{u}$. The set of eigenvalues associated with the matrix A constitute the *spectrum* of the operator.

The reason for finding the eigenvectors and spectrum of an operator is more than a matter of mere curiosity. As noted previously, a set of eigenvectors (once they are orthogonalized and normalized) may serve as an *eigenbasis* for a vector space, and these structures are broadly applied in classical and quantum dynamical systems. As we will see later in the chapter, certain types of operators are more amenable to these applications than others.

Often the physical system is formulated as a differential, rather than algebraic, equation so that the vector space is a function space.[20] The eigenvectors are then called *eigenfunctions*, and the solution follows from an application of the principles of function spaces that were discussed in Chap. 4.

For example, vibrating or oscillatory systems often have eigenfunctions that arise from a coordinate transformation. These *normal modes*, as they are called, serve as a basis in terms of which the behavior of the system may be described via a linear

---

[18] Transformations in complex vector spaces are discussed in Sects. 5.6 and 5.7. Although there are many similarities with transformations in real spaces, there also are important differences.

[19] The German "eigen-" may be interpreted as "(its) own" or "typical." In physics it is usually interpreted as "characteristic." Eigenvectors and eigenvalues are therefore said to be characteristic vectors and values—characteristic of the linear transformation in the given space.

[20] In which case the word "orientation" for eigenvectors is not as illustrative of the main idea.

expansion over all modes. The coefficient of each mode is related to the relative contribution of that mode to the overall motion. Similarly, the state (wavefunction) of a quantum system (e.g., a particle in an infinite square well) may be expanded in terms of the *eigenstates* of a particular operator, such as the Hamiltonian.

With these observations as motivation, let's assume for definiteness that U and V are two-dimensional, so that A is a 2×2 matrix. The first step in solving Eq. 5.35 is to recognize that the term $\lambda\mathbf{u}$ may be written in matrix form as

$$\lambda\mathbf{u} = \lambda \begin{pmatrix} u^1 \\ u^2 \end{pmatrix} = \begin{pmatrix} \lambda u^1 \\ \lambda u^2 \end{pmatrix} = \begin{pmatrix} \lambda & 0 \\ 0 & \lambda \end{pmatrix} \begin{pmatrix} u^1 \\ u^2 \end{pmatrix} = \lambda I\mathbf{u}, \tag{5.36}$$

where I is the unit matrix. The eigenvalue equation may then be written as

$$A\mathbf{u} = \lambda I\mathbf{u} \quad \text{or} \quad (A - \lambda I)\mathbf{u} = 0. \tag{5.37}$$

For nonsingular matrices (det $A \neq 0$) there will be two ket eigenvectors with corresponding eigenvalues. We will distinguish between them with subscripts, such as $\mathbf{u}_1$ and $\mathbf{u}_2$, while using superscripts for their components. Equation 5.35 then becomes

$$(A\mathbf{u}_1, A\mathbf{u}_2) = A \underbrace{(\mathbf{u}_1, \mathbf{u}_2)}_{\in U} = \underbrace{(\lambda_1\mathbf{u}_1, \lambda_2\mathbf{u}_2)}_{\in V} = (\mathbf{v}_1, \mathbf{v}_2), \tag{5.38}$$

where we have indicated the spaces in which the column vectors in each matrix reside (in a coordinate transformation, U and V would be the same space). Expanding Eq. 5.38 gives

$$A \underbrace{\begin{pmatrix} u_1^1 & u_2^1 \\ u_1^2 & u_2^2 \end{pmatrix}}_{\equiv S} = \underbrace{\begin{pmatrix} \lambda_1 u_1^1 & \lambda_2 u_2^1 \\ \lambda_1 u_1^2 & \lambda_2 u_2^2 \end{pmatrix}}_{=AS} = \underbrace{\begin{pmatrix} u_1^1 & u_2^1 \\ u_1^2 & u_2^2 \end{pmatrix}}_{\equiv S} \underbrace{\begin{pmatrix} \lambda_1 & 0 \\ 0 & \lambda_2 \end{pmatrix}}_{\equiv J}, \tag{5.39}$$

or, more succinctly,

$$AS = SJ. \tag{5.40}$$

The columns of matrix S are the eigen*vectors* of A, and the matrix J is a diagonalized matrix whose entries are the eigen*values* of A. As such, both S and J are nonsingular, which means they have inverses. Multiplying Eq. 5.40 on the right by S^{-1} gives

$$A = SJS^{-1}, \tag{5.41}$$

while multiplying on the left gives

$$J = S^{-1}AS. \tag{5.42}$$

Equations 5.41 and 5.42 are examples of *similarity transformations* between A and J whose effect is to change the standard coordinate basis vectors in U to a new set of basis vectors in V that coincides with the eigenvectors of A, or *vice versa*.

The matrix J represents a *diagonalization* of A, it is structurally equivalent[21] to A and it is easier to apply. An $n \times n$ matrix A may be similar to a diagonal matrix J only if the eigenvectors of A are linearly independent.

In considering Eq. 5.37 we see that the vector $\mathbf{u}$ is the kernel of the matrix $(A - \lambda I)$. Recalling our earlier discussion around Eq. 5.30, there are three possibilities[22]:

1. $A - \lambda I = 0$ and therefore has rank $= 0$. In this case λ is known immediately and the kernel $\mathbf{u}$ would be the entire space U—any vector will satisfy the equation[23];
2. $(A - \lambda I)$ is of rank $= 2$, and the kernel of $(A - \lambda I)$ would just be the zero vector;
3. $(A - \lambda I)$ is of rank $= 1$. As a 2×2 matrix of rank 1, its determinant is zero.

Our problem involves option 3. We set $\det (A - \lambda I) = 0$, solve for λ and then substitute the result back into Eq. 5.37 to find $\mathbf{u}$. The two-dimensional case has an advantage in that finding the eigenvalues reduces to solving a quadratic equation. Writing $(A - \lambda I)$ as

$$(A - \lambda I) = \begin{pmatrix} a - \lambda & b \\ c & d - \lambda \end{pmatrix}, \tag{5.43}$$

setting $\det (A - \lambda I) = 0$ and solving for λ gives

$$\lambda = \frac{\text{Tr } A \pm \sqrt{(\text{Tr } A)^2 - 4\det A}}{2}, \tag{5.44}$$

as you may show. This presents three possibilities:

$$(\text{Tr } A)^2 - 4\det A > 0 \rightarrow \text{two real and distinct roots,}$$
$$(\text{Tr } A)^2 - 4\det A = 0 \rightarrow \text{a double real root,}$$
$$(\text{Tr } A)^2 - 4\det A < 0 \rightarrow \text{a conjugate pair of roots.}$$

Example 5.11 Consider the case of distinct real roots and let

$$A = \begin{pmatrix} 0 & 1 \\ -1 & -10/3 \end{pmatrix}.$$

Solving Eq. 5.44 gives $\lambda_1 = -1/3$ and $\lambda_2 = -3$. To find the first eigenvector, we substitute λ_1 into Eq. 5.35 or Eq. 5.37 and write

---

[21]"Structurally equivalent" here means that the determinant and the trace are invariant under similarity transformations. This is considered further in the end-of-chapter problems.

[22]Although this line of reasoning is specific to the two-dimensional case ($n = 2$), it can be extended for $n > 2$, where $(A - \lambda I)$ would then be a singular matrix with $0 < \text{rank} < n$.

[23]We will find this option relevant to our consideration of the degenerate eigenvectors of a 2×2 Hermitian matrix in Example 5.12.

$$\begin{pmatrix} 0 & 1 \\ -1 & -10/3 \end{pmatrix} \begin{pmatrix} u_1^1 \\ u_1^2 \end{pmatrix} = -\frac{1}{3} \begin{pmatrix} u_1^1 \\ u_1^2 \end{pmatrix}.$$

This gives two equations for the two unknown components u_1^1 and u_1^2 of the vector $\mathbf{u}_1$. Recall that although A itself is non-singular, $(A - \lambda I)$ is singular with rank equal to one. Consequently, because this rank is less than the number of unknowns, the best we can do is solve for one of the unknowns in terms of the other.[24] We then have the two equations

$$0u_1^1 + u_1^2 = -\frac{1}{3}u_1^1$$

$$-u_1^1 - \frac{10}{3}u_1^2 = -\frac{1}{3}u_1^2,$$

from which we find that $u_1^1 = -3u_1^2$. Setting $u_1^2 = 1$ (a convenient and perfectly-allowed choice) means that

$$\lambda_1 = -\frac{1}{3} \rightarrow \mathbf{u}_1 \equiv \begin{pmatrix} u_1^1 \\ u_1^2 \end{pmatrix} = \begin{pmatrix} -3 \\ 1 \end{pmatrix} \rightarrow \hat{\mathbf{u}}_1 = \frac{1}{\sqrt{10}} \begin{pmatrix} -3 \\ 1 \end{pmatrix},$$

where $\mathbf{u}_1$ is normalized as $\hat{\mathbf{u}}_1$. Similarly, you can show that

$$\lambda_2 = -3 \rightarrow \mathbf{u}_2 \equiv \begin{pmatrix} u_2^1 \\ u_2^2 \end{pmatrix} = \begin{pmatrix} 1 \\ -3 \end{pmatrix} \rightarrow \hat{\mathbf{u}}_2 = \frac{1}{\sqrt{10}} \begin{pmatrix} 1 \\ -3 \end{pmatrix}.$$

The two vectors in V are $\mathbf{v}_1 = \lambda_1 \mathbf{u}_1$ and $\mathbf{v}_2 = \lambda_2 \mathbf{u}_2$, but when normalized they are (as they must be) the same as the normalized vectors in U.

The two eigenvectors $\mathbf{u}_1$ and $\mathbf{u}_2$ are linearly independent but are *not* orthogonal; you can check this either by direct calculation of their inner product (it's not zero) or by graphing them. It is also straightforward to show that Eq. 5.40 is satisfied. We leave these and a few additional features of this example as exercises.   ▲

5.6 Hermitian Operators

We now turn our attention to complex spaces, but before discussing transformations we describe Hermitian operators (where $A^\dagger = A$) because of their importance to applications in physics. This importance arises in part because even though they are complex matrices their eigenvalues are real and can lend themselves to interpretation as measurable physical parameters. Another reason pertains to the nature of their eigenvectors, which we discuss in Definition 5.7.

---

[24]When using matrix methods to solve a system of n linear algebraic equations, the rank of the matrix of coefficients must equal the number of unknown variables for there to be n independent solutions. Indeed, those methods rest on turning each equation in the linear system into a vector.

We will give a general derivation of the fact that "Hermitian eigenvalues" are real in Sect. 5.7, but we can demonstrate it easily for a 2×2 matrix by using the results of Eq. 5.44. From its definition, we know that we can write the general form of a 2×2 Hermitian matrix as

$$A = \begin{pmatrix} a & b \\ b^* & d \end{pmatrix}, \tag{5.45}$$

where because $A^\dagger = A$ we must have $a, d \in \mathbb{R}$. If the eigenvalues were complex, Eq. 5.44 would require $(a + d)^2 < 4(ad - b^2)$, where $b^2 \equiv b^*b \in \mathbb{R} > 0$. This reduces to $(a - d)^2 < -4b^2$, which is not possible, Therefore, λ must be real.

In the event $(a + d)^2 > 4(ad - b^2)$, we have two distinct real eigenvalues, and we would proceed as we did in Example 5.11. A particularly interesting case, however, arises when $(a + d)^2 = 4(ad - b^2)$. In this instance, λ is a double real root of the eigenvalue equation and is called a *degenerate* eigenvalue—an important result inasmuch degeneracies appear frequently in the spectra of physical systems.

Example 5.12 Consider the Hermitian matrix

$$A = \begin{pmatrix} a & b \\ b^* & d \end{pmatrix},$$

and assume the eigenvalues are degenerate. Then Eq. 5.44 requires $(a + d)^2 = 4(ad - b^2)$, or $(a - d)^2 = -4bc$, which is satisfied only by $b = c = 0$, which implies $a = d$. Therefore, A is diagonal and $(A - \lambda I) = 0$, so *any* vector satisfies the eigenvalue equation (i.e., the entire domain of A is the kernel). Far from being a problem, this circumstance means that we are free to choose a convenient pair of orthogonal basis vectors corresponding to the degenerate eigenvalues:

$$A = \begin{pmatrix} a & 0 \\ 0 & a \end{pmatrix} \quad \rightarrow \quad \mathbf{u_1} = \begin{pmatrix} 1 \\ 0 \end{pmatrix} \quad \text{and} \quad \mathbf{u_2} = \begin{pmatrix} 0 \\ 1 \end{pmatrix},$$

and our problem is solved by having taken advantage of the dimensions associated with the kernel and the image of transformations (Eq. 5.30). $\blacktriangle$

In Sect. 5.7 we will see that a Hermitian operator is just one member of a larger family of operators whose eigenvectors are orthogonal if their eigenvalues are distinct.

5.7 Unitary, Orthogonal and Hermitian Transformations

The main ideas behind transformations in complex vector spaces may be illustrated by considering the transformation $|v\rangle = A|u\rangle$ in two dimensions:

$$|v\rangle = A|u\rangle \rightarrow v^i = A^i_j u^j \rightarrow \begin{pmatrix} v^1 \\ v^2 \end{pmatrix} = \begin{pmatrix} A^1_1 & A^1_2 \\ A^2_1 & A^2_2 \end{pmatrix} \begin{pmatrix} u^1 \\ u^2 \end{pmatrix} = \begin{pmatrix} a & b \\ c & d \end{pmatrix} \begin{pmatrix} u^1 \\ u^2 \end{pmatrix}.$$ (5.46)

The ket vector $|v\rangle$ is then the column vector

$$|v\rangle = \begin{pmatrix} v^1 \\ v^2 \end{pmatrix} = \begin{pmatrix} au^1 + bu^2 \\ cu^1 + du^2 \end{pmatrix},$$ (5.47)

and its dual bra vector $\langle v|$ as the complex conjugate of the corresponding row vector,

$$\langle v| = (v^*_1, v^*_2) = (a^*u^*_1 + b^*u^*_2, c^*u^*_1 + d^*u^*_2).$$ (5.48)

Equation 5.48 may be written more compactly as

$$\langle v| = (u^*_1, u^*_2) \begin{pmatrix} a^* & c^* \\ b^* & d^* \end{pmatrix} = \langle u|A^\dagger \rightarrow v^*_i = u^*_j A^{j*}_i$$ (5.49)

where $A^\dagger$ is the transpose conjugate of A. Equation 5.48 gives the appearance of $\langle u|$ operating on $A^\dagger$, but is more often interpreted as $A^\dagger$ operating to the left on $\langle u|$; the result is the same. Of course, A operates to the right on the ket vector $|u\rangle$.

Now, given a vector space X, we consider the effect of a linear transformation A on the inner product $\langle u|v\rangle$ of two vectors $|u\rangle, |v\rangle \in X$. Because both vectors are subject to the transformation, we can write $A\langle u|v\rangle = \langle Au|Av\rangle = \langle u|A^\dagger Av\rangle$. The consequences of this transformation clearly depend on the specific nature of A.

In particular, for complex spaces and a unitary matrix ($A^\dagger A = I$) we conclude that inner products are invariant under *unitary transformations*: $\langle Au|Av\rangle = \langle u|v\rangle$. It follows directly from Definition 5.6 that for real vector spaces inner products are invariant under *orthogonal transformations*.[25]

Another important result comes from considering two nonsingular operators A and B. If their product $C = AB$ is nonsingular, then C must have an inverse such that $CC^{-1} = I$. This requires $C^{-1} = B^{-1}A^{-1}$ so that $CC^{-1} = ABB^{-1}A^{-1} = A(BB^{-1})A^{-1} = AA^{-1} = I$. Although this is true for all nonsingular matrices, by Definition 5.6 this otherwise general result becomes $(AB)^\dagger = B^\dagger A^\dagger$ for unitary matrices, and $(AB)^T = B^T A^T$ for orthogonal matrices.

We now turn our attention to Hermitian operators, the eigenvalues of which are real (as shown by example in Sect. 5.5.3). A more direct and general way of showing this is to apply the eigenvalue equation $A|u\rangle = \lambda|u\rangle$ for $|u\rangle \neq 0$, which gives

$$\langle u|Au\rangle = \langle u|\lambda u\rangle = \lambda\langle u|u\rangle$$ (5.50)

where λ is the scalar eigenvalue that may be complex. In the event A is Hermitian, we also have

---

[25] Because of these properties, unitary and orthogonal transformations are called *isometries*.

$$\langle u|Au\rangle = \langle u|A^{\dagger}u\rangle = \langle Au|u\rangle = \langle \lambda u|u\rangle = \lambda^{*}\langle u|u\rangle. \tag{5.51}$$

Comparing Eqs. 5.50 and 5.51 shows $\lambda = \lambda^{*}$, so λ must be real.

Another property of Hermitian operators is that if two eigenvectors of a Hermitian transformation have distinct eigenvalues, then those eigenvectors are orthogonal. This clearly is not a general property of all matrices (see Example 5.11 for a counterexample), and we will show this is true for Hermitian matrices. First, however, we note that a Hermitian operator is only one member of a much larger family of operators with this property.

Definition 5.7 A linear operator A that satisfies the condition $AA^{\dagger} = A^{\dagger}A$ is said to be a *normal* operator. Further, if A is normal, then eigenvectors with distinct eigenvalues are orthogonal.   ■

Clearly, unitary and orthogonal transformations fall into this family because of the equivalence of left and right inverses, but so do Hermitian operators, and so do *all* of the operators listed in Definition 5.6.

We will show this explicitly only for Hermitian matrices.[26] Let $A|u\rangle = \lambda_1|u\rangle$ and $A|v\rangle = \lambda_2|v\rangle$, and stipulate that $\lambda_1 \neq \lambda_2$ are both non-zero. Generally, we can write

$$\langle u|Av\rangle = \langle u|\lambda_2 v\rangle = \lambda_2\langle u|v\rangle, \tag{5.52}$$

but in the particular case for $A = A^{\dagger}$ (where the eigenvalues are real) we also have

$$\langle u|Av\rangle = \langle u|A^{\dagger}v\rangle = \langle Au|v\rangle = \langle \lambda_1 u|v\rangle = \lambda_1^{*}\langle u|v\rangle = \lambda_1\langle u|v\rangle. \tag{5.53}$$

Comparison of Eqs. 5.52 and 5.53 shows $\langle u|v\rangle$ must be zero, and therefore the vectors $|u\rangle$ and $|v\rangle$ are orthogonal. The demonstration of this property for the other matrices in Definition 5.6 follows similar lines.

Consequently, if *all* the eigenvalues of a normal operator are distinct, then their corresponding eigenvectors form a basis. However, for Hermitian operators we can add to this the result from Sect. 5.6 and Example 5.12 that if a portion of the spectrum is degenerate, those eigenvectors also can be included in the basis.

We now turn to an important consequence of the fact that the columns of the similarity transformation matrix S are eigenvectors of the matrix A. Recall, from Eq. 5.42 that $S^{-1}AS = J$, where J is a diagonal matrix of the eigenvalues of A. For n normalized eigenvectors that span the vector space (that is, they are linearly independent, but not necessarily orthogonal), we write the $n \times n$ matrix $S = S(\hat{\mathbf{u}}_1, \hat{\mathbf{u}}_2, \ldots \hat{\mathbf{u}}_n)$ as

$$S = \begin{pmatrix} S_1^1 & \cdots & S_n^1 \\ \vdots & & \vdots \\ S_1^n & \cdots & S_n^n \end{pmatrix}, \tag{5.54}$$

---

[26]More general proofs may be found in [3], Sect. 4.6 and [7], Sect. 8.5.

where the first column is $\hat{\mathbf{u}}_1 = |\hat{u}_1\rangle = S_1^i$, and the kth column is $\hat{\mathbf{u}}_k = |\hat{u}_k\rangle = S_k^i$. The scalar product of two arbitrary eigenvectors $\hat{\mathbf{u}}_k$ and $\hat{\mathbf{u}}_l$ is then

$$\langle \hat{u}_k | \hat{u}_l \rangle = S_i^{k*} S_l^i. \tag{5.55}$$

Now, if in addition to spanning the space and being normalized the eigenvectors are orthogonal—as they are guaranteed to be in a Hermitian transformation with either distinct or degenerate eigenvalues—then the scalar product in Eq. 5.55 is just δ_l^k, the Kronecker delta. In matrix form, this is $S^\dagger S = I$, which means S is unitary.

In summary, we have derived three important and related properties of Hermitian operators—properties which have wide application in complex inner product spaces:

1. any Hermitian matrix A may be diagonalized via a unitary transformation S whose columns are the eigenvectors of A;
2. if an operator is Hermitian, we can always find a complete set of orthonormal eigenvectors, regardless of whether the eigenvalues are distinct, degenerate or some combination of both; and
3. the eigenvalues of a Hermitian operator are real.

Example 5.13 Consider the matrix

$$A = \begin{pmatrix} 1 & 1+i \\ 1-i & 0 \end{pmatrix},$$

which by inspection is Hermitian. We find $\operatorname{Tr} A = 1$, $\det A = -2$ and the distinct eigenvalues of A (from Eq. 5.44) to be $\lambda_1 = -1$ and $\lambda_2 = 2$.

Considering each eigenvalue in turn, and following the same procedure as in Example 5.11, we find:

$$\lambda_1 = -1 \quad \rightarrow \quad \mathbf{u}_1 = -\frac{1}{2}\begin{pmatrix} 1+i \\ -2 \end{pmatrix} \quad \rightarrow \quad \hat{\mathbf{u}}_1 = -\frac{1}{\sqrt{6}}\begin{pmatrix} 1+i \\ -2 \end{pmatrix}.$$

Similarly,

$$\lambda_2 = 2 \quad \rightarrow \quad \mathbf{u}_2 = \begin{pmatrix} 1+i \\ 1 \end{pmatrix} \quad \rightarrow \quad \hat{\mathbf{u}}_2 = \frac{1}{\sqrt{3}}\begin{pmatrix} 1+i \\ 1 \end{pmatrix}.$$

We are now in a position to construct the similarity transformation matrix S from the normalized eigenvectors:

$$S = S(\hat{\mathbf{u}}_1, \hat{\mathbf{u}}_2) = \frac{1}{\sqrt{3}}\begin{pmatrix} -\frac{(1+i)}{\sqrt{2}} & 1+i \\ \frac{\sqrt{2}}{\sqrt{2}} & 1 \end{pmatrix} \quad \rightarrow \quad S^\dagger = \frac{1}{\sqrt{3}}\begin{pmatrix} \frac{(i-1)}{\sqrt{2}} & \sqrt{2} \\ 1-i & 1 \end{pmatrix}.$$

A straightforward calculation shows that $S^\dagger S = I$, and that

$$AS = SJ = \frac{1}{\sqrt{3}}\begin{pmatrix} \frac{(i+1)}{\sqrt{2}} & 2(1+i) \\ -\sqrt{2} & 2 \end{pmatrix}, \quad \text{with} \quad J = \begin{pmatrix} -1 & 0 \\ 0 & 2 \end{pmatrix}.$$

Finally, we see that in fact $\text{Tr } A = \text{Tr } J = 1$, and $\det A = \det J = -2$. ▲

Finally, there are two additional results regarding Hermitian operators, although they are essentially another way of stating the orthonormalization and completeness properties in real and complex function spaces as we discussed them in Chap. 4.

First, and by way of an example, let A be a nonsingular Hermitian operator in a three-dimensional complex space U. The eigenvalue equation $A|\psi\rangle = a|\psi\rangle$ will yield a complete orthonormal set of three basis functions, which we take to be $|\phi_1\rangle$, $|\phi_2\rangle$ and $|\phi_3\rangle$. Then, any function $\psi \in U$ may be expressed as the linear combination

$$|\psi\rangle = c_1|\phi_1\rangle + c_2|\phi_2\rangle + c_3|\phi_3\rangle, \tag{5.56}$$

where the c_i are, in general, complex coefficients to the $|\phi_i\rangle$. The dual space bra vector $\langle\psi|$ is, of course,

$$\langle\psi| = c_1^*\langle\phi_1| + c_2^*\langle\phi_2| + c_3^*\langle\phi_3|. \tag{5.57}$$

Normalizing $|\psi\rangle$ yields

$$\langle\psi|\psi\rangle = |c_1|^2\langle\phi_1|\phi_1\rangle + |c_2|^2\langle\phi_2|\phi_2\rangle + |c_3|^2\langle\phi_3|\phi_3\rangle = \sum_i |c_i|^2 \equiv 1, \tag{5.58}$$

where we have applied the orthonormality property of the $|\phi_i\rangle$.

In applying this otherwise general result to quantum mechanics, the probabilistic interpretation of Eq. 5.58 is that if the vector $|\psi\rangle$ represents the wavefunction of the system, then $|c_i|^2$ is the probability that, upon measurement, the system will be found in eigenstate $|\phi_i\rangle$ from among the three possible eigenstates. The associated eigenvalue is a_i. For example, if A is an energy operator, then the $|\phi_i\rangle$ are the energy eigenstates, and the a_i (found as part of our solution for the $|\phi_i\rangle$) are the corresponding energies of those states.[27]

A second consequence of A being Hermitian emerges when we operate on the sum itself with the goal of extracting, or projecting, one term in the sum and not the others. Letting $|\psi\rangle = c_1|\phi_1\rangle + c_2|\phi_2\rangle + c_3|\phi_3\rangle$ as before, we act on the sum with the (admittedly odd-looking) operator

$$\left(\sum_{i=1}^{3}|\phi_i\rangle\langle\phi_i|\right)|\psi\rangle = |\phi_1\rangle\langle\phi_1|\psi\rangle + |\phi_2\rangle\langle\phi_2|\psi\rangle + |\phi_3\rangle\langle\phi_3|\psi\rangle$$

$$= |\phi_1\rangle c_1 + |\phi_2\rangle c_2 + |\phi_3\rangle c_3 = |\psi\rangle,$$

---

[27] Another application of this result appears in a Fourier series expansion of a function. There, the $|c_i|^2$ would be the amplitude of the contribution of each basis function to the function in question.

which we can see is really just an *identity operator*,

$$\left(\sum_{i=1}^{n}|\phi_i\rangle\langle\phi_i|\right) = I, \tag{5.59}$$

and in which any single term serves as a *projection operator* when acting on ψ.

5.8 Functions of Operators

The idea that a linear operator A can serve as the argument of a function—as in $\exp A$, $\sin A$ or $\cos A$—might seem odd at first, but such functions are allowed if the operators form an algebra. The result is another operator. If the function can be expanded as an infinite sum, then the sum must converge for the specified domain in order for the resulting operator to be well-defined.

A simple example is a polynomial of an operator. Given an operator A (for the sake of argument, assume it is an $n \times n$ matrix), then powers of A can be formed, and terms of different order may be summed. For example, the polynomial

$$f(A) = 3 + 2A + 6A^2 - 3A^3 \tag{5.60}$$

is an expression that makes perfectly good sense, provided we interpret "3" as "$3I$," where I is the identity matrix.

Of course, it is possible to imagine some special cases. For example, if $A = I$ in Eq. 5.60, then $f(I) = 8I$. On the other hand, if A is a nilpotent operator, then $f(A)$ might terminate before the final term. The nilpotent matrix in Fig. 5.7 is an example of this, where

$$A = \begin{pmatrix} -1/2 & 1/2 \\ -1/2 & 1/2 \end{pmatrix} \rightarrow A^2 = 0 \quad \rightarrow \quad f(A) = 3I + 2A = \begin{pmatrix} 2 & 1 \\ -1 & 4 \end{pmatrix}. \tag{5.61}$$

If we replace A in Eq. 5.60 with $A + B$ or AB, then we need to be certain that addition and multiplication between A and B are defined before $f(A + B)$ or $f(AB)$ makes any sense. If A and B are related to each other, that also must be taken into account.

In order to understand the meaning of expressions like e^A, $\sin A$ and $\cos A$, it helps to write these functions as infinite sums. However, although matrix addition commutes, matrix multiplication does not necessarily commute. Therefore, the order of the arguments in a function of operators must be respected.

For example, by expanding e^A and e^B to second-order and carrying out the multiplications, we have

$$e^A = I + A + \frac{1}{2}A^2 + \cdots$$

$$e^B = I + B + \frac{1}{2}B^2 + \cdots$$

$$e^A e^B = I + (A + B) + \frac{1}{2}(A^2 + 2AB + B^2) + \cdots$$

$$e^B e^A = I + (B + A) + \frac{1}{2}(B^2 + 2BA + A^2) + \cdots ,$$

from which we find $e^A e^B - e^B e^A = AB - BA = [A, B]$, their commutator (Sect. 5.3). Therefore, $e^A e^B = e^B e^A$ only if A and B commute.

Another example comes from asking whether $e^{(A+B)} = e^A e^B$. Again we will find that it depends on whether A and B commute. Writing (again to second order)

$$e^{(A+B)} = I + (A + B) + \frac{1}{2}(A^2 + AB + BA + B^2) + \cdots ,$$

and comparing this expression with our earlier expression for $e^A e^B$, we have

$$e^{(A+B)} - e^A e^B = AB + BA - 2AB = -[A, B]. \tag{5.62}$$

Next, consider the fact that a differential equation is essentially a function of operators. For example, consider Eq. 5.60 where $A = d/dx$. We then have

$$F(\psi) = \left(3 + 2\frac{d}{dx} + 6\frac{d^2}{dx^2} - 3\frac{d^3}{dx^3}\right)\psi = 0. \tag{5.63}$$

If you have studied differential equations, you may have come across a solution technique where you let $d/dx = D$, and then proceed to do algebra with D before moving on to a solution.[28] What you are doing by this method is essentially the reverse of what we have just done here—going from the differential equation back to the finite polynomial. That technique works because the operators form an algebra.

Finally, operators may simply be multiplicative functions, and they may appear in combination with other operators. This "simple" case often turns out to be one of the trickiest to handle. For example, consider the operator combination

$$[A, B] \quad \text{for} \quad A = \frac{d}{dx} \quad \text{and} \quad B = f(x).$$

If we let the operator $[A, B]$ act on some function $\psi(x)$, then we have

---

[28] The operator F in Eq. 5.63 would be written as $F = 3 + 2D + 6D^2 - 3D^3$.

$$[A, B]\psi = \left[\frac{d}{dx}, f\right]\psi = \left(\frac{d}{dx}f - f\frac{d}{dx}\right)\psi$$
$$= \frac{d}{dx}(f\psi) - f\frac{d\psi}{dx} = \psi\frac{df}{dx},$$

which is often interpreted to mean that

$$\left[\frac{d}{dx}, f\right] = \frac{df}{dx}. \tag{5.64}$$

It is easy to misinterpret Eq. 5.64 if we forget the two different roles played by ψ—first as the argument of a commutator operator, and then as a multiplicative function.

We will consider some functions of operators in the problems that follow.

Problems

5.1 Let

$$A = \begin{pmatrix} 1 & -1 \\ 0 & 1 \end{pmatrix}, \qquad B = \begin{pmatrix} 1 & 0 \\ -1 & 1 \end{pmatrix}, \qquad C = \begin{pmatrix} 0 & 1 \\ 1 & -1 \end{pmatrix}.$$

Find (a) $A + B$; (b) $2A + C$; (c) AB; (d) BC; (e) $(AB)C$; (f) $A(BC)$.

5.2 Let

$$A = \begin{pmatrix} x & y & 1 \\ -1 & 2 & 1 \\ 1 & 0 & 1 \end{pmatrix},$$

and show that $\det A = 0$ represents the straight line $x + y = 1$.

5.3 Let

$$A = \begin{pmatrix} 1 & -2 & 1 \\ 2 & 1 & -3 \\ 0 & 1 & 1 \end{pmatrix}.$$

Find (a) $\det A$; (b) Adj A; (c) A^{-1}; (d) Verify that $AA^{-1} = I$.

5.4 In addition to their representations given in Chap. 3, complex numbers may be written in matrix form. For example, $z = a + ib$ may be written as $z = aI + bi$, or

$$z = \begin{pmatrix} a & -b \\ b & a \end{pmatrix}, \qquad \text{where} \qquad I = \begin{pmatrix} 1 & 0 \\ 0 & 1 \end{pmatrix} \quad \text{and} \quad i = \begin{pmatrix} 0 & -1 \\ 1 & 0 \end{pmatrix}.$$

(a) Verify this expression for z by showing that the matrix product $i^2 = -1$;
(b) Show that $\det z = |z|^2$;
(c) Find the matrix expression for z^{-1};
(d) Applying the matrix formulation for z, show $|z_1 z_2| = |z_1||z_2|$. [*Hint*: This reflects a property of determinants whereby $\det (AB) = (\det A)(\det B)$.]

5.5 The rotation matrix in the two-dimensional plane (Fig. 5.4) is

$$R(\theta) = \begin{pmatrix} \cos\theta & -\sin\theta \\ \sin\theta & \cos\theta \end{pmatrix}.$$

(a) Referring to Problem 5.4, show that $R(\theta)$ is the matrix representation of $e^{i\theta}$;
(b) Find $R^2(\theta)$, first by direct matrix multiplication and then by applying part (a).
[*Note:* The expression in part (a) will be of central importance when we discuss Lie groups in Sect. 8.5.]

5.6 Regarding the definitions of the Lie and Poisson bracket:
(a) Show that the Lie bracket $[A, B]$ satisfies the Jacobi identity;
(b) Show that the Poisson bracket $\{A, B\}$ satisfies the Poisson condition in Eq. 5.18;
(c) Show that if the Poisson bracket $\{A, B\}$ equals the Lie bracket, then Eq. 5.18 reduces to the Jacobi identity.

5.7 The *Pauli matrices* appear in several different contexts within mathematical physics. Among these are: (a) the transformations of spin-1/2 particles, which are formulated as two-component objects known as *spinors*[29]; (b) as contributors to the *Dirac γ matrices*, which appear in Dirac's equation for the dynamics of relativistic spin-1/2 particles; (c) as elements in the isospin transformations between a proton and neutron; and (d) the description of rotations in two and three dimensions.

The purpose of this problem is to introduce the algebra associated the Pauli matrices, whose standard representation is

$$\sigma_1 = \begin{pmatrix} 0 & 1 \\ 1 & 0 \end{pmatrix}, \sigma_2 = \begin{pmatrix} 0 & -i \\ i & 0 \end{pmatrix} \text{ and } \sigma_3 = \begin{pmatrix} 1 & 0 \\ 0 & -1 \end{pmatrix},$$

and where each σ_i is a traceless Hermitian matrix.
(a) Show that each Pauli matrix is a unitary matrix by verifying that $\sigma_i(\sigma_i)^\dagger = 1$;
(b) Show that $\sigma_1\sigma_2 = i\sigma_3$ and that $[\sigma_1, \sigma_2] = 2i\sigma_3$;
(c) Evaluate $[\sigma_1, \sigma_3]$;
(d) Using your results above, show that $[\sigma_i, \sigma_j] = 2i\epsilon_{ij}^k\sigma_k$, where $2i\epsilon_{ij}^k$ represents the set of structure constants for this algebra (note that multiplication in this algebra is defined by the Lie product, making this a non-associative algebra);
(e) Show that the product of two Pauli matrices may be written as $\sigma_i\sigma_j = \delta_{ij} + i\epsilon_{ij}^k\sigma_k$, where δ_{ij} is the Kronecker delta.

5.8 Referring to Problem 5.7, part (d):
(a) Let $\sigma_i/2 \equiv S_i$, and show that $[S_i, S_j] = i\epsilon_{ij}^k S_k$. [*Note:* Within a factor of i, this is the same Lie algebra as for the vector cross product in Example 5.3 and Sect. 5.3.1.];
(b) Show that the following 3×3 matrices satisfy the same algebra as that followed by the 2×2 matrices in part (a):

---

[29] Just FYI—this is pronounced so as to rhyme with "winners."

$$J_1 = \begin{pmatrix} 0 & 0 & 0 \\ 0 & 0 & -i \\ 0 & i & 0 \end{pmatrix}, \ J_2 = \begin{pmatrix} 0 & 0 & i \\ 0 & 0 & 0 \\ -i & 0 & 0 \end{pmatrix} \quad \text{and} \quad J_3 = \begin{pmatrix} 0 & -i & 0 \\ i & 0 & 0 \\ 0 & 0 & 0 \end{pmatrix}.$$

[*Note:* The elements of the two Lie algebras described in Problems 5.7 and 5.8 are the *generators* for two Lie groups—the *special unitary group* $SU(2)$ and the *special orthogonal group* $SO(3)$, respectively. Precisely *how* these algebras generate these groups will be discussed in Sect. 8.5. There we will look at the role of $SU(2)$ in the isospin transformations, and how $SO(3)$ describes rotations in three-dimensional space. In quantum mechanics $SO(3)$ appears in the context of orbital angular momentum.]

5.9 Let

$$A = \begin{pmatrix} 0 & 1 \\ 0 & 0 \end{pmatrix} \quad \text{and} \quad B = \begin{pmatrix} 0 & 0 \\ 1 & 0 \end{pmatrix}.$$

Show that A and B are nilpotent, but that their product AB is not.

5.10 Show that the matrices A and B in Problem 5.9 are similar by finding the matrix S such that $AS = SB$.

5.11 This problem is a continuation of Example 5.11.
(a) Show

$$AS = SJ = \frac{1}{\sqrt{10}} \begin{pmatrix} 1 & -3 \\ -1/3 & 9 \end{pmatrix},$$

where $S = (\hat{\mathbf{u}}_1, \hat{\mathbf{u}}_2)$ and J is diagonal with the two eigenvalues as entries.
(b) Show that $S^T S \neq I$ (A is not orthogonal, so neither is S).
(c) Verify that the determinant and the trace are invariant under a similarity transformation, with $\det A = \det J = 1$, and $\text{Tr } A = \text{Tr } J = -10/3$.

5.12 Referring to Example 5.13:
(a) Complete the details of the calculations and find the two unit vectors $\hat{\mathbf{u}}_1$ and $\hat{\mathbf{u}}_2$;
(b) Show that S is a unitary matrix.

5.13 Fundamentally, the *Heisenberg Uncertainty Principle* is a statement of an incompatibility between position and momentum—a characteristic of Nature that is apparent within quantum mechanics but which has no counterpart in classical (non-quantum) physics.

In mathematical terms, this incompatibility follows from the form and function of the momentum and position operators. These lead to corresponding position-space and momentum-space wave functions for a particle, from which we get a probability distribution for each parameter.

Each probability distribution has its own dispersion, or "uncertainty," about its peak. The product of these two dispersions has a fixed (constant) minimum. Consequently, as one dispersion narrows the other widens (as one parameter is more sharply defined, the other becomes less so), and the two "uncertainties" cannot both be made arbitrarily small simultaneously.[30]

Given a wavefunction $\psi(x)$, we first consider the commutator of the momentum operator $\hat{p}$ and the position operator $\hat{x}$, respectively, which are given as

$$\hat{p}\psi = -i\hbar\frac{\partial\psi}{\partial x} \quad \text{and} \quad \hat{x}\psi = x\psi.$$

We then apply a general result for the dispersions σ_a and σ_b of expectation values for two parameters whose quantum operators are $\hat{A}$ and $\hat{B}$:

$$\sigma_a{}^2\sigma_b{}^2 \geq \left(\frac{1}{2i}[\hat{A}, \hat{B}]\right)^2.$$

The statement of the uncertainty principle—which you are now being asked to show in this problem—follows directly as $\sigma_p\sigma_x \geq \hbar/2$. Often this is written more casually as $\Delta p\Delta x \geq \hbar/2$, but the first form is preferred.

5.14 Find an invertible 2×2 matrix A such that

$$\begin{pmatrix} 0 & 1 \\ 0 & 0 \end{pmatrix} A \begin{pmatrix} 0 & 0 \\ 1 & 0 \end{pmatrix} = \begin{pmatrix} 0 & 0 \\ 1 & 0 \end{pmatrix} A \begin{pmatrix} 0 & 1 \\ 0 & 0 \end{pmatrix}.$$

5.15 Consider the algebra of all 3×3 matrices over the field $F = \mathbb{R}$. Show that matrices of the form

$$\begin{pmatrix} a & b & c \\ 0 & a+c & 0 \\ c & b & a \end{pmatrix},$$

where $a, b, c \in \mathbb{R}$, form a subalgebra.

5.16 Let A be an $n \times m$ matrix and B be a $p \times q$ matrix, both over the same field F. Their direct sum is the $(n + p) \times (m + q)$ matrix

$$A \oplus B = \begin{pmatrix} A & 0 \\ 0 & B \end{pmatrix},$$

a matrix with $(n + p)$ rows and $(m + q)$ columns, where the 0's represent zero matrices of the appropriate dimensions.
(a) Show that this example satisfies Definition 5.5 for the direct sum of algebras;
(b) Let A_1 and A_2 be $n \times n$ matrices, and let B_1 and B_2 be $p \times p$ matrices. Find the expression for $(A_1 \oplus B_1)(A_2 \oplus B_2)$, in matrix form.

---

[30]Our treatment here follows the approach taken in [5], Sect. 3.5, which proceeds from a more general principle. As such, and because it would otherwise take us too far afield into quantum mechanics, we will provide an important intermediate step without further rationale. However, see also the development in [10], Sect. 1.7, for an approach that treats a particle more explicitly as a traveling wave packet with a Gaussian profile near its peak.

5.17 Consider two masses (each of mass m) and three springs (each with spring constant k). Their equilibrium positions are shown below. At $t = 0$, the masses are

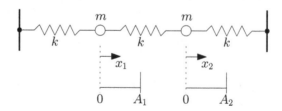

displaced from their equilibrium positions so that $x_1 = A_1$ and $x_2 = A_2$, and are then released. Let $x_1 = A_1 \cos \omega t$ and $x_2 = A_2 \cos \omega t$, where ω is the angular frequency of oscillation. An application of Newton's Second Law (or Lagrange's equations—see [8]) yields two coupled equations:

$$(2k - m\omega^2)A_1 - kA_2 = 0$$
$$-kA_1 + (2k - m\omega^2)A_2 = 0.$$

(a) By the methods of this chapter, find the two eigenvalues of ω (variously known as the eigen-, normal or characteristic frequencies) associated with this system. [*Hint:* Start by defining a vector A with components A_1 and A_2.]
(b) The normal frequencies found in part (a) represent oscillations of the two masses when their motions are "synchronized." Those motions are called the *normal modes of oscillation*. Describe them, either qualitatively or pictorially.

5.18 We wish to show that quaternions, as we defined them in Sect. 3.4.3, may be written in matrix form. Consider a quaternion $\mathbf{q} = \mathbf{1}q_0 + \mathbf{i}q_1 + \mathbf{j}q_2 + \mathbf{k}q_3$, where $q_i \in \mathbb{R}$, for $0 \le i \le 3$.
(a) Write the quaternion $\mathbf{q}$ as a 2×2 matrix, where

$$\mathbf{1} = \begin{pmatrix} 1 & 0 \\ 0 & 1 \end{pmatrix}, \quad \mathbf{i} = \begin{pmatrix} 0 & -1 \\ 1 & 0 \end{pmatrix}, \quad \mathbf{j} = \begin{pmatrix} 0 & -i \\ -i & 0 \end{pmatrix}, \quad \mathbf{k} = \begin{pmatrix} i & 0 \\ 0 & -i \end{pmatrix}.$$

Note the distinction between the unit quaternion (bold-faced $\mathbf{i}$) and $i = \sqrt{-1}$.
(b) Show that the unit quaternions given here as matrices satisfy the relations given in Definition 3.7.

Guide to Further Study
As noted in the Guide to Further Study at the end of Chap. 4, these two chapters fall largely within the purview of linear algebra. Of the references listed in the previous chapter, the texts of Hoffman and Kunze [7], Geroch [4] and Roman [9] are especially relevant to this chapter, particularly for their discussions around subalgebras, quotients and direct sums.

The presentation of matrix algebra in this chapter was premised on the understanding (see the Preface) that you were already familiar with matrix multiplication. More examples and important nuances are found in the standard methods texts, such as those of Arfken et al. [1] and Byron and Fuller [3].

The quantum mechanics text by Griffiths and Schroeter [5] is widely used and highly recommended. Every physics professor seems to have their favorite text for classical mechanics; for upper division and beginning graduate students mine is Landau and Lifshitz [8].The text by Thornton and Marion [11] is an excellent place to start if you are new to classical mechanics.

References

1. Arfken, G.B., Weber, H.J., Harris, F.E.: Mathematical Methods for Physicists—A Comprehensive Guide, 7th edn. Academic Press, Waltham, MA (2013)
2. Birkhoff, G., MacLane, S.: A Survey of Modern Algebra, 5th edn. A.K. Peters Ltd, Wellesley, MA (1997)
3. Byron, Jr., F.W., Fuller, R.W.: Mathematics of Classical and Quantum Mechanics (Two volumes bound as one). Dover, New York (1992); an unabridged, corrected republication of the work first published in two volumes by Addison-Wesley, Reading, MA (Vol. 1 (1969) and Vol. 2 (1970))
4. Geroch, R.: Mathematical Physics. In: Chicago Lectures in Physics. University of Chicago Press, Chicago (1985)
5. Griffiths, D.J., Schroeter, D.F.: Introduction to Quantum Mechanics, 3rd edn. Cambridge University Press, Cambridge (2018)
6. Hirsch, M.W., Smale, S.: Differential Equations, Dynamical Systems, and Linear Algebra. Academic Press, San Diego (1974)
7. Hoffman, K., Kunze, R.: Linear Algebra, 2nd edn. Prentice-Hall, Englewood Cliffs, NJ (1971)
8. Landau, L.D., Lifshitz, E.M.: Mechanics, 3rd edn. Pergamon Press, 1988 Printing, New York (1976)
9. Roman, P.: Some Modern Mathematics for Physicists and Other Outsiders, 2 Volumes. Pergamon Press, Elmsford, NY (1975)
10. Sakurai, J.J.: Modern Quantum Mechanics. Addison-Wesley, Redwood City, CA (1985)
11. Thornton, S.T., Marion, J.B.: Classical Dynamics of Particles and Systems, 5th edn. Cengage Learning, Boston (2003)

Chapter 6
Fundamental Concepts of General Topology

When a layperson thinks of topology, they most likely imagine a geometric shape that can morph into other shapes which, though different in detail from the original, are at once also similar in some broad and qualitative sense. *General topology—* also called *point set topology—*is the study of the properties of spaces, and of the features and structures that remain invariant under smooth transformations between them.[1] When formulated as continuous bijective maps between sets or spaces, these topological transformations are called *homeomorphisms—*not to be confused with the *homomorphisms* we studied earlier, under which particular features of algebraic structures are preserved.[2]

Much of the current research literature in mathematical physics presumes at least a basic understanding of topological concepts. Topological considerations appear across a wide range of fields—from the study of dynamical systems (particularly nonlinear and/or integrable Hamiltonian systems), gauge theories (such as electrodynamics, gravitation and quantum field theory) and current studies in quantum optics, to name just a few. The goal of this chapter is to introduce the essential ideas of general topology and offer a few basic examples. A Guide to Further Study is provided at the end of the chapter.

---

[1] A different approach to topology is distinctly combinatorial in nature. Combinatorial methods are prominent in the theory of networks, for example, and they overlap with general topology in the field of algebraic topology where algebraic methods are used to explore topological properties of spaces. In many respects, algebraic topology is where the field of topology had its historical beginnings in the late ninteenth century (see [3]; also [7], Chap. 50).

[2] From the Latin: *homeo-*, meaning *similar*; versus *homo-*, the *same*. See also the comments at the end of Sect. 1.3.

© Springer Nature Switzerland AG 2021
S. P. Starkovich, *The Structures of Mathematical Physics*,
https://doi.org/10.1007/978-3-030-73449-7_6

6.1   General Topology in a Geometric Context

We begin our discussion of topology by comparing topological transformations with others that are more familiar to us. As discussed earlier,[3] isometric (orthogonal and unitary) transformations preserve inner products (and hence vector norms) in real and complex spaces, respectively. The constraints imposed on these transformations are strict—the distances between points and the angles between lines are invariant under orthogonal and unitary transformations. Examples of orthogonal transformations are rigid rotations, reflections and uniform translations in two- or three-dimensional space, such as the rotation shown in Fig. 5.4b.

Affine transformations satisfy a less stringent set of conditions. These transformations map straight lines to straight lines, and parallel lines remain parallel. However, distances between points and angles between lines are not invariant under affine maps. For example, triangles may be mapped to arbitrary triangles, rectangles to general quadrilaterals, and so forth. The shearing transformation shown in Fig. 5.4a is an example of an affine transformation.

Not discussed previously is the *projective transformation*, one example of which— a *central projection*—is well known to artists and photographers, and to anyone who must occasionally draw a three-dimensional object on a two-dimensional sheet of paper. In a central projection (Fig. 6.1), the point of view (the eye or camera lens) is at the focal point (the "center") of a set of rays that originate from points that lie along an object line. The central projection describes how those rays intersect an image line (the retina of eye, or the film (detector) in the camera). This is a bijective mapping. Note that if the object line is very distant, the central projection becomes a parallel projection and all points on the image line are mapped toward infinity.

A second type of projective transformation (Fig. 6.2) is the *stereographic projection* of the surface of a unit sphere onto a plane. When the plane is the equatorial plane of the sphere, there is a bijective mapping of one hemisphere onto the plane.

Fig. 6.1 The central projection, where the cross-ratio of the distances between points is invariant

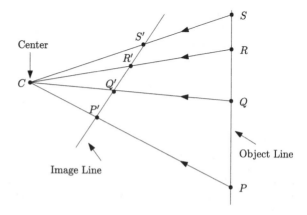

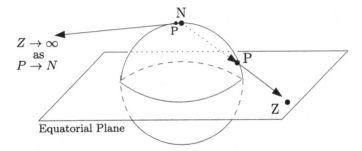

Fig. 6.2 The stereographic projection whose center is the north pole N, and whose plane of projection is the equatorial plane

This projection is often used in complex analysis where the complex plane is mapped onto a hemisphere, and a complex number Z is assigned the spherical coordinates of a point P on the sphere's surface. Points near the center of projection (the north pole, N, in Fig. 6.2) are mapped toward infinity.

Although projective transformations resemble affine transformations in some respects, one important difference is that previously parallel lines may no longer be parallel.

All four of the transformations mentioned above lend themselves to algebraic formulation. Orthogonal and unitary transformations preserve the corresponding quadratic form that defines the inner product. A two-dimensional affine transformation from X to X' may be described in Cartesian coordinates by $x' = a_1x + b_1y + c_1$ and $y' = a_2x + b_2y + c_2$.

In a central projection, the mapping of each point x_o on the object line to the corresponding point x_i on the image line is carried out via the linear fractional transformation $x_i = (ax_o + b)/(cx_o + d)$. The invariant quantity is the *cross-ratio*, defined from Fig. 6.1 as $r = [(PR)(QS)]/[(QR)(PS)]$. Proof that a linear fractional transformation preserves the cross-ratio, and a demonstration of the stereographic mapping of a complex number, are left as exercises for the reader.

We see, then, how a transformation may be described by the quantities it leaves invariant. Significantly, there is a progression from the isometric transformations— where distances between points are invariant—to the projective transformations, where only straight lines are preserved. However, all of the above-mentioned transformations rely on the presence of a coordinate system for their description, which is not required for a topological transformation; for the latter, all we need are elementary concepts from the theory of sets and the properties of maps (see Chap. 1).

6.2   Foundations of General Topology

The word "topology" is used to describe both a particular mathematical structure as well as a field of study in mathematics. As a field of study, its provisions are largely a generalization of many of the properties of the real line, and its goals are to study the structures and invariances associated with topological spaces.

As a mathematical structure, general topology is rooted in set theory.

Definition 6.1 Given a set X, a *topology* on X is a set τ of subsets such that

1. The union of any number of elements of τ is an element of τ;
2. The intersection of any two elements of τ is an element of τ;
3. The set X and the empty set $\emptyset$ are elements of τ.

The combination (X, τ) is called a *topological space*. The elements of τ are, by definition, called *open sets*.   ∎

Example 6.1 Consider the finite set $X = \{a, b, c, d, e\}$.

1. Let $\tau = \{\{a, b, c\}, \{d, e\}, \emptyset, X\}$. The four elements of τ form a topology on X because the union of any number of elements and the intersection of any two elements belongs to τ.
2. Let $\tau = \{\{a\}, \{c, d\}, \{a, c, d\}, \{b, c, d, e\}, \emptyset, X\}$. The six elements of τ form a topology on X.
3. Let $\tau = \{\{a\}, \{c, d\}, \{a, c, d\}, \{b, c, d\}, \emptyset, X\}$. These six elements of τ do not form a topology on X. To show this, we need to find at least one instance in which the terms of Definition 6.1 are not satisfied. Here, $\{a, c, d\} \cup \{b, c, d\} = \{a, b, c, d\} \notin \tau$.
4. Let $\tau = \{\{a\}, \{c, d\}, \{a, c, d\}, \{a, b, c, e\}, \emptyset, X\}$. These six elements of τ do not form a topology on X. Note that $\{a, c, d\} \cap \{a, b, c, e\} = \{a, c\} \notin \tau$.   ▲

Example 6.1(1) and (2), with four and six elements, respectively, show that a given set may have more than one topology defined on it. Consequently, the topologies of a set X may be arranged into a partial ordering according to their number of elements (partial, because different topologies may have the same number of elements).

At the extremes of this partial ordering, the *smallest (coarsest)* topology would contain only X and $\emptyset$. This topology is called the *trivial topology* on X. The *largest (finest)* topology would contain all the subsets (the power set) of X and is called the *discrete topology* on X.

Definition 6.1 specifies that the elements of a topology τ are open sets. From these open sets, we can define closed sets.

Definition 6.2 A set in the topological space (X, τ) is *closed* if it is a relative complement of an open set in (X, τ).   ∎

Because the definition of a closed set in (X, τ) depends on the open sets that comprise the topology, we have the curious situation where a subset of X may be open, closed, neither or both! For a different topology on X, these attributes of a given subset may change. *When it comes to topological spaces, the attributes of "open" and "closed" are neither mutually exclusive nor unique.*

Example 6.2 Let $X = \{a, b, c, d, e\}$ as in Example 6.1, and consider the topological spaces defined in 6.1(1) and (2).

1. Given that $\tau = \{\{a, b, c\}, \{d, e\}, \emptyset, X\}$ forms a topology on X, the complements that correspond to each of the four elements of τ are $\{d, e\}, \{a, b, c\}, X$ and $\emptyset$. These are the closed sets in X. In this case we see that every open set in τ is also closed. This will not typically be the case and is a consequence of how this finite set is partitioned. The other subsets of X, such as $\{b, d\}$, are neither open nor closed; they are not in τ so they are not open in this topology, and they are not complements of the elements of τ, so they are not closed.
2. Another topology on X is $\tau = \{\{a\}, \{c, d\}, \{a, c, d\}, \{b, c, d, e\}, \emptyset, X\}$. These are the open sets in this topology. The corresponding complements are $\{b, c, d, e\}$, $\{a, b, e\}, \{b, e\}, \{a\}, X$ and $\emptyset$. Four of these complements are both open and closed. The other two—$\{a, b, e\}, \{b, e\}$—are closed but not open. All other subsets of X are neither open nor closed in this topology. Note that the subset $\{a, b, c\}$ is both open and closed in (1), but is neither open nor closed in (2).   ▲

It should be apparent that the sets X and $\emptyset$ will always be both open and closed in any topology. In a discrete topology, every set will be both open and closed.

The concept of a neighborhood in a topological space is closely related to that of an open set.

Definition 6.3 Let A be an element of the topology τ (i.e., A is an open set) in the topological space (X, τ), and consider the point $x \in A$. A *neighborhood* of x is any set $B \in X$ such that $B \supset A$. This allows for the possibility that $B = A$, so every open set in a topology is a neighborhood of each of its points.   ■

Note that neighborhoods are defined with respect to points (not sets) in the topological space (X, τ). Conventions differ as to whether a neighborhood, as a set, must be open or whether it may be closed. We adopt the convention in [6] that a neighborhood may be open or closed. Of course, if B is a neighborhood of x and $C \supset B$, then C is a neighborhood of x.

In the trivial topology the only neighborhood of x would be the entire space X. In the discrete topology, every set containing x is a neighborhood of x. An intermediate example of a neighborhood appears in Example 6.2(2). If we let $A = \{c, d\}$, which is open in this topology, and $B = \{b, c, d, e\}$, which is both open and closed, then the fact that $B \supset A$ means that B is a neighborhood of the points $c, d \in A$. Again, a different τ may cause this to change.

Given a topological space (X, τ), a point $x \in X$ may serve as an *accumulation point* of some subset $A \subset X$ even if $x \notin A$. This concept should remind us of how we define the real numbers from the set of rational numbers, and how the limit of a sequence in $\mathbb{Q}$ may not be in $\mathbb{Q}$, i.e., is an irrational number. The concept of an accumulation point is an abstraction and generalization of this limit concept that applies to topological spaces beyond that associated with the real line. We will have more to say about the topology of the real line later in this section, but the real line offers a topology in which several distinct concepts merge. Consequently, we will first define and illustrate accumulation points with finite point sets.

Definition 6.4 Consider a topological space (X, τ), an arbitrary subset $A \subset X$ and a point $x \in X$. Now consider all the neighborhoods of x. If *every* neighborhood of x contains at least one point in A *other than* x (i.e. every neighborhood of x intersects A), then x is an *accumulation point* of A. Note that

1. the accumulation point is with respect to A, not X, and
2. it is not necessary that $x \in A$.   ∎

We recall from Definition 6.3 that a neighborhood of a point x is a superset of an open set containing x. However, the relation $B \supset A$ allows for the possibility that $B = A$. Upon reflection (and in the following example), we see that the definition of an accumulation point comes down to whether *every* open set in the topology τ that contains x has points in A other than x. Therefore, in the discrete topology, the presence of the singleton set (a set with one element) $\{x\}$ for each $x \in X$ means that no point is an accumulation point of *any* set. However, in the trivial topology $\tau = \{\emptyset, X\}$ with only two elements, every point x is an accumulation point of every set other than $\emptyset$ and $\{x\}$.

Example 6.3 Consider the space (X, τ) in Example 6.2(2), where $X = \{a, b, c, d, e\}$ and $\tau = \{\{a\}, \{c, d\}, \{a, c, d\}, \{b, c, d, e\}, \emptyset, X\}$. Let $A = \{a, b, c\}$ which (as we saw in Example 6.2(2)) is neither open nor closed in this topology. We wish to assess each of the five points in X as to whether they are accumulation points of A.

1. The open sets containing the point a are $\{a\}$, $\{a, c, d\}$ and X. The neighborhood that contains only (equals) the singleton open set $\{a\}$ contains no other elements of A (i.e., that neighborhood will contain neither b nor c). Therefore, a is not an accumulation point of $A = \{a, b, c\}$, even though $a \in A$.
2. The open sets containing the point b are $\{b, c, d, e\}$ and X. The neighborhood that contains only (equals) the open set $\{b, c, d, e\}$ contains at least one point in A other than b. Therefore, b is an accumulation point of $A = \{a, b, c\}$.
3. The open sets that contain the point c are $\{c, d\}$, $\{a, c, d\}$, $\{b, c, d, e\}$ and X. The open set $\{c, d\}$ also contains d, but $d \notin A$. Therefore, c is not an accumulation point of $A = \{a, b, c\}$.
4. The open sets that contain the point d are the same as for point c, namely, $\{c, d\}$, $\{a, c, d\}$, $\{b, c, d, e\}$ and X. Here, however, unlike the case for the point c, every open set contains at least one point in A other than d. Therefore, d is an accumulation point of $A = \{a, b, c\}$.
5. The open sets that contain the point e are $\{b, c, d, e\}$ and X, both of which contain at least one point in A other than e. Therefore e is an accumulation point of $A = \{a, b, c\}$.   ▲

Example 6.3 shows that neither membership in A (cf. b with a and c) nor membership in the open sets of the topology τ (cf. c and d) is solely determinative of whether a point is an accumulation point of a set in a topological space. All factors delineated in Definition 6.4 must be taken into account.

With the definitions of open sets, closed sets, neighborhoods and accumulation points now in hand, we are in a position to define the several closely related concepts of interiors, closures and boundaries of sets.

Definition 6.5 Consider a set $A \subset X$ in the topological space (X, τ).

1. The *derived set* of a set A, denoted as A', is the set of accumulation points of A;
2. The *closure* of a set A, denoted as $\bar{A}$, is the union of A with its derived set: $\bar{A} = A \cup A'$. Equivalently,[4] the closure may be described as the intersection of all the closed sets that contain A;
3. The *interior* of a set A, denoted as $A°$, is the union of all open sets in A;
4. The *boundary* of a set A, denoted as ∂A, is the intersection of the closure of A with the closure of its complement: $\partial A = \bar{A} \cap (\bar{A^c})$, where A^c is the complement of A. One consequence is that each neighborhood of a point in the boundary of A will intersect A and A^c. ∎

Example 6.4 Continuing with Example 6.3, we have the following:

1. $A' = \{b, d, e\}$ is the derived set of A;
2. The closure of A is $\bar{A} = A \cup A' = \{a, b, c\} \cup \{b, d, e\} = \{a, b, c, d, e\} = X$, in this case. Equivalently, the closure is the intersection of those closed sets that each contain A. In this case, there is only one closed set that contains A, and that is X;
3. In this topology, the largest open set in $A = \{a, b, c\}$ is $\{a\}$, so $A° = \{a\}$;
4. We leave it as an exercise (same procedure as Example 6.3) to show $(A^c)' = \{b, c, e\}$. Then $(\bar{A^c}) = A^c \cup (A^c)' = \{d, e\} \cup \{b, c, e\} = \{b, c, d, e\}$. From this we can find the boundary of A as $\partial A = \bar{A} \cap (\bar{A^c}) = X \cap \{b, c, d, e\} = \{b, c, d, e\}$. ▲

As noted earlier, the principles of topology are a generalization of the properties of the real line. As it happens, many of the topological properties of $\mathbb{R}$ are "intuitive" (by which we mean "the formal terminology aligns with the vernacular") and familiar, even if we don't recognize these ideas as "topological."

For example, the open sets in $\mathbb{R}$ correspond to open intervals (a, b) on the real line. The topology τ on $\mathbb{R}$ that consists of the set of *all* open intervals is called the *usual topology* on $\mathbb{R}$. In this usual topology, closed sets correspond to closed intervals $[a, b]$, and the only sets in $\mathbb{R}$ that are *both* open *and* closed are $\emptyset$ and $\mathbb{R}$ itself. A neighborhood of a point x in $\mathbb{R}$ is any set that encloses an open set A where $x \in A$.

All points of the open set $A = (a, b)$ in $\mathbb{R}$ are accumulation points of A, as are the points a and b themselves. Therefore, in this case, the derived set and the closure of A are identical to the closed set $[a, b]$. The interior of $[a, b]$ is (a, b), and its boundary is comprised of the two endpoints a and b.

---

[4]As an aside, we note that if $\bar{A} = A$, then A is said to be *everywhere dense* in X. Alternatively, if the complement of the closure of A is everywhere dense, that is, if $A = (\bar{A})^c$, then A is said to be *nowhere dense* in X.

We see that the manifestations of several different concepts merge and become indistinguishable in $\mathbb{R}$. Generally, though, this will not be the case, and by developing the structures necessary to discern differences among the characteristics of sets we will develop a means of describing a wide range of spaces of different structural types. These structural types are described via the *separation axioms* in Sect. 6.4.2.

6.3 Bases and Generators of a Topology

Given a topological space (X, τ), we consider the *base* of the space in terms of which each open set $A \in \tau$, and therefore τ itself, may be expressed.

Definition 6.6 The *base* of a topology τ is a subset $\beta \subset \tau$ of open sets B_i such that every open set $A \in \tau$ may be expressed as the union of elements $B_i \in \beta$. Equivalently, if β forms a base of τ, then every point $x \in A$ belongs to a set $B_i \in \beta$.

A base β is a subset $\beta \subset \tau$ in which

1. X and $\emptyset$ are in β;
2. Every element $A \in \tau$ is the union of elements of β;
3. The intersection of any two elements in β is the union of other elements of β. ∎

All $A \in \tau$ and $B_i \in \beta \subset \tau$ are opens sets, and the B_i are therefore among the open sets defining τ. Further, we would not expect an arbitrary subset $\gamma \subset \tau$ to be a base of τ. However, γ may be used to generate a base and, thereby indirectly, a topology on X. In this circumstance, the subset γ is called then *generator* of τ.

Example 6.5 Consider the sets $X = \{a, b, c, d, e\}$ and $\gamma = \{\{a\}, \{b, c\}, \{a, b, d\}\}$, and ask whether γ (supplemented with $\emptyset$ and X) serves as a base β for some topology τ on X. If it does, then we wish to find τ. If γ itself is not a base, then we can use it to generate a base and a subsequent topology.

One "hit-and-miss" approach would be to assume that γ is a base so that τ would be the set of all possible unions in γ plus γ itself. Here that would give

$$\tau \overset{?}{=} \{\{a\}, \{b, c\}, \{a, b, d\}, \{a, b, c\}, \{a, b, c, d\}, \emptyset, X\},$$

which is *not* a topology because $\{b, c\} \cap \{a, b, d\} = \{b\} \notin \tau$, and so γ is not a base.

A more systematic approach is to apply Definition 6.6(3) to verify that the intersection of any two elements in γ equals the union of other elements. Within γ we see that $\{b, c\} \cap \{a, b, d\} = \{b\} \notin \gamma$. The only "union of other elements" that gives $\{b\}$ is just $\{b\}$, so we must include it in the base. In this way, γ generates the base β:

$$\beta = \{\{a\}, \{b\}, \{b, c\}, \{a, b, d\}, \emptyset, X\}.$$

Taking all possible unions of the $B_i \in \beta$ gives a topology τ:

$$\tau = \{\{a\}, \{b\}, \{a, b\}, \{b, c\}, \{a, b, c\}, \{a, b, d\}, \{a, b, c, d\}, \emptyset, X\}.$$

We get the same result if we consider that any point x in the intersection of two elements of the base β must be included in some other $B_i \in \beta$. From this perspective, the point b in the intersection $\{b, c\} \cap \{a, b, d\}$ must be included in some other B_i. No such set is in γ, so we must append the set $\{b\}$ to generate β. ▲

A basis for a topology is provided by open intervals in $\mathbb{R}$ and open discs in $\mathbb{R}^2$ (see Fig. 6.8 in Sects. 6.4.2 and 6.6). The base for a finite space X is a *countable base*, as is a base in $\mathbb{R}$ (or $\mathbb{R}^2$) if each element in the base is associated with a rational (or ordered pair of rational) numbers.

6.4 Separation and Connectedness

The definitions for the properties of topological spaces described in this and subsequent sections were formulated gradually, and through a process of careful and consistent refinement, primarily during the first half of the twentieth-century. They are essential for constructing a taxonomy of topological spaces.

6.4.1 Separated and Connected Sets and Spaces

We begin with the definitions of a separated set, a disconnected set and a connected space. These results are important for describing the separation axioms.

Definition 6.7 Two sets A and B of a topological space (X, τ) are said to be *separated* if (a) they are disjoint, *and* (b) neither set contains an accumulation point of the other. Note, in particular, that the condition $A \cap B$ is *not* sufficient to say the sets are separated. Combining the effects of these two conditions gives

$$A \cap \bar{B} = \emptyset \quad \text{and} \quad \bar{A} \cap B = \emptyset$$

as the definition of two separated sets, where $\bar{A} = A + A'$ is the closure of A, with A' as the derived set of A. ∎

Example 6.6 Let $X = \mathbb{R}$ with the usual topology, and let $A = (0, 1)$, $B = [1, 2)$, and $C = (2, 3)$. Note that A and C are open, while B is neither open nor closed (some authors would call B "half-open" or "clopen"). Naively (i.e., without proper consideration of the accumulation points), we might conclude that any pair of these sets is separated because they don't appear to "touch."

However, because $\bar{A} = [1, 2]$, we have $\bar{A} \cap B = \{1\} \neq \emptyset$. Therefore, A and B are not separated. Contrast this with B and C. Here, because $\bar{B} = [1, 2]$ and $\bar{C} = [2, 3]$ we see that $B \cap \bar{C} = \bar{B} \cap C = \emptyset$. Therefore, the sets B and C are separated.   ▲

Next we examine the issue of connectedness. When we say that a set is "connected" we wish to convey the idea that the set is, at least in some general sense, in "one piece." We approach this question by first defining when a set is *disconnected*.

Definition 6.8 Given (X, τ), a set $S \subset (X, \tau)$ is *disconnected* if there exist (is at least one example of) two non-empty open sets $A, B \in (X, \tau)$ such that

1. $S \cap A \neq \emptyset$;
2. $S \cap B \neq \emptyset$;
3. $(S \cap A) \cap (S \cap B) = \emptyset$; and
4. $(S \cap A) \cup (S \cap B) = S$.

If S is disconnected, then the union $A \cup B$ is said to form the "disconnection of S." Definition 6.8(3) states that the two sets $(S \cap A)$ and $(S \cap B)$ must be disjoint, but $A \cap B = \emptyset$ is not a necessary condition; what's important is how A and B relate to S. A set that is not disconnected is said to be *connected*.   ■

Example 6.7

1. Let $X = \mathbb{R}$ with the usual topology, $A = \{x : x < 1\}$, $B = \{x : x > 1\}$ and $S = \{x : 0 < x < 2; x \neq 1\}$. Then $S \cap A = (0, 1)$, $S \cap B = (1, 2)$ and S is disconnected. That is, we have shown S to be the union of two non-empty disjoint open sets, the sets $S \cap A$ and $S \cap B$, in X.
2. Consider the previous example, but now with $B = \{x : x \geq 1\}$. Then $S \cap A = (0, 1)$ and $S \cap B = [1, 2)$. In this case Definition 6.8(4) is not satisfied—the union does not equal S because it contains $\{1\}$. Therefore, by *this* choice of A and B, S would *appear to be* connected. However, in (1) above we have shown an A and B to exist that satisfy Definition 6.8 (we only need one such example), so S is disconnected.
3. Let $X = \mathbb{R}$ with the usual topology, and let $S = \{x : |x| \geq 1\} \subset (\mathbb{R}, \tau)$. If we choose $A = \{x : x < -1/2\}$ and $B = \{x : x > 1/2\}$ as two open sets in X, then $S \cap A = \{x : x \leq -1\}$ and $S \cap B = \{x : x \geq 1\}$, and Definition 6.8 is satisfied:

$$(S \cap A) \cap (S \cap B) = \emptyset,$$

$$(S \cap A) \cup (S \cap B) = \{x : |x| \geq 1\} = S.$$

Therefore, S is a disconnected set.
4. Let $X = \{a, b, c, d, e\}$ and let $(X, \tau) = \{\{a, b, d\}, \{c, d, e\}, \{d\}, \emptyset, X\}$. In this topological space, consider the set $S = \{a, c, e\}$. From among the open sets in (X, τ) let $A = \{a, b, d\}$ and $B = \{c, d, e\}$. Note that unlike in the first three examples above, here A and B are *not* disjoint. We have $S \cap A = \{a\}$ and $S \cap B = \{c, e\}$ so that

$$(S \cap A) \cap (S \cap B) = \emptyset,$$

$$(S \cap A) \cup (S \cap B) = \{a, c, e\} = S$$

and S is a disconnected set.   ▲

The empty set $\emptyset$ and a singleton set $\{x\}$ consisting of a single point are always connected. We leave the proofs of these assertions as end-of-chapter exercises.

Thus far we have considered the connectedness of a set S in (X, τ). Whether the space (X, τ) itself is connected is a straightforward application of the same definition, but now with $S = X$. Referring to Definition 6.8, a space X is disconnected if open sets A and B in the topological space (X, τ) can be found such that $(X \cap A)$ and $(X \cap B)$ are non-empty disjoint sets whose union equals X. That is, X is disconnected if

$$(X \cap A) \cap (X \cap B) = \emptyset$$

$$(X \cap A) \cup (X \cap B) = X.$$

However, $X \cap A = A$ and $X \cap B = B$ so that $A \cap B = \emptyset$ and $A \cup B = X$, and we can now give the definition of a connected space as it is usually stated in the literature:

Definition 6.9 A topological space is *connected* if it is *not* the union of two non-empty disjoint open sets.   ■

6.4.2 Separation Axioms and Metric Spaces

The purpose of the separation axioms is to provide a means of classifying topological spaces, with the central criterion in this classification scheme being the extent to which open sets serve to separate points or closed sets within a given space. The classification takes the form of a hierarchy—a progression of spatial types, where each subsequent type represents a space with more properties than the one before.

As seen from a topological perspective, we are most interested in those spaces that possess enough open sets to make it possible to define such things as the completeness of a space, the continuity of a function, the convergence of a sequence and a distance function in the manner to which we have become accustomed in mathematical physics.[5] However, before delineating their content, we make three observations about the separation axioms.

First, it is important to note that they are independent of the axioms defining a topological space (Definition 6.1). As it happens, there are topological spaces that fall outside the hierarchy described by the separation axioms, although these will not be of interest to us in this text. Second, a metric space does not follow logically from

---

[5]The important concepts of completeness, continuity and convergence—as well as compactness—are discussed in Sect. 6.5.

Fig. 6.3 In a T_0 space, points are identified by the open set to which they belong

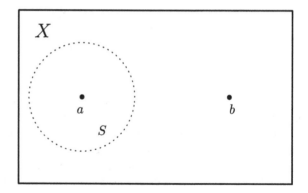

either the topology axioms or the separation axioms but requires the additional feature of a distance function,[6] in which case we say that a distance function "induces" the topology of a metric space.

Finally, we can do a lot of mathematical physics while knowing little or nothing of the hierarchy of topological spatial types. This is because physics is done primarily within the context of a "T_2 space," of which a metric space is a special case. The principal goal of this section is to describe what that means, and to mark more precisely where metric spaces fit in the hierarchy of topological spaces.

We will use Venn diagrams and some examples to illustrate the essential differences among the first five spatial types (labeled T_0 thru T_4). In the figures that follow, areas enclosed by dotted lines ($\cdots$) denote open sets, and solid lines (—) closed sets. The space X is both open and closed. Topology textbooks describe the separation axioms to varying degrees of detail and formality, and the Guide to Further Study at the end of the chapter offers guidance if you wish to explore them beyond this text. Our approach here is motivated primarily by the accounts in [2, 6, 11].

T_0 (Kolmogorov) Spaces: We can think of the T_0 axiom as giving us a "first degree of separation" of points within a space such that points may be identified by the open sets to which they belong. Consider the space (X, τ) with points $a, b \in X$ (Fig. 6.3). The space is a T_0 space if there exists an open set $S \in \tau$ such that *either* (1) $a \in S$ and $b \notin S$, or (2) $a \notin S$ and $b \in S$.

An example of a finite T_0 space for a two-element set $X = \{a, b\}$ would be $(X, \tau) = \{\{a\}, \emptyset, X\}$, or $(X, \tau) = \{\{b\}, \emptyset, X\}$. Note that the trivial topology $(X, \tau) = \{\emptyset, X\}$ is *not* a T_0 space; the only non-empty set in this topology (X itself) contains both a and b.

T_1 (Fréchet) Spaces: Fundamentally, a T_1 space is one in which each set that contains a single point (a *singleton*) is closed. Consider the space (X, τ) with points $a, b \in X$ (Fig. 6.4). A space is a T_1 space if there exist two open sets $S_a, S_b \in \tau$ such that (1)

---

[6]See the definitions and discussion of norms, metrics and distance functions in Sect. 4.2.3.

Fig. 6.4 The essential
property of a T_1 space is that
each individual point is a
closed set

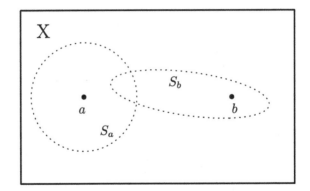

Fig. 6.5 T_2 spaces are the
context in which we frame
virtually all of mathematical
physics

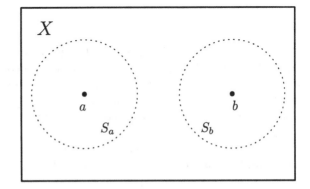

$a \in S_a$ and $b \notin S_a$, and (2) $a \notin S_b$ and $b \in S_b$. For finite T_1 spaces this implies the discrete topology, where all sets in τ are both open and closed.

An example of a T_1 space for a two-element set $X = \{a, b\}$ would be $(X, \tau) = \{\{a\}, \{b\}, \emptyset, X\}$. Each singleton is open by virtue of it being a set in the topology, but each singleton is also closed since it is the complement of some other set that is open (another view is to recognize that a singleton contains "all" of its accumulation points and is therefore closed). If $X = \{a, b, c\}$, then $(X, \tau) = \{\{a\}, \{b\}, \{c\}, \{a, b\}, \{a, c\}, \{b, c\}, \emptyset, X\}$ is a T_1 topology, and for the same reasons the singletons are both open and closed as are the other sets in the topology. For comparison, note that the T_0 space $(X, \tau) = \{\{a\}, \emptyset, X\}$ is not a T_1 space for $X = \{a, b\}$. Although X contains $\{b\}$ it also contains $\{a\}$, contrary to the requirements of a T_1 space.

T_2 (Hausdorff) Spaces: Consider the space (X, τ) with points $a, b \in X$ (Fig. 6.5). The space is a T_2 space if there exist two *disjoint* open sets $S_a, S_b \in \tau$ such that $a \in S_a$ and $b \in S_b$.

Essentially, a T_2 space is T_1 with disjoint (rather than simply distinct) open sets S_a and S_b. A T_2 space is therefore a special case of a T_1 space, and by extension is a special case of T_0. As such, the logical relationship among these three spaces is

Fig. 6.6 The points in the closed set A in a T_3 (regular) space are necessarily closed only by including the T_1 axiom in the definition

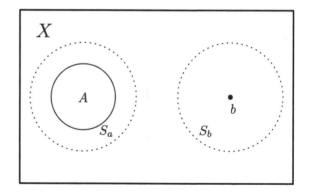

$$T_2 \Longrightarrow T_1 \Longrightarrow T_0. \qquad\qquad (6.1)$$

When a space meets the T_2 criteria (as does the real line $\mathbb{R}$; see Problem 6.11) it becomes possible to define a notion of "convergence" upon which rests most of what we do in physics. Consequently, and as we have noted previously, the topological spaces in which we do virtually all of mathematical physics fit within the definition of a T_2 space, though usually with additional specifications (e.g., a metric).

T_3 *(Regular T_1) Spaces:* Consider the space (X, τ) with closed set $A \in X$ and point $b \notin A$. The space is a T_3, or *regular*[7] T_1, space if

1. the space is a T_1 space, and
2. there exist two disjoint open sets $S_a, S_b \in \tau$ such that $A \subset S_a$ and $b \in S_b$.

Essentially, the point a (a closed set) in T_2 is replaced by a more general closed set A in T_3 (Fig. 6.6).

Note that we specify the T_1 property, namely, that singletons in X are closed. To see why this is necessary, let $X = \{a, b, c\}$, and $(X, \tau) = \{\{b\}, \{a, c\}, \emptyset, X\} = \{\{b\}, A, \emptyset, X\}$. The complements of the four open sets in τ are the same four sets, which makes the sets both open and closed. The set $\{a\}$, however, is not among them and is therefore neither open nor closed. Consequently, even though A is closed, not all points in A are necessarily closed singletons without including the T_1 axiom.

T_4 *(Normal T_1) Spaces:* Consider the space (X, τ) with two disjoint closed sets $A, B \in X$ (Fig. 6.7). The space is a T_4, or *normal*[8] T_1, space if

1. the space is a T_1 space, and
2. there exist two disjoint open sets $S_a, S_b \in \tau$ such that $A \subset S_a$ and $B \subset S_b$.

---

[7]Terminological nuances vary among authors. Our convention is "regular = criterion (2)" so that "$T_3 = T_1 +$ regular." This follows the convention in [2, 6, 11]. Another convention is to say $T_3 =$ criterion(2), in which case "regular $= T_1 + T_3$." This latter convention is followed in [12].

[8]Here, our convention is "normal = criterion (2)" so that "$T_4 = T_1 +$ normal." This follows [2, 6, 11]. In [12] the convention is $T_4 =$ criterion(2) so that "normal $= T_1 + T_4$."

Fig. 6.7 For the points in a T_4 (normal) space, we must include the T_1 axiom

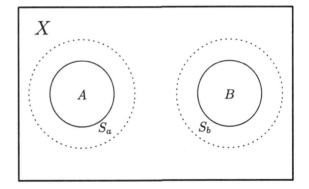

The considerations here are the same as for T_3 in that A and B could be closed yet contain singletons that are not closed. In order for the space to be T_4 we must include the T_1 axiom.

The classification of topological spaces may be continued beyond (and refined within) what we have outlined here (see [12]), but we now turn our attention to metric spaces and describe how they connect to the separation axioms. We construct this *metric topology* by defining a base, and then showing that the elements (open sets) of the base are consistent with a distance function and with the T_4 axioms.

That is, metric spaces form a proper subset of all possible T_4 (normal T_1) spaces. Further, different metrics may correspond to equivalent spaces.

For example, each of the open sets in Fig. 6.8 is a prototype of a basis element for a topology in $\mathbb{R}^2$ (see Sect. 6.6), and we would denote the corresponding equivalent metric spaces as (X, d_i) for $i = 1, 2, 3$. Heuristically[9] we can argue that within the basis elements we can locate disjoint closed sets and require that all points (as sets, i.e., as singletons) are closed. In this way, a metric topology is a T_4 (normal T_1) space.

In summary, the hierarchy of

$$\text{Metric} \Longrightarrow T_4(\text{normal } T_1) \Longrightarrow T_3(\text{regular } T_1) \Longrightarrow T_2 \Longrightarrow T_1 \Longrightarrow T_0, \quad (6.2)$$

describes the logical relationship of the spaces discussed in this section, where the goal has been to place metric spaces in their proper topological context. With this context in hand, we can now move on to examine other topological properties.

---

[9]For a formal proof that a metric space is T_4, see, for example, [11], pp. 30–31.

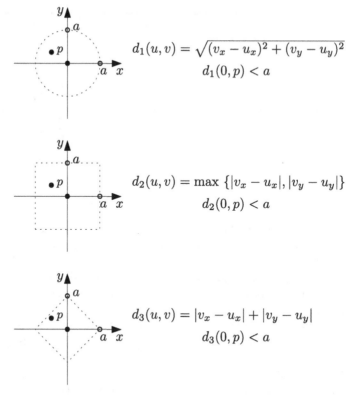

$$d_1(u,v) = \sqrt{(v_x - u_x)^2 + (v_y - u_y)^2}$$
$$d_1(0,p) < a$$

$$d_2(u,v) = \max\{|v_x - u_x|, |v_y - u_y|\}$$
$$d_2(0,p) < a$$

$$d_3(u,v) = |v_x - u_x| + |v_y - u_y|$$
$$d_3(0,p) < a$$

Fig. 6.8 Three different metrics in $\mathbb{R}^2$ and their corresponding sets for $d(0,p) < a$ (*after* [11])

6.5   Compactness, Continuity, Convergence and Completeness

In introducing the foundations of general topology in Sect. 6.2, we noted that topology as a field of study is "...largely a generalization and abstraction of many of the properties of the real line." This is somewhat apparent in our definition of connectedness (Definition 6.8)—a definition that reflects our intuition that a connected set should be thought of as "being in one piece" (like the real line), and does so using language that is applicable to arbitrary sets and spaces.

In Sect. 1.3.2 we showed how the concept of a continuous function may be considered at three different levels of sophistication, where the last of these (expressed in terms of maps and inverse maps of open sets) is relevant to our discussion here. Similarly, the convergence of sequences and the completeness of a space were discussed in Sect. 3.4.1, but in this case we framed the discussion in the context of the real line or complex plane.

In this section we further develop the concepts of continuity, convergence and completeness in language that takes us beyond the real line and is applicable to arbitrary sets and spaces. First, however, we consider the issue of the *compactness* of a topological space.

Compactness. When we hear that a topological space (X, τ) is "compact" we imagine it to be (in at least some vague sense) "closely and firmly united"—a layperson's description that might apply equally well to a vase of flowers, a crowd of people or a cluster of stars.

In defining compactness for topological spaces, we want to think of each element of the space X as belonging to at least one open set in a collection of open sets. This collection of sets is called an *open covering* of X. A subset of this collection that includes every element of X constitutes an open *subcovering* of X. We formalize this picture in the following definition, starting from the definition of a power set (the set of all subsets) of X, denoted as 2^X (see Sect. 1.1.1).

Definition 6.10 Consider a collection of sets $\mathcal{U} \subset 2^X$ whose elements are the sets U_i such that the union of all U_i equals X. Then $\mathcal{U}$ is a *covering* of X.

If the union of a finite number of U_i equals X, this finite subset of $\mathcal{U}$ is a *finite subcovering* of X. The topological space (X, τ) is said to be *compact* if *every* open covering has a finite subcovering. ∎

Example 6.8

1. The definition makes it clear that any finite space of n elements is compact.
2. A *closed* interval in $\mathbb{R}^1$ is compact. Regardless of how we choose to cover the interval with open sets, every subcovering—no matter how small its sets or how many sets are necessary to cover the interval—must eventually include every point (including the endpoints) of the interval. Hence, the number of sets in the subcovering is finite, and a closed interval is compact.
3. Alternatively, the space $\mathbb{R}^1$ (or any *open* interval in $\mathbb{R}^1$) is not compact. Open coverings of an open interval $(x_1, x_2) \subset \mathbb{R}^1$ can easily be constructed, but not *every* open covering has a finite subcovering. ▲

Continuity and homeomorphisms. One motivation for defining a continuous function as we did in Sect. 1.3.2 (Definition 1.4) was to move away from the $\epsilon - \delta$ formulation from elementary calculus, and instead allow for the eventuality of defining continuity without reference to a distance function (and thereby outside the context of a metric space). That eventuality has now arrived.

In that earlier development, we established the more general criterion that a function $f : A \to B$ is *continuous* if, given the open sets $U \subset A$ and $V \subset B$, it is the case that for all V there exists a U such that $f^{-1}(V) = U$.

The question before us here is how the existence of a continuous function on a space depends on the topological properties of that space as articulated by the separation axioms; that is, where in the hierarchy of topological spaces can continuous functions fit? The answer is that the space must be Hausdorff.

The proof (which we will not do here) depends on the bijective nature of topological maps called *homeomorphisms*, and on the above-mentioned definition of a continuous function (map) that does not rely on a distance function.

Definition 6.11 A *homeomorphism* is a continuous bijective map $\phi : A \to B$, and as such the inverse $\phi^{-1} : B \to A$ also is a continuous map. Two spaces related to each other by a homeomorphism are called *homeomorphs*. ∎

When extending a function from a subset to the entire space, the first step in the procedure (regarding functions on closed subsets) is the substance of Urysohn's theorem, while the second step (extending the function to the entire space via homeomorphisms) is the substance of the Tietze extension theorem. The proofs of both theorems are well beyond the scope of our text, but may be found in most topology texts.[10]

Convergence, directed sets, nets and completeness. Recall that we relied on the presence of a distance function when we defined a convergent sequence in a field F and assessed whether F is complete (see Sect. 3.4.1). In that context we defined a sequence $\{x_n\}$ in F as a *Cauchy sequence* if, for every $\epsilon > 0$, there is an integer N such that $d(x_n, x_m) < \epsilon$ for all $n, m \geq N$. That is, the sequence converges eventually, though not necessarily monotonically. This makes the $d(x_n, x_m)$ the elements of a partially-ordered (as opposed to totally-ordered) set (see the discussion and examples in Sect. 1.2.1). We also recall that the point to which the sequence converges is called the *limit* of the sequence.[11] Further, if for every Cauchy sequence in F the limit x is in F, then F is said to be *complete*.

Absent a distance function, however, we must generalize the concepts of a sequence and convergence if we wish to apply these ideas to more general topological spaces. The generalization of a sequence is called a *net* which is closely related to the idea of a *directed set*.[12] A directed set is a partially-ordered set which, in the present context, may be thought of as a set whose elements are neighborhoods of points. A net is a mapping of that directed set into the topological space X. We then inquire as to what it means for the net to converge to a point.[13]

Ultimately, we will conclude that the space of greatest and most general interest with regard to the convergence of nets is the T_2 space. This property of "convergence of nets," perhaps more than any other, is the reason T_2 spaces are the starting point for applications of topology to mathematical physics.

Figure 6.9 shows how we might represent this Moore-Smith convergence theory on the real line. The points in (a) and the open sets in (b) are intentionally not in

---

[10] See, for example, [11], Sect. 2.2.

[11] Our treatment of continuous functions in Sect. 1.3.2 was essentially expressed in terms of convergent sequences: if $f(x_n)$ converges to $f(x)$ as x_n converges to x, then f is continuous.

[12] In many texts convergence is discussed solely in the context of metric spaces. The more generalized theory of convergence involving directed sets and nets is called *Moore-Smith convergence theory*. We outline the basic ideas here. See [2], Sect. I.6 and [6], Chap. 2 for more robust treatments.

[13] The term "limit point" tends to be used in the context of metric spaces, whereas "accumulation point" is used for sets. The term "cluster point" is discussed below.

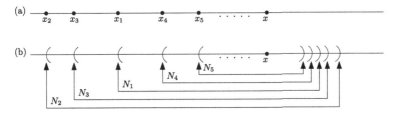

Fig. 6.9   a In metric spaces, sequences converge via partial ordering of the distance between points.
b In more general spaces, the sequence of points is replaced with a net of open sets. Convergence
to a unique point requires at least a T_2 space

numerical order (as read left-to-right). We do this so as to emphasize that partial-
ordering is sufficient, and that convergence does occur eventually. Whether by a
distance function in (a) or a net of open sets in (b), we can define convergence.

These considerations of directed sets, nets and eventual convergence are expressed
more formally[14] in the several definitions that follow.

Definition 6.12 A binary relation $\succeq$ *directs* a non-empty set D with elements m, n
and p if the following criteria are met:

1. if $m \succeq n$ and $n \succeq p$, then $m \succeq p$ (transitive, or partial-ordering, property);
2. $m \succeq m$ (reflexive property);
3. there is a $p \in D$ such that $p \succeq m$ and $p \succeq n$ (necessary for convergence).

A set D on which such a binary relation exists is called a *directed set*. ∎

Definition 6.13 A *net* S is a map $S : D \rightarrow X$ of a directed set D into a topological
space X. ∎

For example, if $X = \mathbb{R}^1$ (where, of course, we have a metric) and $D = \mathbb{N}$, then
the net may be thought of as either the set of points $x_n \in X$ as indexed by $n \in \mathbb{N}$, or
as the set of distances $d(x_n, x_m)$ between points (see Fig. 6.9). We recognize this net
as a sequence in $\mathbb{R}^1$.

It is not necessary for the convergence of a sequence in $\mathbb{R}^1$ to be monotonic;
all we require is that the sequence converge to its limit eventually. This *eventual
convergence* of a sequence is to be contrasted with *frequent convergence*, where the
limit is approached only from time to time.

These two concepts, though reasonably intuitive, are formalized for more general
spaces beyond $\mathbb{R}^1$ in Definition 6.14 and illustrated in Fig. 6.10. However, if you have
any experience in applying convergence tests to infinite sequences, then you know
that some sequences might appear to converge (perhaps frequently, as the sequence
proceeds) but do not converge eventually.

---

[14]We primarily follow the treatment in [6], pp. 65–67, but see also [2], pp. 14–15.

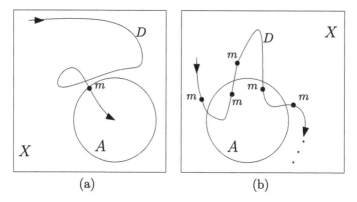

Fig. 6.10 **a** Eventually in A, for all $n \succeq m$; **b** frequently in A for each m

Definition 6.14 Consider a topological space (X, τ), a subset $A \subset X$, a directed set D with binary operation $\succeq$, and a net $S : D \rightarrow X$. Let $m, n \in D$, and let $S(n)$ denote the mapping of the element $n \in D$ into X.

If there exists an element $m \in D$ such that $S(n) \in A$ for *all* $n \succeq m$, then we say the net S is *eventually* in A. On the other hand, if the most we can say is that for *each* element $m \in D$ there *exists* an element $n \succeq m$ such that $S(n) \in A$, then S is said to be *frequently* in A.   ∎

In Fig. 6.10 the directed set D is depicted as a directed path[15] in X, the net S having mapped D into X. In Fig. 6.10a there is a point m after which all points in D are in A. In Fig. 6.10b, each point m has a subsequent point in A regardless of whether $m \in A$ or $m \notin A$.

With this context in hand, we are now in a position to define convergence, irrespective of any metric.

Definition 6.15 Given a topological space (X, τ) and a net S, the net is said to *converge* to a point $x \in X$ if S is eventually in each neighborhood of x.   ∎

Because neighborhoods are defined relative to a topology τ ("τ-neighborhoods"), the convergence of S to a point in X is with respect to that topology. The point x in Definition 6.15 is called an accumulation point. However, if the net S is only frequently in every neighborhood of x, then x is called a *cluster point*.

We close this discussion with a comment as to why a T_2 space is essential for convergence. Consider the case where a net (via a directed set) appears in two neighborhoods, A and B, of two distinct points. This means the net appears in the intersection $A \cap B$. However, if a space is a T_2 (Hausdorff) space, then it becomes possible to establish disjoint neighborhoods for distinct points. Indeed, a Hausdorff space may

---

[15]If there were a distance function, we could think of D as the set of all distances from the set A for each point along the path. The binary relation $\succeq$ in D would then give a partial ordering to those distances.

be defined as a space in which each net converges to at most one point. As noted earlier, this is one of the more important properties of T_2 spaces in terms of applications to mathematical physics. In very practical terms, it is highly advantageous when convergence occurs to unique points in the space, so that functions may then be defined unambiguously.

As for the issue of completeness, the basic ideas are roughly similar to those associated with Hilbert spaces, namely, that a space is complete when sequences converge to points that lie within the space. The discussion of uniform spaces in [6] should be consulted if you are interested in pursuing this further.

6.6 Product Spaces

In one respect, we have anticipated the results of this section (at least as they apply to $\mathbb{R}^2 = \mathbb{R}^1 \times \mathbb{R}^1$) in the discussion of metric spaces at the end of Sect. 6.4.2 and in presenting sketches such as those shown in Fig. 6.8. Our purpose here is to generalize those ideas to the Cartesian product[16] of arbitrary sets and define a *product topology* on that product. The combination of a Cartesian product of sets with a product topology gives a topological space called a *product space*.

Consider two topological spaces (X_1, τ_1) and (X_2, τ_2), with open sets $U_i, V_i \in \tau_i$ for $i = 1, 2$. For graphical purposes only, Fig. 6.11 shows these two spaces, along with their corresponding U_i, V_i, as two separate one-dimensional axes. However, it is important to remember that we are dealing here with general sets of an arbitrary nature on which neither a coordinate system nor a metric is yet necessarily defined, and which in fact may be of any dimension.

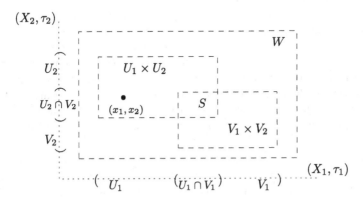

Fig. 6.11 Two topological spaces X_1 and X_2, each with corresponding open sets U_i and V_i. These subsets intersect to form the open sets $U_i \cap V_i \subset X_i$. The Cartesian products $U_1 \times U_2$ and $V_1 \times V_2$ intersect and form the set S

---

[16]For a review of Cartesian products of sets and projection maps, see Sect. 1.4.

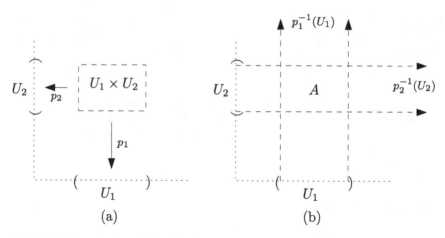

Fig. 6.12 a The projections of $U_1 \times U_2$. **b** The inverse projections intersect at $A = U_1 \times U_2$

The Cartesian product $X_1 \times X_2$ is represented in Fig. 6.11 as the two-dimensional area bounded by X_1 and X_2. The pairwise Cartesian products of all opens sets in τ_1 with those in τ_2 form open sets in $X_1 \times X_2$, only two of which—$U_1 \times U_2$ and $V_1 \times V_2$—are shown in Fig. 6.11. We wish to ask whether these Cartesian products form a topology on $X_1 \times X_2$ in accordance with Definition 6.1.

First, we see that the intersection of these two Cartesian products is the set $S = (U_1 \times U_2) \cap (V_1 \times V_2) = (U_1 \cap V_1) \times (U_2 \cap V_2)$, which is contained in the product set $X_1 \times X_2$. Next, it is plain to see that the union of these two Cartesian products is likewise in $X_1 \times X_2$. Therefore, together with the null set and the set $X_1 \times X_2$ itself, the Cartesian products $U_1 \times U_2$, $V_1 \times V_2$ and all the others in $X_1 \times X_2$ form a product topology on the set $X_1 \times X_2$, which is then a product space.

Although we have framed this product topology without reference to coordinates, consider the case where such coordinates have been assigned to X_1 and X_2. Given a point in U_1 with coordinates[17] x_1 and another point in U_2 with coordinates x_2, this product topology identifies a point in the open set $U_1 \times U_2$ with coordinates (x_1, x_2).

This is familiar to us in the event X_1 and X_2 are in fact one-dimensional coordinate spaces, where we then speak of the first and second coordinates of the point in the product space as the ordered pair (x_1, x_2). The set $U_1 \times U_2$ forms a neighborhood of this point in the product space, as does any other set, e.g., the set W, that contains the point.

As we recall from Sect. 1.4, the projection operator essentially "unwraps" the Cartesian product. Figure 6.12a shows how projecting the product set $U_1 \times U_2$ back to X_1 and X_2 recovers the original open sets $U_1 \subset X_1$ and $U_2 \subset X_2$. The inverse projections are shown in Fig. 6.12b as "infinite strips," and it is their intersection that gives the open set $U_1 \times U_2$.

---

[17]If X_1 and X_2 are multi-dimensional, the "coordinate" x_1 as written here has multiple components incorporated within it, and similarly for x_2. Hence, we use the plural, "coordinates."

This is a reminder of an important point discussed in Sect. 1.3 with respect to inverses and inverse maps. Given any map $f : A \rightarrow B$, we can generate an inverse by just "reversing the arrow," and we write this inverse as f^{-1}. However, for f^{-1} to be a *map*, then f must be bijective, i.e., both one-to-one and onto.

The projections in Fig. 6.12a are not bijective maps (they are, in fact, neither one-to-one nor onto), and the inverses are "one-to-man" which disqualifies them as maps. This asymmetry between, say, p_1 and p_1^{-1} is apparent when we note that $p_1 : (U_1 \times U_2) \rightarrow U_1$, while $p_1^{-1} : U_1 \rightarrow (U_1 \times X_2)$.

6.7 Quotient Spaces

Quotient sets and the various maps associated with them were first described in Sect. 1.5 and depicted in Fig. 1.9. In effect, a quotient set X/R is a partition of a set X into disjoint equivalence classes, pursuant to some equivalence relation R. We then proceeded over the course of several chapters to define a quotient for various algebraic structures, as summarized in Table 5.1.

We now apply the same concept to topological spaces. By analogy with product sets and spaces, the combination of a quotient set with a *quotient topology* will yield a *quotient space*. Given a mapping $\phi : X \rightarrow Y$ between two topological spaces and an equivalence relation R on the domain X, our goal is to identify τ_q, the quotient topology on the quotient set X/R.

Figure 6.13a shows the surjective canonical map (also called a *quotient map*) $p : X \rightarrow X/R$ and the induced map ϕ', which is a bijection from X/R to Y. The map $\phi = \phi'p$ (implemented right-to-left) is the composition of these two maps. If ϕ and p are continuous, then so is ϕ'. Two open sets $A_1, A_2 \subset X$ are equivalent if they are in the same equivalence class $[A]$ in X/R, that is if $p : A_1, A_2 \rightarrow [A]$. Consequently, we can think of the maps in Fig. 6.13a as being between the topologies of the various spaces, i.e., between the open sets in those respective topologies.

Definition 6.16 Consider the topological spaces (X, τ_x) and (Y, τ_y), with the sets $U_x \in \tau_x$ and $U_y \in \tau_y$. The quotient topology τ_q on X/R is defined as the set of open sets $U_q \subset X/R$ such that the sets $p^{-1}(U_q) \subset X$ are open sets in τ_x. More formally,

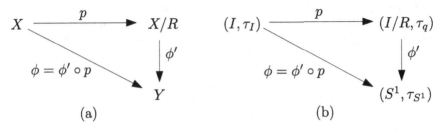

Fig. 6.13 A quotient structure for topological spaces

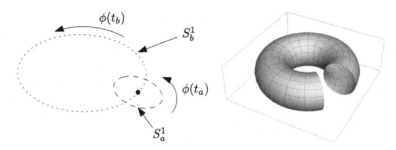

Fig. 6.14 A torus may be thought of as the space $T = S^1 \times S^1$

$$\tau_q = \{U_q \subset X/R : p^{-1}(U_q) \in \tau_x\}.$$

■

Equivalently, and because of the bijective nature of ϕ', we can also say that $\tau_q = \{U_y \subset Y : \phi^{-1}(U_y) \in \tau_x\}$. This latter formulation for τ_q is often more transparent than that given in Definition 6.16, as shown in the following example.

Example 6.9 Let $X = I = [0, 1] \subset \mathbb{R}^1$, $Y = S^1$ and $\phi : I \to S^1$, as shown below. That is, ϕ maps the unit interval in $\mathbb{R}$ to a unit circle in the complex plane. The quotient diagram is shown in Fig. 6.13b. An equivalence relation R for two points

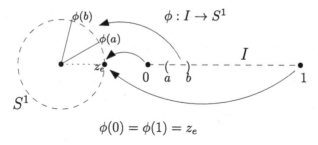

$$\phi(0) = \phi(1) = z_e$$

in I may be expressed as "at the same location when mapped to the unit circle."

Letting $t \in I$, it is clear that the points $t = 0$ and $t = 1$ are equivalent; each is mapped to the identity $z_e = 1 \equiv (1, 0)$ in the complex plane. Each of the other $t \in I$ is an equivalence class unto itself. Consequently, the quotient set may be written either in terms of the equivalence classes, as $I/R = \{[t], [1]\}$, or in terms of its open sets, as $I/R = \{\{t\}, \{0, 1\}\}$, where $t \neq 0, 1$. Hence, we have the quotient topology t_q, thereby making I/R a quotient space. Finally, the bijective induced map may be written as $\phi' : X/R = \{\{t\}, \{0, 1\}\} \to \{e^{2\pi i t}, z_e\}$.   ▲

The product space and quotient space concepts merge in the example of a hollow torus $T = S^1 \times S^1$ (Fig. 6.14). By the quotient structure above, we can interpret this as the mapping of the unit square $I \times I \to T$.

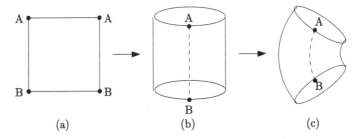

Fig. 6.15 Mapping a unit square to a torus

Geometrically, if we identify the opposite sides of the unit square as equivalent, then a succession of two continuous maps takes us first from a unit square to a cylinder, and then from a cylinder to a torus, a portion of which is shown in Fig. 6.15c.

The quotient diagram would be the same as Fig. 6.13b, but with $I \times I$ and $S^1 \times S^1$ in place of I and S^1, respectively.

6.8 Topological Invariants

We began this chapter with a discussion of those properties of spaces (or subsets of those spaces) that remain invariant under various transformations. The purpose of that account was to show that there is a hierarchy of transformations—from Euclidean to projective—where at each step fewer properties are left invariant.

Least constraining of all are homeomorphisms, and you may now be asking which from among the topological properties we have been discussing in this chapter remain invariant under these "topological transformations." The answer[18] that applies to the widest collection of all spaces is "very few." For example, connectedness is invariant under homeomorphisms, and it is occasionally said that the only property preserved under homeomorphisms is a sense of "nearness" between two points. However, this language implies the presence of a distance function, and it is more precise to say "connectedness." For example, $\mathbb{R}^1$ is homeomorphic to any open interval $(a, b) \subset \mathbb{R}$, so "nearness" becomes a matter of opinion.

The T_1 and T_2 properties are invariant under homeomorphisms, and this is particularly advantageous for those of us in physics or engineering who rely on T_2 spaces for our daily work. Beyond that, however, the best that can be done is to ask whether two particular spaces are topologically equivalent by virtue of their sharing a robust set of invariant properties. Answering these kinds of questions is a central mission of topology, a field of study which we now leave for your future exploration.

---

[18] We base our account here on the brief summary in [6], p. 88.

Problems

6.1 Referring to Fig. 6.1, show that the cross-ratio $r = [(PR)(QS)]/[(QR)(PS)]$ is invariant under central projections (i.e., compare the ratio as determined on the image line with that as found on the object line).

6.2 The essence of a stereographic projection of a spherical surface into the complex plane is the projection of a great circle onto a single coordinate axis. In the figure, the circle is of unit radius ($R = 1$), the angle α is the "latitude," and the circle represents the "prime meridian" so that the x-axis is along the "equator" and points along $\theta = 0°$ of "longitude." The point P is the "north pole," the point A is on the

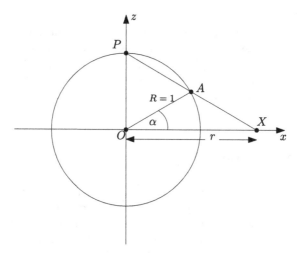

surface of the sphere and X marks the spot where the projection $P - A - X$ intersects the equator. The origin of the x-axis is at the center of the sphere (circle). Find r, the x-coordinate of the point X, in terms the latitude α. Because $\theta = 0$, this coordinate represents a real number. Now find the latitude and longitude that correspond to a general complex number $z = re^{i\theta}$.

6.3 List all possible topologies on the set $X = \{a, b\}$.

6.4 A set X is partitioned into three non-empty subsets, A, B and C. Can $\tau = \{A, B, C, \emptyset, X\}$ be a topology on X? If yes, please give an example. If no, then please explain your answer.

6.5 Given the set $X = \{a, b, c\}$, find at least two topologies other than the trivial and discrete topologies.

6.6 Given the set $X = \{a, b, c, d\}$, find at least four topologies other than the trivial and discrete topologies.

6.7 Find the derived set for each of the following sets:
(a) $A = \{x : 0 \le x < 2, \text{or} x = 2\}$, in the usual topology on $\mathbb{R}$;
(b) $B = \{(x, y) : x, y \in \mathbb{N}\}$, in the usual topology on $\mathbb{R}$;
(c) $C = \emptyset$, for any topology.

6.8 Find the closure of each of the sets in Problem 6.7.

6.9 Which, if any, of the sets in Problem 6.7 are connected?

6.10 In a discrete topology for the set $X = \{a, b, c\}$, is the empty set $\emptyset$ connected? What about the singleton set $\{a\}$?

6.11 The usual topology on $\mathbb{R}^1$ consists of open intervals (a, b), but other topologies are possible. An equally valid topology can be defined by intervals $(a, b]$. Show that this topology on $\mathbb{R}$ preserves the T_2 nature of the real line. [*Hint:* This is not difficult; you need to show that two distinct points can be located in two distinct open sets, and that these intervals satisfy the definition of a topology.]

6.12 Show that each of the expressions for d in Fig. 6.8 satisfies the definition of a distance function (Definition 4.4).

6.13 Referring to Figs. 6.11 and 6.12, find the following:

$$(a)p_1(S); \; (b)p_2(S); \; (c)p_1^{-1}(X_1); \; (d)p_2^{-1}(X_1); \; (e)p_2(\emptyset).$$

6.14 Find a homeomorphism ϕ such that:
(a) $\phi : (-3, 3) \subset \mathbb{R}^1 \rightarrow (-1, 1) \subset \mathbb{R}^1$;
(b) $\phi : \mathbb{R}^1 \rightarrow (-1, 1) \subset \mathbb{R}^1$.

Guide to Further Study
There are three texts—all relatively short—that you should consider if you wish to (a) resume your study of topology with a review of the material in this chapter; (b) explore at greater depth the concepts we have covered; (c) move beyond the content of this chapter; and (d) do so at a level that is approximately the same as this chapter. In alphabetical order, those texts are Baum [1], McCarty [8] and Patterson [9]. Of these, the first two are readily available. Patterson's book may be difficult to find in print, but it is the most conversational of the three.

A fourth text, that of Flegg [4], deserves special mention. More than the others, it takes the reader from the geometric transformations we discussed at the beginning of this chapter (Flegg's approach motivated ours) to a discussion of networks, upon which many present-day applications of topology are founded. The short historical note near the end of Flegg's text should not be missed.

Having said that, Kelley's text [6] on general topology is unsurpassed for anyone who has a good grasp of the material in the present chapter or in the above-mentioned texts. Although it is considered a graduate text, you will find that it offers many new insights in a reader-friendly way. Also in this category are Section IIB of Roman [10], and over a dozen chapters (26 to 42—they're short) in Geroch [5]. Mathematics majors should consider the texts of Singer and Thorpe [11] (at the undergraduate level) and Bredon [2] (at the graduate level).

Counterexamples are an important tool for anyone engaged in self-study (or taking a class), and for this a good reference is the work by Steen and Seebach [12]. However, beware that some of the definitions pertaining to the separation axioms differ from what is most common (see our footnotes in Sect. 6.4.2).

References

1. Baum, J.D.: Elements of Point Set Topology. Prentice-Hall, Englewood Cliffs, NJ (1964). Republished by Dover, New York (1991)
2. Bredon, G.E.: Topology and Geometry. Springer-Verlag, New York (1993)
3. Dieudonné, J.: A History of Algebraic and Differential Topology, 1900–1960, Reprint of the 1989 Edition. Birkhäuser, Boston (2009)
4. Flegg, H.G.: From Geometry to Topology. Crane, Russack and Company, New York (1974). Republished by Dover, New York (2001)
5. Geroch, R.: Mathematical Physics. In: Chicago Lectures in Physics. University of Chicago Press, Chicago (1985)
6. Kelley, J.L.: General Topology. Springer (1975). Originally Published by Van Nostrand, New York (1955)
7. Kline, M.: Mathematical Thought from Ancient to Modern Times, Published in 3 Volumes. Oxford University Press, Oxford (1990)
8. McCarty, G.: Topology—An Introduction with Application to Topological Groups. McGraw-Hill, New York (1967). Republished by Dover, New York (1988)
9. Patterson, E.M.: Topology. Oliver and Boyd, Edinburgh and London (1956)
10. Roman, P.: Some Modern Mathematics for Physicists and Other Outsiders, 2 Volumes. Pergamon Press, Elmsford, NY (1975)
11. Singer, I.M., Thorpe, J.A.: Lecture Notes on Elementary Topology and Geometry. Scott, Foresman and Company, Glenview, IL (1967). Republished by Springer, New York (1976)
12. Steen, L.A., Seebach, Jr., J.A.: Counterexamples in Topology, Corrected, Revised and Expanded Second Edition. Springer-Verlag, New York (1978). Republished by Dover, New York (1995)

Chapter 7
Differentiable Manifolds

7.1 Differentiation in $\mathbb{R}^n$

When we first study the differentiation of a single-variable, real-valued function in elementary calculus, we depict the function as a graph and define its derivative as the slope of a tangent line at various points along the curve. For a multi-variable function whose graph may be a surface, the depiction of derivatives is essentially the same; we consider each coordinate independently, define the corresponding partial derivative and examine how the function behaves along lines parallel to each coordinate axis. More generally we define the directional derivative of a function as how a function changes along the direction parallel to a particular vector.

This graph-oriented, "slope of a tangent" approach to understanding differentiation has the distinct pedagogical advantage of being easily visualized, and it is not irrelevant to higher-dimensional applications. Nonetheless, in this section we elaborate upon this standard approach toward defining the derivative of a function by emphasizing the foundational role of sets and maps, rather than graphs and slopes. This generalization of the concept of the derivative will expand the range of its application over what is possible with the traditional approach.

If you have already studied multi-variable calculus at length, then this section will be largely a review of that material. Still, you should read this section as preparation for what follows.

7.1.1 Review of Single-Variable Differentiation and Directional Derivatives

Figure 7.1 depicts the essentials involved in defining the derivative of a single-variable, real-valued function $\phi(x)$. Here, $x \in X = \mathbb{R}^1$ is the independent variable, and $\phi(a) \in U = \mathbb{R}^1$ as the value of the function at $x = a$. The set $A \subset X$ is an open neighborhood of the point $a \in \mathbb{R}^1$. We also draw $B \subset U$ as an open neighborhood

© Springer Nature Switzerland AG 2021
S. P. Starkovich, *The Structures of Mathematical Physics*,
https://doi.org/10.1007/978-3-030-73449-7_7

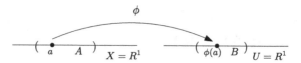

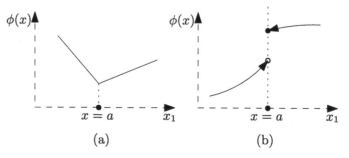

Fig. 7.1 A real-valued function $\phi(x) : A \subset \mathbb{R}^1 \to \mathbb{R}^1$

Fig. 7.2 Two functions $\phi(x)$ that are not differentiable at $x = a$

of $\phi(a)$ and assume ϕ to be continuous. As such, ϕ^{-1} will map every open neighborhood $B \subset U$ into an open set in X. Though necessary, continuity is not sufficient for this function to be differentiable.

Definition 7.1 The derivative $\phi'(x)$ at $x = a \in A$ is defined as

$$\phi'(a) = \lim_{t \to 0} \frac{\phi(a+t) - \phi(a)}{t} \begin{array}{l} \leftarrow \text{ in } B \subset U \\ \leftarrow \text{ in } A \subset X \end{array}, \qquad (7.1)$$

where $t \in X = \mathbb{R}^1$. The function $\phi(x)$ is said to be *differentiable* at $x = a$ if

1. the limit in Eq. 7.1 exists, and
2. the function $\phi(x)$ is continuous at $x = a$.   ∎

All of this should be familiar, but it is worth reminding ourselves again that Definition 7.1(2) is not a sufficient condition for differentiability. Figure 7.2a shows a continuous function that is not differentiable at $x = a$, though we sometimes speak of "two-sided derivatives," where the derivative is discontinuous at $x = a$.

However, without continuity (Fig. 7.2b), the condition in Definition 7.1(1) will not be satisfied; that continuity is necessary follows from a rearrangement of Eq. 7.1:

$$\lim_{t \to 0} [\phi(a+t) - \phi(a)] = \lim_{t \to 0} \frac{\phi(a+t) - \phi(a)}{t} \cdot t = \phi'(a) \cdot \lim_{t \to 0} t = 0. \qquad (7.2)$$

The derivative of a multi-variable ($m > 1$) function $\phi : \mathbb{R}^m \to \mathbb{R}^n$ for $n = 1$ is formulated in analogous fashion. The result is called a *directional derivative*. In

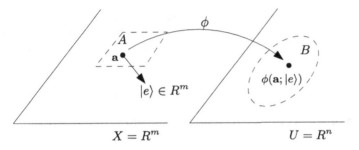

Fig. 7.3 The real-valued function $\phi(\mathbf{x}) : A \subset \mathbb{R}^m \to \mathbb{R}^n$

Fig. 7.3 we take $\mathbf{x} \in X = \mathbb{R}^m$ with coordinates[1] $(x^1, x^2, \ldots x^m)$ as the independent variable, and consider a point $\mathbf{a} = (a^1, a^2, \ldots a^m) \in A \subset \mathbb{R}^m$. The vector $|e\rangle \in \mathbb{R}^m$ is a unit vector that identifies the direction along which we will calculate the derivative. As before, A and B are open sets in their respective spaces.

Not shown in the figure is a parameter $t \in \mathbb{R}^1$. Unlike in Eq. 7.1, t is not an element of $X = \mathbb{R}^m$, but rather is an element of a one-dimensional parameter space. It serves to mark various points along the direction specified by $|e\rangle$, and will approach zero in the limit of the derivative.

We define the directional derivative $\phi'(\mathbf{x})$ at the point $\mathbf{x} = \mathbf{a}$ along the direction specified by $|e\rangle$ as

$$\phi'(\mathbf{a}, |e\rangle) = \lim_{t \to 0} \frac{\phi(\mathbf{a} + t|e\rangle) - \phi(\mathbf{a})}{t} \begin{array}{l} \leftarrow \text{in } B \subset U \\ \leftarrow \text{in } \mathbb{R}^1 \end{array}. \tag{7.3}$$

Example 7.1 Consider $X = \mathbb{R}^2$ with Cartesian coordinates $(x^1, x^2) = (x, y)$, and let $U = \mathbb{R}^1$ with $\phi(\mathbf{x}) = \phi(x, y) = x^2 y$. We wish to find the directional derivative $\phi'(\mathbf{a}, |e\rangle)$ at the point $\mathbf{a} = (a^1, a^2) = (x_0, y_0)$ in a direction that is parallel to the x-axis.

We note first that

$$\mathbf{a} + t|e\rangle = \begin{pmatrix} x_0 \\ y_0 \end{pmatrix} + t \begin{pmatrix} 1 \\ 0 \end{pmatrix} = \begin{pmatrix} x_0 + t \\ y_0 \end{pmatrix},$$

so that $\phi(\mathbf{a} + t|e\rangle) = (x_0 + t)^2 y_0$, with $\phi(\mathbf{a}) = x_0^2 y_0$.

Equation 7.3 then becomes

$$\phi'(\mathbf{a}, |e\rangle) = \lim_{t \to 0} \frac{(x_0 + t)^2 y_0 - x_0^2 y_0}{t} = \lim_{t \to 0} \frac{2t x_0 y_0 + t^2 y_0}{t} = 2x_0 y_0,$$

---

[1] In writing the components with superscripts, we are taking $\mathbf{x}$ and $\mathbf{a}$ as ket (column) vectors.

which is the expected result, since by taking $|e\rangle$ along the x-axis we are just finding the partial derivative $\partial\phi/\partial x$ at $\mathbf{x} = (x_0, y_0)$. Alternatively, had we let $|e\rangle = (1/\sqrt{2})(1, 1)$, we would find $\phi'(\mathbf{a}, |e\rangle) = (2x_0y_0 + x_0^2)/\sqrt{2}$. We leave this as an exercise.[2]   ▲

7.1.2 Multi-variable Differentiation and the Jacobian

The derivative of a function is a local quantity; it is evaluated at a point, as opposed to over a region or a finite segment of a curve. We also recognize the first derivative is a *linear* approximation to the localized rate of change of the function, which is why we speak of the tangent *line*.

Consequently, Eq. 7.1 for $\phi'(a)$ may be recast as

$$.(\phi(a + t) - \phi(a)) \approx \lambda(a)t \tag{7.4}$$

for some constant $\lambda(a)$, the first derivative of ϕ as evaluated at the point $x = a$. Further, a comparison of Eqs. 7.1 and 7.3 shows they are essentially the same equation, but with $|e\rangle = |1\rangle$ in the single-variable case.

Now, when expressing the first derivative for some $\phi : A \subset \mathbb{R}^m \to \mathbb{R}^n$ for m, $n > 1$, we can no longer use $t \in \mathbb{R}^1$ as the linear parameter but must generalize it in such a way that $t|e\rangle = \mathbf{h} = (h^1, h^2, \ldots h^m) \in \mathbb{R}^m$. This allows us to write the derivative in a more general form, which is applicable for all values of m, n, and of which our earlier examples are special cases.

Specifically, we write the first derivative of ϕ as evaluated at $\mathbf{a}$ as

$$\phi'(\mathbf{a}, \mathbf{h}) = \lim_{\|\mathbf{h}\| \to 0} \frac{\phi(\mathbf{a} + \mathbf{h}) - \phi(\mathbf{a})}{\|\mathbf{h}\|} \begin{array}{l} \leftarrow \text{in } B \subset \mathbb{R}^n \\ \leftarrow \text{in } A \subset \mathbb{R}^m \end{array}, \tag{7.5}$$

where $\|\mathbf{h}\|$ is the norm[3] of $\mathbf{h}$.

We again apply the linear approximation and recast Eq. 7.5 as

$$(\phi(\mathbf{a} + \mathbf{h}) - \phi(\mathbf{a})) \approx [D\phi(\mathbf{a})]\mathbf{h}, \tag{7.6}$$

which is analogous to $(\phi(a + t) - \phi(a)) \approx \lambda(a)t$ from before. Because the left-hand side of Eq. 7.6 is in $\mathbb{R}^n$ and $\mathbf{h}$ is in $\mathbb{R}^m$, $[D\phi(\mathbf{a})]$ must be an $n \times m$ matrix.

The matrix $[D\phi(\mathbf{a})]$ in Eq. 7.6 is the *Jacobian matrix* of ϕ as evaluated at $\mathbf{x} = \mathbf{a}$. It is the first derivative of ϕ as evaluated at the point $\mathbf{x} = \mathbf{a}$. The entries in the Jacobian are first-order partial derivatives, all of which must exist and be continuous (as in

---

[2]If you are familiar with elementary vector calculus, then you will recognize the directional derivative as the scalar product $\nabla\phi \cdot |e\rangle$, where $\nabla\phi$ is the gradient of ϕ. See Example 7.3 in Sect. 7.1.2.
[3]See Definition 4.3 in Sect. 4.2.3 for the definition of the norm.

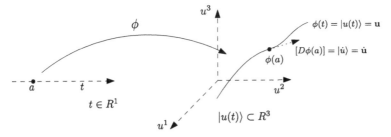

Fig. 7.4 The mapping of $t \in \mathbb{R}^1$ to a parameterized curve $\phi(t) \subset \mathbb{R}^3$

the one-dimensional case). Given $\phi : R^m \to \mathbb{R}^n$, with $\mathbf{x} = (x^1, x^2, \ldots x^m) \in \mathbb{R}^m$ and $\mathbf{u} = (u^1, u^2, \ldots u^n) \in \mathbb{R}^n$, we write the Jacobian of the map ϕ as

$$[D\phi(\mathbf{x})] \equiv \frac{\partial(u^1, u^2, \ldots u^n)}{\partial(x^1, x^2, \ldots x^m)} = \begin{pmatrix} \dfrac{\partial u^1}{\partial x^1} & \dfrac{\partial u^1}{\partial x^2} & \cdots & \dfrac{\partial u^1}{\partial x^m} \\[2mm] \dfrac{\partial u^2}{\partial x^1} & \dfrac{\partial u^2}{\partial x^2} & \cdots & \dfrac{\partial u^2}{\partial x^m} \\[2mm] \vdots & \vdots & \vdots & \vdots \\[2mm] \dfrac{\partial u^n}{\partial x^1} & \dfrac{\partial u^n}{\partial x^2} & \cdots & \dfrac{\partial u^n}{\partial x^m} \end{pmatrix}. \tag{7.7}$$

The Jacobian generalizes and unifies into one structure (an $n \times m$ matrix) the concept of the first derivative of a single-variable function; if $m = n = 1$, then $[D\phi(\mathbf{a})]$ is just the number $\lambda(a)$ in Eq. 7.4. This should not be too surprising, given how we derived the Jacobian by building it up from this special case. Moreover, other combinations of m and n in the Jacobian can be associated with other forms of the derivative which, at least superficially, appear to be wholly unrelated to each other, and which in some cases may be expressed in vector notation. In fact, all these expressions are just special cases of this more general structure[4].

Example 7.2 Consider the case of $m = 1$ and $n = 3$. The range of the map ϕ : $\mathbb{R}^1 \to \mathbb{R}^3$ may be depicted as a parameterized curve in three dimensions (Fig. 7.4). For example, the curve might represent the path of an object through space, with points in time mapped to points along the curve. The Jacobian $[D\phi(t)]$ is a 3×1 column matrix, which here represents the velocity vector at various points along the curve:

$$[D\phi(t)] = \begin{pmatrix} \partial u^1/\partial t \\ \partial u^2/\partial t \\ \partial u^3/\partial t \end{pmatrix} = |\dot{u}\rangle = \dot{\mathbf{u}}(t).$$

▲

---

[4]If you have studied multivariable integration, then you likely know the Jacobian in a different context, viz., as effecting a change of variables for the integrand.

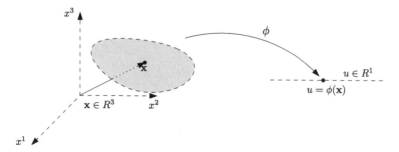

Fig. 7.5 The mapping of a position vector $\mathbf{x} \in \mathbb{R}^3$ to the value of a scalar function $\phi(\mathbf{x}) \in \mathbb{R}^1$

Example 7.3 The map $\phi : A \subset \mathbb{R}^3 \rightarrow B \subset \mathbb{R}^1$ (i.e., $m = 3$ and $n = 1$) is depicted in Fig. 7.5 as the mapping of points in a three-dimensional space to points in $\mathbb{R}^1$.

A typical application involves finding the values of a scalar function[5] $\phi(\mathbf{x})$ (such as a temperature or pressure field) at points defined by $\mathbf{x}$. The action of the mapping $\phi(\mathbf{x})$ is to assign a scalar value to each point in that region.

In this case, the Jacobian is a 1×3 row matrix, which in vector notation we would interpret as the *gradient* of $u = \phi(\mathbf{x})$:

$$[D\phi(\mathbf{x})] = (\frac{\partial u^1}{\partial x^1} \frac{\partial u^1}{\partial x^2} \frac{\partial u^1}{\partial x^3}) = \langle Du| = \nabla u.$$

Thus, the gradient of a scalar function is a one-form (see Sect. 4.3.2), and (as noted earlier) the scalar product $\langle Du|e\rangle$ is the directional derivative of u in the direction specified by the unit vector $|e\rangle$.   ▲

In Fig. 7.4 we depict $[D\phi(t)]$ as tangent vectors to the curve $\mathbf{u}(t) \in \mathbb{R}^3$. At each point along the curve we have a different vector, but we can (at least in principle) sketch them all on the same coordinate system as $\mathbf{u}(t)$. If we want to depict the gradient in Fig. 7.5, we have a problem; in fact, it is the same problem as in Fig. 7.4, but we easily fail to see it there because the tangent vectors are just arrows.

The "problem" is that the tangent vectors and the gradients do not exist in the vector spaces drawn. Instead, they exist in their respective spaces, with a different space at each point. It is very important to remember where (in which space) these various objects exist. As we discussed in Chap. 4, vectors (like a velocity tangent vector) are ket vectors and live in vector spaces, but one-forms (such as gradients) are bra vectors and live in dual spaces. We will return to this when we discuss transformations of forms and vectors in (Sect. 7.7).

Further, in Examples 7.2 and 7.3 it was possible to express the Jacobian in ordinary vector notation because m or n equaled 1 and the other parameter was 3 (or less). For other combinations of m and n we rely on the definition of the Jacobian and on its role as an $n \times m$ matrix operator, and forgo the traditional vector notation.

---

[5]For convenience we usually sketch the region of the scalar field on the same three-dimensional coordinate system as that used for $\mathbf{x}$ (the shaded area in Fig. 7.5).

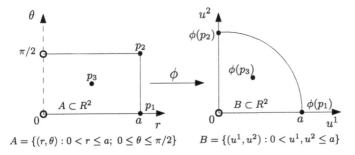

$$A = \{(r, \theta) : 0 < r \leq a; \ 0 \leq \theta \leq \pi/2\} \qquad B = \{(u^1, u^2) : 0 < u^1, u^2 \leq a\}$$

Fig. 7.6 The map $\phi : A \to B$ as defined in Eq. 7.8

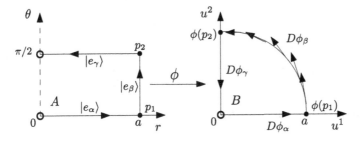

Fig. 7.7 Several unit vector displacements in A and the corresponding $[D\phi]$ in B

Example 7.4 Consider the map $\phi : A \subset \mathbb{R}^2 \to B \subset \mathbb{R}^2$ in Fig. 7.6 (i.e., $m = n = 2$).

In terms of ordered pairs, the map is given as

$$\phi : (r, \theta) \to (u^1, u^2) = (r \cos \theta, r \sin \theta). \tag{7.8}$$

Among the points on the boundaries of A and B, we have $\phi(p_1) = (a, 0)$ and $\phi(p_2) = (0, a)$. Because all points on the θ-axis ($r = 0$) in A are mapped to zero in B, we omit the θ-axis from the domain on the grounds that we wish for ϕ to be non-singular (Sect. 5.5.1). We will have problems at p_1 and p_2 as well.

The Jacobian of this transformation is the 2×2 matrix

$$[D\phi(r, \theta)] \equiv \frac{\partial(u^1, u^2)}{\partial(r, \theta)} = \begin{pmatrix} \dfrac{\partial u^1}{\partial r} & \dfrac{\partial u^1}{\partial \theta} \\[2mm] \dfrac{\partial u^2}{\partial r} & \dfrac{\partial u^2}{\partial \theta} \end{pmatrix} = \begin{pmatrix} \cos \theta & -r \sin \theta \\ \sin \theta & r \cos \theta \end{pmatrix}, \tag{7.9}$$

and its effect is shown along the boundaries of the domain and range in Fig. 7.7.

Although ϕ is continuous over the entire domain, $[D\phi]$ is discontinuous at p_1 and p_2, giving us "two-sided derivatives" we can avoid by redefining the domain. ▲

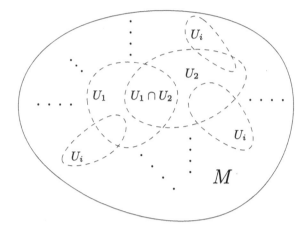

Fig. 7.8 Open sets $U_i \subset M$, where intersections such as $U_1 \cap U_2$ allow for coordinate transformations between U_1 and U_2 provided differentiability between them is established

7.2 Differentiable Manifolds in $\mathbb{R}^n$

A *manifold M* may be defined as a T_2 (Hausdorff) space with a set of maps obeying certain differentiability conditions[6] among the open sets in the space. In physics we tend to think of manifolds as spaces that are *locally* $\mathbb{R}^n$, though not necessarily Euclidean. In either case, for each region (open set) $U_i \subset M$ we can draw an $\mathbb{R}^n$ coordinate system. Where the regions overlap (Fig. 7.8), we carry out smooth (continuously differentiable) coordinate transformations between them.

This practical perspective of a manifold serves us well, but it is worth getting a sense of how it connects to a more formal definition. We can make the connection by referring to the separation axioms, the concept of a topological base and the results of Sect. 7.1.2. This last step is important because only after differentiability is established can we perform smooth coordinate transformations.

Recall from Sect. 6.4.2 that a topological space (X, τ) is a T_2 (Hausdorff) space if, for each pair of points $a, b \in X$, there exist two *disjoint* open sets $S_a, S_b \in \tau$ such that $a \in S_a$ and $b \in S_b$ (see Fig. 6.5). The points (as singleton *sets*) of a T_2 space are closed sets.[7]

Also, given a topological space (X, τ), the base of the topology τ is a subset $\beta \subset \tau$ of open sets B_i such that every open set $A \in \tau$ may be expressed as the union of elements $B_i \in \beta$ (Definition 6.6). Equivalently, if β forms a base of τ, then every point $x \in A$ belongs to a set $B_i \in \beta$. The intersection of any two elements in β is the union of other elements of β.

---

[6]See, for example, [15], p. 110. In physics, we focus our attention on differentiable manifolds rather than general manifolds. See Definition 7.2 below.

[7]Recall that a T_2 space is T_1 with disjoint, rather than simply distinct, open sets S_a and S_b.

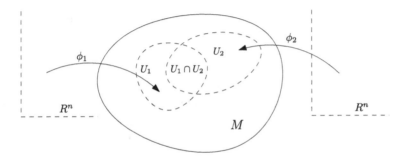

Fig. 7.9 Homeomorphisms $\phi_i : \mathbb{R}^n \rightarrow U_i$ onto the n-manifold M

Definition 7.2 An *n-dimensional manifold* (or *n-manifold*) M is a Hausdorff space with a countable base[8] so that every point $p \in M$ has a neighborhood U_i. We then require that each U_i be homeomorphic to $\mathbb{R}^n$. Each pair (U_i, ϕ_i) forms a *chart* (coordinate patch), and the set of all such pairs is called an *atlas* on M (Fig. 7.9). Because ϕ is a homeomorphism, ϕ^{-1} is also a map, where $\phi_i^{-1} : U_i \rightarrow \mathbb{R}^n$. Consequently, the Jacobian $[D\phi(\mathbf{x})]$ of ϕ will have rank[9] n for all $\mathbf{x} \in \mathbb{R}^n$. A manifold in which the Jacobian has this property is called a *differentiable manifold*. ∎

7.3 Antisymmetric Tensors and *p*-Forms in $\mathbb{R}^n$

In our brief introduction to tensors in Sects. 4.5.3 and 4.5.4, we referred to the fact that a metric tensor is symmetric, i.e., that $g_{ik} = g_{ki}$ for $i \neq k$. In fact, all of the off-diagonal elements in a metric tensor are zero, and the tensor is said to be *diagonal*. We then offered several examples of metric tensors.

We now wish to further explore tensor symmetries, starting with second-rank *antisymmetric tensors*[10]. In particular, antisymmetric second-rank *covariant* tensors are a starting point for *exterior calculus*—a ubiquitous language in modern mathematical physics. As we will see, ordinary vector calculus in three dimensions has several direct correspondences with the exterior calculus, but the latter is significantly broader in scope and gives us the tools to do calculus on arbitrary manifolds (although we will develop it in the context of $\mathbb{R}^n$). These tensors also represent some important quantities in physics, among them being the electromagnetic field tensor, F_{ik}, from which we can find the full relativistic form of Maxwell's equations.

---

[8] Spaces with a countable base are sometimes called *separable*, but we avoid that term so as to allay confusion. These spaces are also called *second-countable* in that they satisfy the *second axiom of countability*, which (along with the *first axiom*) is described in [8], pp. 48–50.

[9] The rank of a matrix is the number of linearly independent columns (or rows). See Sect. 5.5.2.

[10] Also referred to as a *skew-symmetric*, or *alternating*, tensors. A good understanding of this section is key to understanding most of the remainder of this text.

We begin with the definition of the *wedge product* ($\wedge$, read as "wedge" or "hat") by framing it in terms of the tensor product (Definition 4.11). This definition is facilitated by using the *Kronecker tensor*,

$$\epsilon_{\alpha\beta\gamma\ldots\pi}^{123\ldots p} = 0, \text{ if } \alpha\beta\gamma\ldots\pi \text{ is not a permutation of} 123\ldots p;$$
$$= +1, \text{ if} \alpha\beta\gamma\ldots\pi \text{ is an even permutation of} 123\ldots p; \qquad (7.10)$$
$$= -1, \text{ if} \alpha\beta\gamma\ldots\pi \text{ is an odd permutation of} 123\ldots p,$$

where, for example,

$$\epsilon_{\alpha\beta\gamma}^{123} = \epsilon_{\gamma\alpha\beta}^{123} = +1; \epsilon_{\alpha\gamma\beta}^{123} = \epsilon_{\gamma\beta\alpha}^{123} = -1; \epsilon_{\alpha\alpha\gamma}^{123} = 0.$$

The wedge product of p basis one-forms can then be defined in terms of their tensor product as

$$\hat{e}^1 \wedge \hat{e}^2 \wedge \hat{e}^3 \wedge \cdots \hat{e}^p \equiv \epsilon_{\alpha\beta\gamma\ldots\pi}^{123\ldots p} (\hat{e}^\alpha \otimes \hat{e}^\beta \otimes \hat{e}^\gamma \otimes \cdots \hat{e}^\pi), \qquad (7.11)$$

where the sum on the righthand side of Eq. 7.11 is taken over all possible permutations. As an example, letting $p = 2$ and keeping only the non-zero terms gives

$$\hat{e}^1 \wedge \hat{e}^2 = \epsilon_{\alpha\beta}^{12}(\hat{e}^\alpha \otimes \hat{e}^\beta) + \epsilon_{\beta\alpha}^{12}(\hat{e}^\beta \otimes \hat{e}^\alpha) = (+1)(\hat{e}^1 \otimes \hat{e}^2) + (-1)(\hat{e}^2 \otimes \hat{e}^1),$$

or

$$\hat{e}^1 \wedge \hat{e}^2 = (\hat{e}^1 \otimes \hat{e}^2) - (\hat{e}^2 \otimes \hat{e}^1), \qquad (7.12)$$

where the antisymmetry of the wedge product is manifest.

Framing the wedge product this way—by its relationship to the tensor product—allows not only for easy comparison with the tensor product, but also allows for the construction of antisymmetric tensors of any order. Continuing with our example, the wedge product of two one-forms $\boldsymbol{\alpha} = \alpha_i \hat{e}^i$ and $\boldsymbol{\beta} = \beta_j \hat{e}^j$ yields a *two-form*, $\boldsymbol{\omega}$, which may be expressed as

$$\boldsymbol{\omega} = \boldsymbol{\alpha} \wedge \boldsymbol{\beta} = \alpha_i \beta_j (\hat{e}^i \wedge \hat{e}^j)$$
$$= \alpha_i \beta_j [\epsilon_{\alpha\beta}^{ij}(\hat{e}^\alpha \otimes \hat{e}^\beta)]$$
$$= \alpha_i \beta_j (\hat{e}^i \otimes \hat{e}^j) - \beta_j \alpha_i (\hat{e}^j \otimes \hat{e}^i)$$
$$\boldsymbol{\omega} = \boldsymbol{\alpha} \wedge \boldsymbol{\beta} = \boldsymbol{\alpha} \otimes \boldsymbol{\beta} - \boldsymbol{\beta} \otimes \boldsymbol{\alpha}, \qquad (7.13)$$

where again the antisymmetry is clear. We know this is a tensor because in Sect. 4.5.3 (Eq. 4.30) we showed that a second-rank covariant tensor (of any symmetry) is the tensor product of two one forms,

$$\mathbf{t} = \boldsymbol{\alpha} \otimes \boldsymbol{\beta} = \alpha_i \beta_j (\hat{e}^i \otimes \hat{e}^j) = t_{ij}(\hat{e}^i \otimes \hat{e}^j) \in T^*, \qquad (7.14)$$

and because the difference of two tensors is a tensor. As such, we identify ω as an element of a subspace (see Sect. 4.5.1), denoted as $\Lambda^2 \subset T^*$, whose dimension we will define below.

At this point you may be asking *"Yes, but what actually is a two-form? What does it do? What does it* represent?" We know from Definition 4.6 that we can think of a *one*-form as a "machine" that takes a vector and maps it to a scalar. Analogously, a *two*-form is a map that takes *two* vectors as arguments and generates a scalar. The scalar is defined in terms of a determinant whose entries are inner products:

$$\omega(\mathbf{v}_1, \mathbf{v}_2) = (\alpha \wedge \beta)(\mathbf{v}_1, \mathbf{v}_2) \equiv \begin{vmatrix} \langle \alpha | \mathbf{v}_1 \rangle & \langle \alpha | \mathbf{v}_2 \rangle \\ \langle \beta | \mathbf{v}_1 \rangle & \langle \beta | \mathbf{v}_2 \rangle \end{vmatrix} = -\omega(\mathbf{v}_2, \mathbf{v}_1). \qquad (7.15)$$

The determinant structure in Eq. 7.15 serves nicely in dealing with antisymmetric tensors since switching any two columns (vectors) reverses the sign of the determinant. As for what a two-form represents, a geometric interpretation of Eq. 7.15 is related to the concept of an oriented area, which we will discuss in Chap. 8.

Like the tensor product, the wedge product is associative, homogeneous and distributive, but as we have seen these two operations differ regarding commutativity. Whereas the definition of the tensor product makes no mention of commutativity, the antisymmetry of the wedge product is one of its central features. Indeed, the wedge product allows us to associate p-forms with totally antisymmetric (antisymmetric under all inversions) covariant p-rank tensors. Closure of Λ^2 (or, more generally, the space of p-forms Λ^p) follows from Eq. 7.11.

The wedge product, whose properties are defined in Definition 7.3, is a central feature of a non-commutative algebra[11] called an *exterior* or *Grassman algebra*, which is among the more important algebraic structures in mathematical physics. Their principal elements are p-forms[12], of which one-forms and 2-forms are special cases.

Definition 7.3 Let α, β and γ be forms of various orders. The *wedge product* is

1. Associative: $\alpha \wedge (\beta \wedge \gamma) = (\alpha \wedge \beta) \wedge \gamma = (\alpha \wedge \beta \wedge \gamma)$;
2. Homogeneous: $(c\alpha) \wedge \beta = c(\alpha \wedge \beta) = \alpha \wedge (c\beta)$;
3. Distributive (if α and β have the same order): $(\alpha + \beta) \wedge \gamma = \alpha \wedge \gamma + \beta \wedge \gamma$.

4. Graded commutativity, where if α is a p-form and β is a q-form, then

$$\alpha \wedge \beta = (-1)^{pq} \beta \wedge \alpha.$$

---

[11] See Definition 5.1, with $\odot$ changed to $\wedge$. The wedge product is frequently called the *exterior* or *Grassman product*, for Hermann Grassman, (1809–77), a *gymnasium* teacher of mathematics. Following [10], p. 782, we know that Grassman had no university-level education in mathematics, but he nonetheless developed a non-commutative algebra *before* Hamilton. However, he did not publish his work until 1844, one year *after* Hamilton published his work on quaternions. See Sects. 3.4.2 and 3.4.3 for the background on Hamilton's work.

[12] The wedge product may also be applied to vectors, yielding *p-vectors*. An example is the wedge product of two ordinary vectors called a *bivector*, which is the same as the vector cross-product. In this text we focus on the algebra of p-forms.

We will examine this property after defining the dimension and a basis for Λ^p.

5. The wedge product of a p-form and a q-form is a $(p + q)$-form, a result which follows directly from the associative property of the wedge product. ∎

We now turn to a discussion of the basis and dimension of Λ^p and how the dimension of Λ^p is related to $n = \dim X$. We start with an example.

Example 7.5 Consider a two-dimensional ($n = 2$) space X on which we may define a second-rank covariant tensor space T^* (Eq. 4.30). A tensor $\mathbf{t} \in T^*$ and a two-form $\boldsymbol{\omega} \in \Lambda^2 \subset T^*$ may be written as

$$\mathbf{t} = \alpha_1 \beta_1 (\hat{\mathbf{e}}^1 \otimes \hat{\mathbf{e}}^1) + \alpha_1 \beta_2 (\hat{\mathbf{e}}^1 \otimes \hat{\mathbf{e}}^2) + \alpha_2 \beta_1 (\hat{\mathbf{e}}^2 \otimes \hat{\mathbf{e}}^1) + \alpha_2 \beta_2 (\hat{\mathbf{e}}^2 \otimes \hat{\mathbf{e}}^2)$$
$$\boldsymbol{\omega} = \alpha_1 \beta_1 (\hat{\mathbf{e}}^1 \wedge \hat{\mathbf{e}}^1) + \alpha_1 \beta_2 (\hat{\mathbf{e}}^1 \wedge \hat{\mathbf{e}}^2) + \alpha_2 \beta_1 (\hat{\mathbf{e}}^2 \wedge \hat{\mathbf{e}}^1) + \alpha_2 \beta_2 (\hat{\mathbf{e}}^2 \wedge \hat{\mathbf{e}}^2),$$

where we have carried out the summation over all indices in Eqs. 7.13 and 7.14.

At first glance it appears as though $\boldsymbol{\omega} \in \Lambda^2$ is also four-dimensional, but the antisymmetry of the wedge product shows otherwise. Instead, we have

$$\boldsymbol{\omega} = 0 + \alpha_1 \beta_2 (\hat{\mathbf{e}}^1 \wedge \hat{\mathbf{e}}^2) - \alpha_2 \beta_1 (\hat{\mathbf{e}}^1 \wedge \hat{\mathbf{e}}^2) + 0.$$

Further, because of the linear dependence of the two remaining terms we are left with just *one* linearly independent antisymmetric basis tensor, and therefore Λ^2 is a one-dimensional space for $n = 2$. There are several equivalent ways we can express this result:

$$\begin{aligned}
\boldsymbol{\omega} &= (\alpha_1 \beta_2 - \alpha_2 \beta_1)(\hat{\mathbf{e}}^1 \wedge \hat{\mathbf{e}}^2) \\
&= \sum_{1 \le i < j \le n = 2} (\alpha_i \beta_j - \alpha_j \beta_i)(\hat{\mathbf{e}}^i \wedge \hat{\mathbf{e}}^j) \\
&\equiv (\alpha_I \beta_J - \alpha_J \beta_I)(\hat{\mathbf{e}}^I \wedge \hat{\mathbf{e}}^J) \\
&\equiv \omega_{IJ}(\hat{\mathbf{e}}^I \wedge \hat{\mathbf{e}}^J) \in \Lambda^2 \subset T^*,
\end{aligned} \tag{7.16}$$

where we introduced a new "capitalized index" (or, "ordered") notation[13] to indicate that we are summing only over $1 \le I < J \le n = \dim X$. The purpose of this notation is to avoid the double counting that comes from summing over all indices. ▲

It is often the case that $\boldsymbol{\omega}$ will be written so as to carry out the sum over *all* indices in accordance with the usual summation convention. In this circumstance, we get (from the second line in Eq. 7.16)

$$\boldsymbol{\omega} = \frac{1}{2}(\alpha_i \beta_j - \alpha_j \beta_i)(\hat{\mathbf{e}}^i \wedge \hat{\mathbf{e}}^j) \equiv \frac{1}{2}\omega_{ij}(\hat{\mathbf{e}}^i \wedge \hat{\mathbf{e}}^j), \tag{7.17}$$

---

[13]Conventions vary among authors. Equation 7.16 follows the convention in [3], where ordered coordinates are called "strict" coordinates.

where ω_{ij} are the components of an antisymmetric tensor. However, because of the double counting (and the subsequent linear dependence, as before), *the* $(\hat{\mathbf{e}}^i \wedge \hat{\mathbf{e}}^j)$ *do not form a basis for* Λ^2. Therefore, the only basis element of Λ^2 for $n = 2$ is the *ordered* wedge product of the two basis one-forms. Expressions like Eq. 7.17 have their place in tensor analysis, but operations involving forms more often use an ordered product basis, as in Eq. 7.16.

The results in Example 7.5 carry over to higher-order forms and their spaces. Generally, the dimension of Λ^p is the number of linearly independent antisymmetric *ordered* basis tensors. Several immediate consequences are that dim $\Lambda^1 = n$, dim $\Lambda^n = 1$ and $\Lambda^p = 0$ for $p > n$.

For $1 < p < n$, we find the dimension of Λ^p by finding the number of non-zero ordered combinations of $(\hat{\mathbf{e}}^1 \wedge \hat{\mathbf{e}}^2 \wedge \hat{\mathbf{e}}^3 \wedge \cdots \wedge \hat{\mathbf{e}}^p)$ from among the set of n basis one-forms. The result is just the binomial coefficient:

$$\dim \Lambda^p = \binom{n}{p} \equiv \frac{n!}{p!(n-p)!}. \tag{7.18}$$

For example, for $n = 3$ the space Λ^2 of two-forms will be three-dimensional,

$$\omega = \alpha_1 \beta_2 (\hat{\mathbf{e}}^1 \wedge \hat{\mathbf{e}}^2) + \alpha_1 \beta_3 (\hat{\mathbf{e}}^1 \wedge \hat{\mathbf{e}}^3) + \alpha_2 \beta_3 (\hat{\mathbf{e}}^2 \wedge \hat{\mathbf{e}}^3), \tag{7.19}$$

the space Λ^3 will be one-dimensional with basis $(\mathbf{e}^1 \wedge \mathbf{e}^2 \wedge \mathbf{e}^3)$ and dim $\Lambda^1 = 3$. The demonstration of Eq. 7.19 is left as an exercise[14].

An unusual aspect of the wedge product is its graded commutative property (Definition 7.3(4)), where commutativity depends on the order of the forms. That is, in forming the wedge product of a p-form and a q-form, commutativity depends on the total number of inversions that arise from reversing the two sets of basis tensors. If p and q are both odd, then the wedge product anticommutes; otherwise it commutes.

Example 7.6 Let α be a two-form in basis σ^i, and let β be a three-form in basis θ^j. The basis for the five-form $\omega = \alpha \wedge \beta$ may be rearranged through a succession of inversions so as to eventually yield the basis for $\beta \wedge \alpha$. In detail, we have

$$
\begin{aligned}
(\sigma^1 \wedge \sigma^2) \wedge (\theta^1 \wedge \theta^2 \wedge \theta^3) &= (\sigma^1 \wedge \sigma^2 \wedge \theta^1 \wedge \theta^2 \wedge \theta^3) \\
&= (-1)(\sigma^1 \wedge \theta^1 \wedge \sigma^2 \wedge \theta^2 \wedge \theta^3) \\
&= (+1)(\theta^1 \wedge \sigma^1 \wedge \sigma^2 \wedge \theta^2 \wedge \theta^3) \\
&= (-1)(\theta^1 \wedge \sigma^1 \wedge \theta^2 \wedge \sigma^2 \wedge \theta^3) \\
&= (+1)(\theta^1 \wedge \theta^2 \wedge \sigma^1 \wedge \sigma^2 \wedge \theta^3) \\
&= (-1)(\theta^1 \wedge \theta^2 \wedge \sigma^1 \wedge \theta^3 \wedge \sigma^2) \\
&= (+1)(\theta^1 \wedge \theta^2 \wedge \theta^3 \wedge \sigma^1 \wedge \sigma^2) \\
&= (-1)^6 (\theta^1 \wedge \theta^2 \wedge \theta^3) \wedge (\sigma^1 \wedge \sigma^2) \\
\Rightarrow \alpha \wedge \beta &= \beta \wedge \alpha.
\end{aligned}
$$

---

[14] The set of p-forms for all $1 \le p \le n$ constitute a *graded algebra* on the space.

In this case, $pq = 6$ and the wedge product commutes. However, if α were a one-form or a three-form, say, then the wedge product would anticommute.   ▲

Equation 7.19 (where $n = 3$ and $p = 2$) may remind you of the vector cross product, and there is good reason for this. First, however, consider the first term in Eq. 7.19 and ask whether the basis two-form $(\hat{\mathbf{e}}^1 \wedge \hat{\mathbf{e}}^2)$ is related to $\hat{\mathbf{e}}^3$—the basis one-form that is not included in that first term. More generally, given a space Λ^p of p-forms with dim $X = n$, what can we say about the space $\Lambda^{(n-p)}$ of $(n - p)$-forms? The answer is provided by the *Hodge star* ($\star$) *operator*.

Definition 7.4 Let ω be a p-form with basis $(\hat{\mathbf{e}}^{i_1} \wedge \hat{\mathbf{e}}^{i_2} \wedge \hat{\mathbf{e}}^{i_3} \wedge \cdots \hat{\mathbf{e}}^{i_p})$. Then there is an $(n - p)$-form $\star\omega$ (read "star omega") with the same corresponding coefficients (by Eq. 7.18, Λ^p and $\Lambda^{(n-p)}$ have the same dimension) as ω, but with the basis

$$\star(\hat{\mathbf{e}}^{i_1} \wedge \hat{\mathbf{e}}^{i_2} \wedge \hat{\mathbf{e}}^{i_3} \wedge \cdots \hat{\mathbf{e}}^{i_p}) \equiv (-1)^\pi (\hat{\mathbf{e}}^{j_1} \wedge \hat{\mathbf{e}}^{j_2} \wedge \hat{\mathbf{e}}^{j_3} \wedge \cdots \hat{\mathbf{e}}^{j_{n-p}}).$$

Each basis (one labeled with i, the other with j) is ordered, the set of n indices $(i_1, i_2, i_3 \ldots i_p, j_1, j_2, j_3 \ldots j_{n-p})$ is a permutation π of the set $(1,2,3 \ldots n)$, $\pi = 0$ if the permutation is even and $\pi = -1$ if the permutation is odd. s   ■

The plethora of indices and subindices in Definition 7.4 can be very confusing, but in practice finding $\star\omega$ from ω reduces to asking "What's left out?" of the basis in each term in ω, and then determining whether the permutation of the combined n indices of ω and $\star\omega$ is even or odd.

Example 7.7 Let $\omega = \alpha_1\beta_2(\hat{\mathbf{e}}^1 \wedge \hat{\mathbf{e}}^2) + \alpha_1\beta_3(\hat{\mathbf{e}}^1 \wedge \hat{\mathbf{e}}^3) + \alpha_2\beta_3(\hat{\mathbf{e}}^2 \wedge \hat{\mathbf{e}}^3)$ be a two-form for $n = 3$ (see Eq. 7.19). Noting the permutations as either even or odd, we get

$$(\hat{\mathbf{e}}^1 \wedge \hat{\mathbf{e}}^2)\text{in } \omega \Rightarrow (+1)\hat{\mathbf{e}}^3\text{in}\star\omega\text{because } (1,2,3) \text{ is even}$$
$$(\hat{\mathbf{e}}^1 \wedge \hat{\mathbf{e}}^3)\text{in } \omega \Rightarrow (-1)\hat{\mathbf{e}}^2\text{in}\star\omega\text{because } (1,3,2) \text{ is odd}$$
$$(\hat{\mathbf{e}}^2 \wedge \hat{\mathbf{e}}^3)\text{in } \omega \Rightarrow (+1)\hat{\mathbf{e}}^1\text{in}\star\omega\text{because } (2,3,1) \text{ is even.}$$

This gives the one-form

$$\star\omega = \alpha_1\beta_2\hat{\mathbf{e}}^3 - \alpha_1\beta_3\hat{\mathbf{e}}^2 + \alpha_2\beta_3\hat{\mathbf{e}}^1.$$

Again, by virtue of Eq. 7.18, both ω and $\star\omega$ are three-dimensional.

As another example, let $n = 2$, and let $\alpha = \alpha_1\hat{\mathbf{e}}^1 + \alpha_2\hat{\mathbf{e}}^2$ be a two-dimensional one-form. Then $\star\alpha = \alpha_1\hat{\mathbf{e}}^2 - \alpha_2\hat{\mathbf{e}}^1$.   ▲

Example 7.8 Let $n = 4$, and let $\omega = \omega_{12}(\hat{\mathbf{e}}^1 \wedge \hat{\mathbf{e}}^2) + \omega_{14}(\hat{\mathbf{e}}^1 \wedge \hat{\mathbf{e}}^4)$. Then

$$\star\omega = \omega_{12}(\hat{\mathbf{e}}^3 \wedge \hat{\mathbf{e}}^4) + \omega_{14}(\hat{\mathbf{e}}^2 \wedge \hat{\mathbf{e}}^3),$$

since $(1, 2, 3, 4)$ and $(1, 4, 3, 2)$ are even permutations of $(1, 2, 3, 4)$.   ▲

In Sect. 7.4 the wedge and Hodge star operations will be used to establish a general correspondence between p-forms and vector calculus for three-dimensional space. We can already see how a correspondence might arise from Example 7.7 by letting of $(\hat{\mathbf{e}}^1, \hat{\mathbf{e}}^2, \hat{\mathbf{e}}^1)$ be $(\hat{\mathbf{i}}, \hat{\mathbf{j}}, \hat{\mathbf{k}})$ and letting the wedge product $(\wedge)$ be the vector cross product. However, for $n > 3$ (as in Example 7.8), our familiar three-dimensional vector calculus no longer exists. Fortunately, p-forms, the wedge product and the Hodge star operation live on in higher dimensions, and we can think of three-dimensional vector calculus as a special case built around these more fundamental structures.

Finally, Eq. 7.17 establishes the relationship between a two-form and the components of an antisymmetric covariant tensor, and the tensor notation is often very useful. Consider a second-rank covariant tensor $\mathbf{t} = t_{ij}(\hat{\mathbf{e}}^i \otimes \hat{\mathbf{e}}^j)$ of unspecified symmetry. Then using Eqs. 7.11 and 7.17 we may write an antisymmetric tensor as

$$\mathbf{t}_a = \frac{1}{2}t_{ij}(\hat{\mathbf{e}}^i \wedge \hat{\mathbf{e}}^j) = \frac{1}{2}t_{ij}[(\hat{\mathbf{e}}^i \otimes \hat{\mathbf{e}}^j) - (\hat{\mathbf{e}}^j \otimes \hat{\mathbf{e}}^i)].$$

There is also a symmetric tensor (obviously not derived from the wedge product),

$$\mathbf{t}_s = \frac{1}{2}t_{ij}[(\hat{\mathbf{e}}^i \otimes \hat{\mathbf{e}}^j) + (\hat{\mathbf{e}}^j \otimes \hat{\mathbf{e}}^i)],$$

and together their sum is $\mathbf{t}_s + \mathbf{t}_a = t_{ij}(\hat{\mathbf{e}}^i \otimes \hat{\mathbf{e}}^j) = \mathbf{t}$.

In many places it is customary to write this solely in terms of components without reference to the basis tensors:

$$t_{ij} = \frac{1}{2}(t_{ij} + t_{ji}) + \frac{1}{2}(t_{ij} - t_{ji}) \equiv t_{(ij)} + t_{[ij]}. \tag{7.20}$$

Higher-order p-forms are associated with higher-rank *totally antisymmetric* covariant tensors, i.e., covariant tensors that are antisymmetric upon the interchange of any two indices. For example, the component expression for a totally antisymmetric covariant third-rank tensor $t_{[ijk]}$ is

$$t_{[ijk]} = \frac{1}{3!}(t_{ijk} + t_{kij} + t_{jki} - t_{jik} - t_{ikj} - t_{kji}), \tag{7.21}$$

a result which may be found by making full use of Eq. 7.11 and by following the same method that we followed when finding $t_{[ij]}$. We leave this as an exercise. The higher-order version of Eq. 7.17 for a p-form written in terms of a totally antisymmetric tensor of rank p is then seen to be

$$\omega = \frac{1}{p!}\omega_{ij\ldots p}(\hat{\mathbf{e}}^i \wedge \hat{\mathbf{e}}^j \wedge \cdots \hat{\mathbf{e}}^p). \tag{7.22}$$

As an aside, a *totally symmetric* tensor $t_{(ijk)}$ corresponding to Eq. 7.21 is

$$t_{(ijk)} = \frac{1}{3!}(t_{ijk} + t_{kij} + t_{jki} + t_{jik} + t_{ikj} + t_{kji}). \tag{7.23}$$

Importantly, and unlike what we just found for second-rank tensors, $t_{ijk} \neq t_{(ijk)} + t_{[ijk]}$; the sum on the right omits terms of mixed symmetry that are part of t_{ijk}.

The discussion in this section may remind you of the discussion of the symmetric and alternating groups in Chap. 2. For example, in Sect. 2.3.3 we saw that the alternating group A_n is a subgroup of the symmetric group S_n. Here, Λ^p is the space of p-forms (totally antisymmetric tensors) and is a subspace of tensor space T^*. If you are interested in learning more about the symmetries or group properties of tensors, please see the Guide to Further Study at the end of the chapter.

7.4 Differential Forms in $\mathbb{R}^n$

In order for p-forms to be applicable to physics, there must be structures that provide for their differentiation and integration. This requires formulations of the derivative and integral that are more general and inclusive than—but nonetheless consistent with—what we know from elementary calculus about the differentiation and integration of functions. The focus of this section is on differentiation. We introduce integration on manifolds in Chap. 8 where we will derive a generalized version of Stokes's Theorem, of which line and surface integrals are special cases.

A *differential form* is the differential of a p-form. It is a structure that combines three familiar ideas—the differential of a function, the scalar product and the directional derivative—into a single concept.

Consider a function of coordinates $f(x^i)$ (which we take to be a 0-form), its differential df and the scalar product $\langle \alpha | \mathbf{v} \rangle$ of a one-form α with a vector $\mathbf{v}$. We write the differential and the scalar product as

$$df = \frac{\partial f}{\partial x^i} dx^i \text{ and } \langle \alpha | \mathbf{v} \rangle = \langle \alpha_i \, \hat{e}^i | v^j \, \hat{e}_j \rangle. \tag{7.24}$$

If we make the assignments

$$\alpha_i \longrightarrow \frac{\partial f}{\partial x^i} \text{ and } \hat{e}^i \longrightarrow \mathbf{dx}^i, \text{ so that } \alpha \longrightarrow \mathbf{df} = \frac{\partial f}{\partial x^i} \mathbf{dx}^i, \tag{7.25}$$

then the scalar product is reframed accordingly as

$$\langle \alpha | \mathbf{v} \rangle \longrightarrow \langle \mathbf{df} | \mathbf{v} \rangle = \langle \frac{\partial f}{\partial x^i} \mathbf{dx}^i | v^j \hat{e}_j \rangle = v^j \frac{\partial f}{\partial x^i} \langle \mathbf{dx}^i | \hat{e}_j \rangle \equiv \mathbf{v} \cdot \nabla f, \tag{7.26}$$

where $\langle \mathbf{dx}^i | \hat{e}_j \rangle = \delta^i_j$ because $\mathbf{dx}^i$ is a unit basis one-form acting on a unit vector.

As noted previously (Example 7.3), the result in Eq. 7.26 is the directional derivative of the function $f(x^i)$ along the direction specified by the vector $\mathbf{v}$. Therefore, like the gradient of a function, $\mathbf{df}$ is a one-form—in this case, a *differential* one-form—whose components are those of the gradient[15]. Generally, we write a differential one-form as $\boldsymbol{d\alpha} = \alpha_i \mathbf{dx}^i$, where the α_i are functions of the coordinates.

The properties and operations discussed in Sect. 7.3 for p-forms apply to differential forms[16]. This includes the wedge product, the Hodge star operator, the ordered nature of the basis in Λ^p, the graded commutativity property and so forth; we just change notation to reflect that we are now working with differential forms. A more substantive change will be to replace the general basis one-forms $\hat{\mathbf{e}}^i$, $\hat{\mathbf{e}}^j$ and $\hat{\mathbf{e}}^k$ with $\mathbf{dx}^i$, $\mathbf{dx}^j$ and $\mathbf{dx}^k$, representing a *local frame of reference* (see Example 7.10).

Example 7.9 Consider the one-forms $\boldsymbol{\alpha} = \alpha\mathbf{dx}$, $\boldsymbol{\beta} = \beta\mathbf{dy}$ and $\boldsymbol{\gamma} = \gamma\mathbf{dz}$ for $n = 3$ and Cartesian coordinates, and let ω be a three-dimensional two-form given as

$$\omega = \alpha\beta(\mathbf{dx} \wedge \mathbf{dy}) + \alpha\gamma(\mathbf{dx} \wedge \mathbf{dz}) + \beta\gamma(\mathbf{dy} \wedge \mathbf{dz}).$$

Then (compare Example 7.7):

$$\star\omega = \alpha\beta\mathbf{dz} - \alpha\beta\mathbf{dy} + \alpha\beta\mathbf{dx}$$

is a three-dimensional one-form. ▲

We need to more fully explore what the "$\mathbf{d}$" in $\mathbf{df}$ or $\boldsymbol{d\alpha}$ actually means. Equations 7.24–7.26 would seem to suggest that $\mathbf{d}$ is "just like" the gradient operator in vector calculus. Rather, $\mathbf{d}$ is the *exterior differential operator*, the operation of which changes a p-form into a $(p + 1)$-form.

For example, if $\mathbf{f}$ is a function (a 0-form), then $\mathbf{df}$—the differential of a function that we usually just write as df—is a one-form. If $\boldsymbol{\alpha}$ is a one-form, then $\omega = \boldsymbol{d\alpha}$ is a two-form[17], and so forth. Depending on the context in which it is applied, the $\mathbf{d}$ operator can play a role in Λ^p that is analogous to (and corresponds to) the roles played by the gradient, curl or divergence operators in vector spaces.

Definition 7.5 Let f be a function, and consider the p-form $\boldsymbol{\alpha} = \alpha_I \mathbf{dx}^I$ and the q-form $\boldsymbol{\beta} = \beta_J \mathbf{dx}^J$ (note the ordered basis). Let c be a constant. The exterior differential operator is defined such that

---

[15] As we will see shortly, the gradient is defined only for spaces with dimension $n \leq 3$, whereas the concept of a differential one-form may be applied more broadly.

[16] You may be wondering: "Are we going to insist on using the word 'differential' every time when the context is otherwise clear?" The answer is "no," but we have made it a point in this text to define forms more broadly than their differential formulation. From here onward, however, we will usually, though not always, drop "differential" whenever the context is clear.

[17] Note that in Example 7.9 we just wrote down a differential two-form based upon our earlier discussion of p-forms; we did not actually differentiate a one-form to get a two-form.

1. $\mathbf{df} = \frac{\partial f}{\partial x^i}\mathbf{dx}^i$ is a one-form—the ordinary differential of f. See Eq. 7.25;
2. $\boldsymbol{d\alpha} = \boldsymbol{d\alpha}_I \wedge \mathbf{dx}^I$—the central part of the definition;
3. $\mathbf{d}(\alpha + \beta) = \boldsymbol{d\alpha} + \boldsymbol{d\beta}$—linearity;
4. $\mathbf{d}(c\alpha) = c\boldsymbol{d\alpha}$—homogeneity;
5. $\mathbf{d}(\alpha \wedge \beta) = \boldsymbol{d\alpha} \wedge \beta + (-1)^p \alpha \wedge \boldsymbol{d\beta}$—a Leibnitz product rule;
6. $\mathbf{d}(\boldsymbol{d\alpha}) \equiv \mathbf{d}^2\alpha = 0$.

   Properties (3) and (4) are straightforward. Proof of property (5) is left as an exercise in light of the other properties. The proof of property (6) will be given later in this section, and property (2) will be demonstrated by the examples that follow. Two additional concepts will have particular relevance when we discuss integration on manifolds in Chap. 8:

7. If $\boldsymbol{d\alpha} = 0$, then α is said to be *closed*;
8. If $\alpha = \boldsymbol{d\beta}$ for some form β, then α is said to be *exact*. If a form is exact, then by (6) it is closed. The converse is not always true. See Sects. 8.2 and 8.3. ∎

As in Example 7.9, the following examples employ a notation that is frequently used for differential forms in a three-dimensional space with Cartesian coordinates. Rather than writing basis two-forms with arbitrary indices, we write terms like $\boldsymbol{dx} \wedge \boldsymbol{dy}$. A typical one-form is written as $\alpha = \alpha_1\mathbf{dx} + \alpha_2\mathbf{dy} + \alpha_3\mathbf{dz}$, where the α_i are functions.

Example 7.10 One of the simplest functions is simply $f(x^j) = x^j$. In this instance,

$$\mathbf{df} = \frac{\partial f}{\partial x^i}\mathbf{dx}^i = \frac{\partial x^j}{\partial x^i}\mathbf{dx}^i = \delta_i^j\mathbf{dx}^i,$$

and the one-form $\mathbf{df}$ is just the basis one-form $\mathbf{dx}^j$. In Cartesian coordinates, $\mathbf{dx}^1 = \mathbf{dx}$, $\mathbf{dx}^2 = \mathbf{dy}$ and $\mathbf{dx}^3 = \mathbf{dz}$, and among the scalar products we might form are

$$\langle\mathbf{dx}|\hat{\mathbf{i}}\rangle = \langle\mathbf{dy}|\hat{\mathbf{j}}\rangle = \langle\mathbf{dz}|\hat{\mathbf{k}}\rangle = 1,$$

whereas terms like $\langle\mathbf{dx}|\hat{\mathbf{j}}\rangle$, $\langle\mathbf{dy}|\hat{\mathbf{k}}\rangle$ and so forth all equal zero. We asserted these results in Eqs. 7.24–7.26; here they follow from Definition 7.5(1). Again (Eq. 7.25), when $\mathbf{d}$ operates on a function, the components of $\mathbf{df}$ are those of a gradient. ▲

Example 7.11 Consider the one-form

$$\alpha = A(x, y, z)\mathbf{dx} + B(x, y, z)\mathbf{dy} + C(x, y, z)\mathbf{dz}$$

Using Definition 7.5(2) we take the exterior derivative of α to obtain a two-form. Note that terms like $\mathbf{dx} \wedge \mathbf{dx} = 0$, and $\mathbf{dx} \wedge \mathbf{dz} = -\mathbf{dz} \wedge \mathbf{dx}$ by the antisymmetry of the wedge product of two one-forms. Because A, B and C are functions, we find

$$d\alpha = (dA) \wedge dx + (dB) \wedge dy + (dC) \wedge dz$$

$$= \left(\frac{\partial A}{\partial x} dx + \frac{\partial A}{\partial y} dy + \frac{\partial A}{\partial z} dz \right) \wedge dx$$

$$+ \left(\frac{\partial B}{\partial x} dx + \frac{\partial B}{\partial y} dy + \frac{\partial B}{\partial z} dz \right) \wedge dy$$

$$+ \left(\frac{\partial C}{\partial x} dx + \frac{\partial C}{\partial y} dy + \frac{\partial C}{\partial z} dz \right) \wedge dz$$

$$d\alpha = \left(\frac{\partial C}{\partial y} - \frac{\partial B}{\partial z} \right) (dy \wedge dz) + \left(\frac{\partial A}{\partial z} - \frac{\partial C}{\partial x} \right) (dz \wedge dx) + \left(\frac{\partial B}{\partial x} - \frac{\partial A}{\partial y} \right) (dx \wedge dy).$$

Applying the Hodge star operator (see Definition 7.4 and Example 7.7) to $d\alpha$ gives the one-form

$$\star d\alpha = \left(\frac{\partial C}{\partial y} - \frac{\partial B}{\partial z} \right) dx + \left(\frac{\partial A}{\partial z} - \frac{\partial C}{\partial x} \right) dy + \left(\frac{\partial B}{\partial x} - \frac{\partial A}{\partial y} \right) dz.$$

If you are familiar with elementary vector calculus, then you recognize the components of $d\alpha$ and $\star d\alpha$ as those of the curl $\nabla \times \mathbf{v}$ of a vector $\mathbf{v} = A(x, y, z)\hat{\mathbf{i}} + B(x, y, z)\hat{\mathbf{j}} + C(x, y, z)\hat{\mathbf{k}}$. In this sense we say that when $\mathbf{d}$ operates on a one-form, it corresponds to taking the curl of a vector in vector calculus.

Further, the basis two-forms in the expression for $d\alpha$ remind us of differential (infinitesimal) oriented areas, and the basis one-forms in $\star d\alpha$ resemble differential line elements. We'll say more about oriented areas and differential line elements when we discuss integration on manifolds in Chap. 8. ▲

Example 7.12 Drawing on Example 7.9, consider the two-form

$$\omega = P(x, y, z)(dy \wedge dz) + Q(x, y, z)(dz \wedge dx) + R(x, y, z)(dx \wedge dy),$$

where we have reordered the terms and relabeled the coefficients. Again we will apply Definition 7.5(2) and the antisymmetry of the wedge product of two one-forms.

We note that the only surviving terms will be those containing all three basis one-forms, such as terms like $dy \wedge dz \wedge dx$. Terms containing repeated basis one-forms, such as $dy \wedge dz \wedge dy$, will vanish. This follows from the definition of the wedge product (Definition 7.3); see also Eqs. 7.10 and 7.11.

Proceeding as before, we have

$$
\begin{aligned}
d\omega &= (\mathbf{dP}) \wedge (\mathbf{dy} \wedge \mathbf{dz}) + (\mathbf{dQ}) \wedge (\mathbf{dz} \wedge \mathbf{dx}) + (\mathbf{dR}) \wedge (\mathbf{dx} \wedge \mathbf{dy}) \\
&= \left(\frac{\partial P}{\partial x}\mathbf{dx} + \frac{\partial P}{\partial y}\mathbf{dy} + \frac{\partial P}{\partial z}\mathbf{dz} \right) \wedge (\mathbf{dy} \wedge \mathbf{dz}) \\
&\quad + \left(\frac{\partial Q}{\partial x}\mathbf{dx} + \frac{\partial Q}{\partial y}\mathbf{dy} + \frac{\partial Q}{\partial z}\mathbf{dz} \right) \wedge (\mathbf{dz} \wedge \mathbf{dx}) \\
&\quad + \left(\frac{\partial R}{\partial x}\mathbf{dx} + \frac{\partial R}{\partial y}\mathbf{dy} + \frac{\partial R}{\partial z}\mathbf{dz} \right) \wedge (\mathbf{dx} \wedge \mathbf{dy}) \\
&= \frac{\partial P}{\partial x}(\mathbf{dx} \wedge \mathbf{dy} \wedge \mathbf{dz}) + \frac{\partial Q}{\partial y}(\mathbf{dy} \wedge \mathbf{dz} \wedge \mathbf{dx}) + \frac{\partial R}{\partial z}(\mathbf{dz} \wedge \mathbf{dx} \wedge \mathbf{dy}).
\end{aligned}
$$

Carrying out the necessary inversions (as in Example 7.6), or recognizing that the three basis three-forms are in cyclic order and hence of even permutation, we have the three-form

$$
d\omega = \left(\frac{\partial P}{\partial x} + \frac{\partial Q}{\partial y} + \frac{\partial R}{\partial Z} \right) (\mathbf{dx} \wedge \mathbf{dy} \wedge \mathbf{dz}).
$$

Applying the Hodge star operator to $d\omega$ gives a 0-form, i.e., the function

$$
\star d\omega = f(x, y, z) = \left(\frac{\partial P}{\partial x} + \frac{\partial Q}{\partial y} + \frac{\partial R}{\partial Z} \right).
$$

Again comparing this result with those of vector calculus, we recognize the component of $d\omega$ and the "component" of $\star d\omega = f(x, y, z)$ as the divergence $\nabla \cdot \mathbf{v}$ of a vector $\mathbf{v} = P(x, y, z)\hat{\mathbf{i}} + Q(x, y, z)\hat{\mathbf{j}} + R(x, y, z)\hat{\mathbf{k}}$.

Therefore, the operation of $\mathbf{d}$ on a two-form corresponds to the divergence operator in vector calculus. Note also that the basis three-form $(\mathbf{dx} \wedge \mathbf{dy} \wedge \mathbf{dz})$ resembles a differential volume element.   ▲

The property $\mathbf{d}^2\omega = 0$ (Definition 7.5(6))—that a double application of $\mathbf{d}$ to a p-form ω yields zero—has important implications when we consider the integration of forms in Chap. 8. It also helps us to better understand the correspondences between exterior and vector calculus. In order to see how the operator $\mathbf{d}^2 = 0$ comes about, we will first apply it to a function and then use that result to show that this property holds for any p-form.

Given a function f, we take

$$
\mathbf{d}(\mathbf{d}f) = \mathbf{d}\left(\frac{\partial f}{\partial x^i}\mathbf{dx}^i \right) = \left(\frac{\partial^2 f}{\partial x^j \partial x^i}\mathbf{dx}^j \right) \wedge \mathbf{dx}^i = \frac{\partial^2 f}{\partial x^j \partial x^i} \left(\mathbf{dx}^j \wedge \mathbf{dx}^i \right). \quad (7.27)
$$

The summation indices in Eq. 7.27 are *not ordered*. Consequently, when we sum over all indices, the factor $\left(\mathbf{dx}^j \wedge \mathbf{dx}^i \right)$ yields zero either when $i = j$, or through pairwise cancellation due to antisymmetry i.e., $\left(\mathbf{dx}^j \wedge \mathbf{dx}^i\right) = -\left(\mathbf{dx}^i \wedge \mathbf{dx}^j\right)$; the mixed partials are equal when i and j are reversed. Therefore, $\mathbf{d}^2\mathbf{f} = 0$.

Next consider the basis one-form $\mathbf{dx}^i = (1)\mathbf{dx}^i$, where we pulled out the factor of unity to make a point, namely that $\mathbf{d}(\mathbf{dx}^i) = \mathbf{d}(1) \wedge \mathbf{dx}^i = 0 \wedge \mathbf{dx}^i = 0$. We are now in a position to consider $\mathbf{d}^2$ when applied to a p-form $\omega = a_I \mathbf{dx}^I$, where the a_I are functions and the basis is ordered. From Definition 7.5(5) (the proof of which was left as an exercise) we have

$$
\begin{aligned}
\mathbf{d}(d\omega) &= \mathbf{d}(\mathbf{da}_I \wedge \mathbf{dx}^I) \\
&= \mathbf{d}(\mathbf{da}_I) \wedge \mathbf{dx}^I + (-1)^p \mathbf{da}_I \wedge \mathbf{d}(\mathbf{dx}^I) \\
&= 0 \wedge \mathbf{dx}^I + (-1)^p \mathbf{da}_I \wedge 0 \\
\Rightarrow \mathbf{d}(d\omega) &= 0,
\end{aligned}
\tag{7.28}
$$

where we used the results noted previously.

As a final example in this section we evaluate the effect of a two-form acting on two vectors to yield a scalar, as defined earlier in Eq. 7.15 (see also Example 7.10).

Example 7.13 Let $n = 2$ and consider the two-form $\omega = \omega_{12}(\hat{\mathbf{e}}^1 \wedge \hat{\mathbf{e}}^2)$ (Eq. 7.16). In a two-dimensional Cartesian coordinate system we can write this as $\omega = \omega(\mathbf{dx} \wedge \mathbf{dy})$, where $\omega = \omega(x, y)$ is a function and represents the magnitude of the two-form.

Equation 7.15 then becomes

$$
\omega(\mathbf{v}_1, \mathbf{v}_2) = \omega(\mathbf{dx} \wedge \mathbf{dy})(\mathbf{v}_1, \mathbf{v}_2) \equiv \omega \begin{vmatrix} \langle \mathbf{dx}|\mathbf{v}_1\rangle & \langle \mathbf{dx}|\mathbf{v}_2\rangle \\ \langle \mathbf{dy}|\mathbf{v}_1\rangle & \langle \mathbf{dy}|\mathbf{v}_2\rangle \end{vmatrix}.
$$

Letting $\mathbf{v}_1 = x_1\hat{\mathbf{i}} + y_1\hat{\mathbf{j}}$ and $\mathbf{v}_2 = x_2\hat{\mathbf{i}} + y_2\hat{\mathbf{j}}$ gives

$$
\omega(\mathbf{v}_1, \mathbf{v}_2) = \omega \begin{vmatrix} x_1 & x_2 \\ y_1 & y_2 \end{vmatrix} = \omega(x_1 y_2 - x_2 y_1).
$$

We see that the scalar quantity $\omega(\mathbf{v}_1, \mathbf{v}_2)$ is the *signed* magnitude of the vector cross product $\mathbf{v}_1 \times \mathbf{v}_2$, as scaled by the magnitude of ω. Geometrically, this is the oriented area of the parallelogram bounded by $\mathbf{v}_1$ and $\mathbf{v}_2$. If $\omega = \omega(x, y) = 1$, then we have the area of the parallelogram as defined by the two vectors in the plane. In general, however, ω is a function, in which case both the magnitude and orientation of the area formed by these same two vectors will vary with position. A much more interesting example is that of a two-form in a three-dimensional space, and we discuss this as a (guided) end-of-chapter problem in Chap. 8.   ▲

7.5   Correspondences Between Exterior and Vector Calculus in $\mathbb{R}^3$

The correspondences between differential forms and ordinary vectors that arose in Examples 7.10–7.13 are summarized in Eq. 7.29.

$$
\begin{array}{cccc}
\underline{\star\text{form}} \Longleftrightarrow & \underline{\text{form}} & \Longleftrightarrow & \underline{\text{vector or function}} \\
3 - \text{form} \Longleftrightarrow & 0 - \text{form} & \Longleftrightarrow & f\,(\text{function}) \\
\uparrow \mathbf{d} \Longleftrightarrow & \downarrow \mathbf{d} & \Longleftrightarrow & \nabla(\text{gradient}) \\
2 - \text{form} \Longleftrightarrow & 1 - \text{form} & \Longleftrightarrow & \nabla f\,(\text{vector}) \\
\uparrow \mathbf{d} \Longleftrightarrow & \downarrow \mathbf{d} & \Longleftrightarrow & \nabla \times (\text{curl}) \\
1 - \text{form} \Longleftrightarrow & 2 - \text{form} & \Longleftrightarrow & \nabla \times \mathbf{v}(\text{vector}) \\
\uparrow \mathbf{d} \Longleftrightarrow & \downarrow \mathbf{d} & \Longleftrightarrow & \nabla \cdot (\text{div}) \\
0 - \text{form} \Longleftrightarrow & 3 - \text{form} & \Longleftrightarrow & \nabla \cdot \mathbf{v}(\text{function})
\end{array}
\tag{7.29}
$$

The vertical arrows represent single-steps; if we attempt to apply a succession of steps, we get zero. For example, if we start with a 0-form f and apply $\mathbf{d}$, we get $\mathbf{df}$, but applying it again gives $\mathbf{d}^2\mathbf{f} = 0$, by Definition 7.5(6). The vector equivalent would be to first form the gradient ∇f and then take its curl, which yields the vector identity $\nabla \times \nabla f = 0$.

Similarly, if we start with a one-form α and apply $\mathbf{d}$, we get a two-form; applying it again gives $\mathbf{d}^2\alpha = 0$. The vector equivalent would be to start with a vector $\mathbf{v}$ and then take the curl followed by the divergence, which gives the identity $\nabla \cdot \nabla \times \mathbf{v} = 0$.

Figure 7.10 replicates a portion of the pattern shown in Eq. 7.29.

Starting with a one-form, Fig. 7.10a illustrates the process we described in Example 7.11, where we first applied the $\mathbf{d}$ operator and then the $\star$ operator. Figure 7.10b

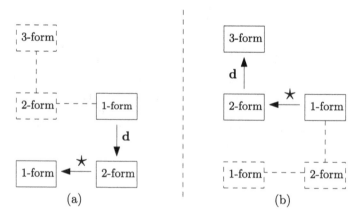

Fig. 7.10 **a** Further elaboration of Example 7.11; **b** the effect of reversing the **d** and $\star$ operators

shows the process with the **d** and $\star$ operations having been reversed. Clearly, the **d** and $\star$ operations do not commute.

The process of reversing the $\star$ operation itself (going from $\star\alpha$ back to α) is simplified by our choice of metric[18]. The reversal stems from a second application of the $\star$ operator. For a p-form α the result is

$$\star\star(\alpha) = (-1)^{p(n-p)}\alpha, \tag{7.30}$$

which follows directly from Definition 7.4. The derivation is left as an exercise.

The correspondences developed in the previous pages will reappear in Chap. 8 in the context of the integration of p-forms on manifolds, and they will assist in our development of a generalized Stokes's theorem. It is important to remember, however, that these correspondences (like the vector operations themselves) are defined only for $\mathbb{R}^3$; for $\mathbb{R}^n$ with $n > 3$ the correspondences are with tensor (not vector) calculus.

7.6 Hamilton's Equations and Differential Forms

The study of classical mechanics is a revelatory journey from simple applications of Newton's laws of motion to the elegant formulation of Hamilton's equations by way of the *Lagrangian*, and it is in the study of Hamilton's equations that you are likely to first encounter differential forms. In broad outline (Fig. 7.11), Newtonian and Lagrangian dynamics are framed in the context of space and time coordinates in a coordinate-space manifold, whereas Hamiltonian dynamics is framed in the context of a closely-related *phase space*. A structure of central importance within the phase space is a closed differential two-form, which is known as a *symplectic structure*. A manifold on which such a structure exists is called a *symplectic manifold*.

The transition from Lagrange's to Hamilton's equations is carried out via a *Legendre transformation*,[19] and the variables in phase space (q, p) form a *conjugate pair* (as defined by how they are related to one another in Hamilton's equations). Our purpose in this section is to give a sense of what all this means without undertaking a full exposition of classical mechanics—a goal we can accomplish by limiting our consideration to single-particle one-dimensional motion.

---

[18]We have been limiting our presentation in this section to the Euclidean metric, which simplifies things considerably while still conveying the essential structural concepts pertaining to differential forms. A more general treatment may be found in [3] and [5]. A few of our results (such as Eq. 7.30) would need to be modified to accommodate different metrics.

[19]Another instance where Legendre transformations appear is in the study of classical thermodynamics. In that setting we start with the differential form of the first law of thermodynamics (energy conservation) and, through various Legendre transformations, find differential expressions for the enthalpy, and the Gibbs and Helmholtz free energies.

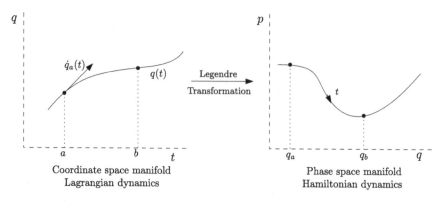

Fig. 7.11 A Legendre transformation from coordinate space (t, q) to phase space (q, p)

7.6.1 Lagrange's Equation, Legendre Transformations and Hamilton's Equations

The Lagrangian ($\mathcal{L}$) of a dynamical system is defined as the *difference* between a "kinetic" term $\mathcal{T}$ and a "potential" term $\mathcal{U}$:

$$\mathcal{L} \equiv \mathcal{T} - \mathcal{U}.$$

Genarally, $\mathcal{L}$, $\mathcal{T}$ and $\mathcal{U}$ are functions of (a) time; (b) a generalized coordinate q (such as a linear or angular displacement, or a field variable); and (c) a corresponding generalized velocity $\dot{q}$ (linear, angular or a time rate-of-change of the field variable). A relativistic treatment places coordinates and time on an equal footing.

In the most familiar cases from elementary dynamics where we deal with one or a few particles, each of fixed mass, $\mathcal{T}$ is the kinetic energy of the system of particles. Further, if $\mathcal{U}$ depends solely on the configuration of the system as represented by the coordinate q, then $\mathcal{U}(q)$ represents the potential energy of the system.

The integral of the Lagrangian between two points in time is called the *action*,

$$S = \int_{t_0}^{t_1} \mathcal{L} dt,$$

and it is a quite remarkable fact that Nature behaves in such a way that the action is minimized (formally, it achieves an extremal value) for the realized time evolution of the system. That this is so is called the *action principle*.

The process of minimizing (extremizing) the action is among the seminal problems in the *calculus of variations*,[20] and the solution that arises from this procedure is a

---

[20]The calculus of variations is a general method by which we find stationary values of integrals as the integrand is varied. With these methods we can find such things as the shortest distance between

differential equation—*Lagrange's equation*—which is then solved for the evolution of the system given the particulars of the problem at hand.

For a non-relativistic single-particle dynamical system, Lagrange's equation is

$$\frac{d}{dt}\left(\frac{\partial \mathcal{L}}{\partial \dot{q}}\right) - \frac{\partial \mathcal{L}}{\partial q} = 0, \tag{7.31}$$

and the generalized momentum is defined as $p \equiv \partial \mathcal{L}/\partial \dot{q}$. The simplest application of Eq. 7.31 is to let $\mathcal{L} = \frac{1}{2}m\dot{q}^2 - \mathcal{U}(q)$, in which case Lagrange's equation yields Newton's Second Law:

$$\frac{dp}{dt} - F = 0, \quad \text{where } p = m\dot{q} \text{ and the force } F = -\frac{\partial \mathcal{U}}{\partial q}.$$

There is no particular advantage to using Lagrange's equation over Newton's Second Law when solving for the motion of a single-particle system. However, for systems with multiple particles (such as several masses on springs, or a double pendulum), Newton's approach is usually intractable while the Lagrangian approach (with one Lagrange equation for each generalized coordinate) yields a coupled system of ordinary differential equations, which can then be solved for each $q(t)$.

Given $\mathcal{L}(q, \dot{q}, t)$, a *Legendre transformation* replaces $\mathcal{L}(q, \dot{q}, t)$ with a new function $\mathcal{H}(q, p, t)$. This new function is called the *Hamiltonian* of the system. In modern mathematical physics—including areas as diverse as the study of chaotic behavior in simple systems, and quantum gravity—it is the Hamiltonian, more than the Lagrangian, that is the focus of the inquiry.

In order to see how a Legendre transformation works, consider a function $f(x, y, t)$ whose differential is

$$df = \frac{\partial f}{\partial x}dx + \frac{\partial f}{\partial y}dy + \frac{\partial f}{\partial t}dt \equiv udx + vdy + \frac{\partial f}{\partial t}dt. \tag{7.32}$$

Replacing the independent variable y with the variable v (replacing x with u is equally valid, but gives a slightly different result) and defining a new function $g(x, v, t) \equiv vy - f$, we get

$$dg = vdy + ydv - df = ydv - udx - \frac{\partial f}{\partial t}dt, \tag{7.33}$$

which gives

$$y = \frac{\partial g}{\partial v} \quad \text{and} \quad u = -\frac{\partial g}{\partial x}. \tag{7.34}$$

---

two points on any surface, Snell's law of elementary optics (shortest travel time for a light ray), the shape of a suspended cable (configuration of lowest energy) or (in its full relativistic formulation) the principal equations of mathematical physics, including those of Dirac, Maxwell and Einstein.

Because of their relationship to one another through the function g in Eq. 7.34, the variables x and v are said to form a conjugate pair; the same is said of the equations themselves.

Returning to our Lagrangian, we carry out the following steps:

1. Make the corresponding substitutions $\mathcal{L}(q, \dot{q}, t) \leftrightarrow f(x, y, t)$ in Eqs. 7.32–7.34;
2. Apply the definition $p \equiv \partial\mathcal{L}/\partial\dot{q}$;
3. Apply Lagrange's equation (7.31), which gives $\dot{p} = \partial\mathcal{L}/\partial q$.

In this context, the function g becomes the Hamiltonian $\mathcal{H}$, and Eq. 7.34 becomes the conjugate pair we call *Hamilton's equations*. That is,

$$\mathcal{H} \equiv p\dot{q} - \mathcal{L}, \qquad \dot{q} = \frac{\partial\mathcal{H}}{\partial p}, \qquad \dot{p} = -\frac{\partial\mathcal{H}}{\partial q}, \qquad \frac{\partial\mathcal{H}}{\partial t} = -\frac{\partial\mathcal{L}}{\partial t}, \qquad (7.35)$$

where it is straightforward to show that $\mathcal{H} = \mathcal{T} + \mathcal{U}$, the total mechanical energy of the system. The details of the development above are left as exercises.

Equation 7.35 constitute a Hamiltonian system and describe the motion of the particle in phase space with coordinates (q, p). Paths in phase space are called *orbits*, along which the time t is a parameter. If $\mathcal{L}$ is independent of time then so, too, is $\mathcal{H}$, and the total energy along a given orbit is constant. If $\mathcal{H}$ is independent of a coordinate q (known in this context as an "ignorable" coordinate), then the corresponding momentum is conserved. Thus, the symmetries associated with the time- or position-independence of the equations of motion lead to the conservation of energy and momentum, respectively.

Figure 7.12 shows examples of phase space diagrams for several familiar single-particle systems. By writing Hamilton's equations in terms of generalized coordinates and momenta as we have done, we can speak of both linear and angular displacements and their corresponding momenta via the same formulation.

7.6.2 Hamiltonian Phase Space as a Symplectic Manifold

The goal of this section is to gain a global sense of the dynamics associated with a Hamiltonian system. Central to the discussion are two differential forms that characterize Hamiltonian phase space: the exact one-form $\boldsymbol{\alpha} = p\mathbf{dq}$, and the closed two-form $\omega = d\boldsymbol{\alpha} = \mathbf{dp} \wedge \mathbf{dq}$ (see Definition 7.5). The variables $\mathbf{p}$ and $\mathbf{q}$ form an ordered basis in the phase space.

We first consider the exterior differential of the time-independent Hamiltonian $\mathcal{H}(q, p)$. Because $\mathcal{H}$ is a function, the differential is the one-form

$$\mathbf{dH} = \frac{\partial\mathcal{H}}{\partial x^i}\mathbf{dx}^i = \frac{\partial\mathcal{H}}{\partial q}\mathbf{dq} + \frac{\partial\mathcal{H}}{\partial p}\mathbf{dp}. \qquad (7.36)$$

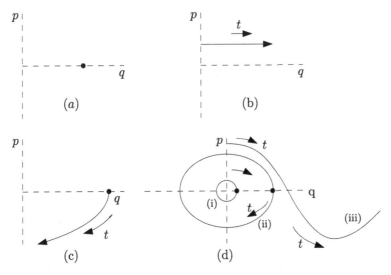

Fig. 7.12 Phase space diagrams for **a** a particle at rest; **b** a particle moving with constant momentum in one dimension; **c** a particle dropped from rest under the influence of gravity; **d** a simple pendulum for (i) a small angular displacement from equilibrium (resulting in simple harmonic motion), (ii) a larger displacement (and energy) and (iii) enough energy to cause the pendulum to go over-the-top and continue rotating rather than oscillate

Applying the **d** operator a second time (Definition 7.5(6) and Eq. 7.28) yields

$$\omega = \mathbf{d}^2 \mathbf{H} = \left(\frac{\partial^2 \mathcal{H}}{\partial p \partial q} - \frac{\partial^2 \mathcal{H}}{\partial q \partial p} \right) (\mathbf{dp} \wedge \mathbf{dq}) = 0, \qquad (7.37)$$

as written with respect to the ordered basis two-form $(\mathbf{dp} \wedge \mathbf{dq}) = d\alpha$.

Equation 7.37 conveys the substance of *Liouville's theorem*, which is essentially a "law of conservation of phase space volume" for conservative systems.[21] This interpretation is more transparent if we approach the problem from a vector calculus perspective and consider the divergence theorem.

The divergence theorem is a special case of a generalized Stokes's theorem that we will discuss in Chap 8. In substance, it tells us that the volume integral of the divergence of some vector field **F** equals the area integral of the flux over the boundary of that volume. That is,

$$\int_V \nabla \cdot \mathbf{F} dV = \int_A \mathbf{F} \cdot \hat{\mathbf{n}} dA \Longrightarrow \nabla \cdot \mathbf{F} = \lim_{\Delta V \to 0} \int_A \mathbf{F} \cdot \hat{\mathbf{n}} dA,$$

---

[21] Here, "volume" is a generic term. For a two-dimensional phase space it is an area; for higher even-dimensional spaces it is a "hypervolume." Liouville's theorem also applies to transformations of phase space coordinates that preserve the form of Hamilton's equations. Such transformations are called *canonical transformations* See, for example, [12], Sect. 45.

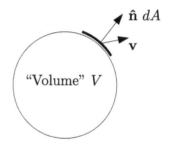

Fig. 7.13 A change in the volume arises from the "flux" of the boundary itself

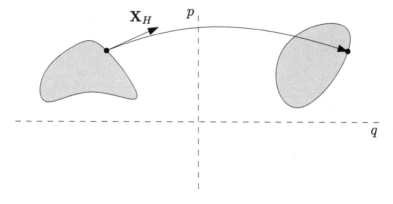

Fig. 7.14 A phase flow, where $\mathbf{X}_H$ is the Hamiltonian vector field

from which the divergence can be interpreted as a "source strength" of $\mathbf{F}$ per volume, as evaluated at a point.

One application of the divergence theorem is to let $\mathbf{F}$ be a velocity vector $\mathbf{v}$ of the boundary (Fig. 7.13). In this case, $(\mathbf{v} \cdot \hat{\mathbf{n}}) dA$ is the time-rate-of-change of a differential volume element itself. By implication, if $\nabla \cdot \mathbf{v} = 0$, then the enclosed volume will be constant, though it may change its shape.

We now apply this formulation to phase space volumes by writing the velocity of the boundary as

$$\mathbf{X}_H = \dot{q}\hat{\mathbf{q}} + \dot{p}\hat{\mathbf{p}} = \frac{\partial H}{\partial p}\hat{\mathbf{q}} - \frac{\partial H}{\partial q}\hat{\mathbf{p}},$$

where we have applied Hamilton's equations. The vector field $\mathbf{X}_H$ is called the *Hamiltonian vector field* and represents a velocity vector of the surface enclosing a volume in phase space (Fig. 7.14).

Writing ∇ in terms of its phase space coordinates and evaluating the divergence of $\mathbf{X}_H$ yields

$$\nabla \cdot \mathbf{X_H} = \left(\hat{\mathbf{q}}\frac{\partial}{\partial q} + \hat{\mathbf{p}}\frac{\partial}{\partial p}\right) \cdot \mathbf{X}_H = \left(\frac{\partial^2 \mathcal{H}}{\partial q \partial p} - \frac{\partial^2 \mathcal{H}}{\partial p \partial q}\right) = 0,$$

which is the same result as in Eq. 7.37. Consequently, a conservative system in which Hamilton's equations hold—either in their original form or as a result of a canonical transformation—maintains a constant volume in phase space.

Liouville's theorem has applications beyond classical mechanics, as for example in statistical mechanics. This should not be too surprising since in both instances a point in phase space represents the state of a system. In statistical mechanics the number of states available to a system is related (logarithmically) to the system's entropy, from which (with the help of partition functions) the macro-level relations of classical thermodynamics may be derived.[22]

7.7   Transformations of Vectors and Differential Forms

Transformations between manifolds that carry resident algebraic structures along with them are particularly important in physics. Indeed, classical tensor analysis[23] involves the construction of tensorial structures and tensor equations that remain invariant under transformations of the underlying coordinate system.

The focus of this section is on the transformations of vectors and p-forms, and in the process we will re-introduce the notion of *tangent spaces* (see the discussion following Example 7.3 in Sect. 7.1.2) and expand on our earlier discussions of dual spaces. The results of this section will be needed in Chap. 8 when we discuss integration on manifolds and develop the generalized Stokes's theorem. We will also take a brief look at transformations as described in traditional tensor analysis.

We start with transformations of functions (zero-forms). Figure 7.15a shows the manifolds U and V with subspaces A and B, respectively. There is a map $\phi : A \to B$, and a function f that is defined on all of V. Our goal is to evaluate f at the point $p \in A \subset U$, even though f is not yet defined there. What makes this even remotely possible is the map ϕ. The challenge we face is to make the connection between the function f—as it is evaluated at some $q \in B$—and the function as we might evaluate it at p. The point r in Fig. 7.15 is outside the range of ϕ and is not relevant to achieving our goal.

There are two equivalent ways of evaluating f at p: (1) apply ϕ to map the point p to the point q and then evaluate the function f at $q \in B$ to give $f(q) \in \mathbb{R}$, and (2) devise a new map called ϕ^* and use it to map f back to $A \subset U$ so as to evaluate it at the point $p \in A$. The equivalence of these two approaches is shown in Figure 7.15(b), and it is plain to see that all we have really described is a direct substitution of variables. The map ϕ^* goes by several names—an *induced* or *reciprocal* map are two formal terms—but most often the map ϕ^* is called the *pullback transformation* of the function f to $A \subset U$.

---

[22] See, among other places, the extensive treatment in [9].

[23] In the older literature, tensor calculus is called "the absolute differential calculus," of which the work of Levi-Civita is a classic. For an English translation see [13]. The Guide to Further Study offers additional references along these classical, as well as more modern, lines.

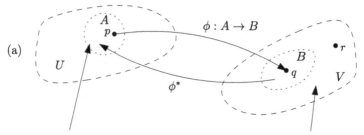

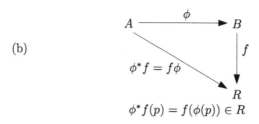

Fig. 7.15 The relationship between a map ϕ and the reciprocal map ϕ^* (a "pullback")

Example 7.14 Let $U = V = \mathbb{R}$.

1. Let $\phi(p) = q = p^2$ and $f(q) = 4q$. Then $\phi^* f(p) = f(\phi(p)) = 4p^2$. Here, A and B are the non-negative real numbers.
2. Let $\phi(p) = q = \sin p$; $f(q) = 1 - q^2$. Then $\phi^* f(p) = f(\phi(p)) = 1 - \sin^2 p = \cos^2 p$. Here, we would have $A = [0, 2\pi]$ and $B = [-1, 1]$.   ▲

In effect, we can say that ϕ "pushes points forward" to B, while ϕ^* "pulls functions back" to A, and indeed, the map ϕ is often referred to as a *push-forward*.[24]

If we consider a point as a zero-vector and a function as a zero-form, the framework described above (which admittedly is a bit overwrought for a simple substitution of variables) suggests that vectors are pushed forward while forms are pulled back. In fact, this is the case, provided we define ϕ^* appropriately. As for ϕ, when operating on vectors instead of points, it is no longer a direct coordinate substitution but a linear transformation of the kind we discussed in Sect. 7.1.2.

Consider, then, a vector **v** defined at a point $p \in A$, and the map $\phi : A \to B$, or $\phi : p \mapsto q$, as in Fig. 7.16.

The set of all vectors (vectors of all magnitudes, in all available directions as allowed by A) at p forms a vector space that is tangent to A at the point p. This vector space is called a *tangent space at p in A* and is denoted as $T_p(A)$; we can then write $\mathbf{v} \in T_p(A)$. Then ϕ (or, as a notational alternative, ϕ_*) maps $\mathbf{v} \in T_p(A)$ to a vector $\phi_* \mathbf{v} \in T_q(B)$ at $q = \phi(p) \in B$. Also in Fig. 7.16 (though not shown) is a one-form $\boldsymbol{\alpha}(q)$ at the point q, which we may write as $\boldsymbol{\alpha}(\phi(p))$. It exists in a *dual tangent vector space*, denoted as $T_q^*(B)$.

---

[24]In this context, a push forward is often denoted as ϕ_* so as to contrast it with the pullback ϕ^*.

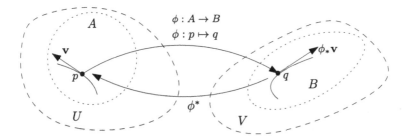

Fig. 7.16 The map ϕ_* pushes forward the vector $\mathbf{v}$ from p to a vector $\phi_*\mathbf{v}$ at q. The map ϕ^* pulls back a one-form $\boldsymbol{\alpha}$ (not shown) at q to a one-form $\phi^*\boldsymbol{\alpha}$ at p

An underlying principle is that the inner product of a one-form with a vector is defined only if the vector and form are located at the same point. Therefore, in a fashion similar to our treatment of points and forms, we have two choices: (1) push the vector $\mathbf{v}$ forward to $\boldsymbol{\alpha}$, or (2) pull the one-form $\boldsymbol{\alpha}$ back to $\mathbf{v}$. These operations will be linear maps via a Jacobian, and not just variable substitutions.[25]

These two paths to the inner product may be constructed as follows:

1. Apply the map $\phi_* : \mathbf{v}(p) \in T_p(A) \mapsto \phi_*\mathbf{v} \in T_{\phi(p)}(B)$. This new vector is now at the same point $(q = \phi(p))$ as the one-form $\boldsymbol{\alpha} \in T_q^*(B)$. The inner product at q is then $\langle \boldsymbol{\alpha} | \phi_*\mathbf{v} \rangle$, often written as $\boldsymbol{\alpha}\phi_*(\mathbf{v})$.
2. Apply the map $\phi^* : \boldsymbol{\alpha}(q) \in T_q^*(B) \mapsto \phi^*\boldsymbol{\alpha}(q) \in T_p^*(A)$. This new one-form is now at the same point (p) as the vector $\mathbf{v} \in T_p(A)$. The inner product at p is then $\langle \phi^*\boldsymbol{\alpha} | \mathbf{v} \rangle$, often written as $\phi^*\boldsymbol{\alpha}(\mathbf{v})$.

These two formulations of the inner product are equivalent because scalar quantities are invariant under coordinate transformations, therefore yielding

$$\phi^*\boldsymbol{\alpha}(\mathbf{v}) = \boldsymbol{\alpha}\phi_*(\mathbf{v}). \tag{7.38}$$

Equation 7.38 serves as a definition of ϕ^*, but before we can put it into practice we need to know the nature of the spaces A and B. Following the pattern in Sect. 7.1.2, we let $A = \mathbb{R}^m$ and $B = \mathbb{R}^n$. Further, let the coordinates of a point in A be written as $\mathbf{x} = (x^1, x^2, \ldots x^m) \in \mathbb{R}^m$, and those in B be written in "barred" coordinates $\bar{\mathbf{x}} = (\bar{x}^1, \bar{x}^2, \ldots \bar{x}^n) \in \mathbb{R}^n$.

Recall that the Jacobian (Eq. 7.7) of a transformation is a set of partial derivatives and represents the first derivative of a multivariable function in $\mathbb{R}^n$. When the Jacobian acts on a differential line element, it serves to transform that line element to new coordinates or to a new space. This is how we wish to use the Jacobian here, i.e., to carry out the linear transformation of the velocity vectors from A to B. We will simplify the notation by writing the Jacobian of the transformation evaluated at p as

---

[25]Essentially, we are saying that at each point in a manifold a tangent space and dual space intersect, thereby allowing the inner product to be evaluated at that point. This puts us at the precipice of a discussion of fibres and fibre bundles, from which, in this text, we step back.

$D\phi_p$; when applied to $\mathbf{v}$, we write it as $D\phi_p(\mathbf{v})$. We can now proceed to calculate $\phi_*(\mathbf{v})$ and $\phi^*\boldsymbol{\alpha}(\mathbf{v})$ in Eq. 7.38.

It helps to keep a specific example in mind. Let $A = B = \mathbb{R}^2$ and $\mathbf{x}(t) = (x^1, x^2)$ be a path in A that intersects point p. Let the time t be a parameter along the path, so the time derivative gives the velocity $\mathbf{v}(t) = (v^1, v^2)$. We also have the linear transformation $\phi : \mathbf{x} \rightarrow \bar{\mathbf{x}}$, where $\bar{\mathbf{x}} = (\bar{x}^1, \bar{x}^2)$ and $\bar{\mathbf{v}} = (\bar{v}^1, \bar{v}^2)$. Functionally we can write $\bar{\mathbf{x}}(\mathbf{x}) = \phi(\mathbf{x})$.

Applying $\bar{\mathbf{v}} = \phi_*(\mathbf{v}) = D\phi_p(\mathbf{v})$ we have a 2×2 matrix transformation of velocity components:

$$\bar{\mathbf{v}} = [\phi_*(\mathbf{v})] = [D\phi_p(\mathbf{v})] = \begin{pmatrix} \dfrac{\partial \bar{x}^1}{\partial x^1} & \dfrac{\partial \bar{x}^1}{\partial x^2} \\[2ex] \dfrac{\partial \bar{x}^2}{\partial x^1} & \dfrac{\partial \bar{x}^2}{\partial x^2} \end{pmatrix} \begin{pmatrix} v^1 \\[1ex] v^2 \end{pmatrix}, \tag{7.39}$$

where the ith component of $\bar{\mathbf{v}}$ may be written as

$$\bar{v}^i = \frac{\partial \bar{x}^i}{\partial x^j} v^j. \tag{7.40}$$

Equation 7.40 is indicative of the kind of tensor component equation that is part of the "origin story" of tensors and of the Einstein summation convention itself. It is also the source of the term "*contra*variant" when describing tensors like the velocity, i.e., tensors with upper indices. Writing the equation in a finite form as $\bar{v}^i \delta x^j = v^j \delta \bar{x}^i$, we interpret this as v changing in a sense "opposite to," or "contrary to," the change in the coordinates.

Of course, Eq. 7.40 focuses on *components*; the vector itself is $\bar{\mathbf{v}} = \bar{v}^i \bar{\mathbf{e}}_i$, where the $\bar{\mathbf{e}}_i$ are basis vectors in B. That is,

$$\phi_*(\mathbf{v}) = \frac{\partial \bar{x}^i}{\partial x^j} v^j \bar{\mathbf{e}}_i. \tag{7.41}$$

Next, we write a one-form in B as $\bar{\boldsymbol{\alpha}} = \bar{\alpha}_i d\bar{\mathbf{x}}^i \in B$, and then pull it back to a one-form $\boldsymbol{\alpha} \in A$:

$$\boldsymbol{\alpha} = \phi^*\bar{\boldsymbol{\alpha}} = \bar{\alpha}_i \frac{\partial \bar{x}^i}{\partial x^j} d\mathbf{x}^j = \alpha_j d\mathbf{x}^j. \tag{7.42}$$

Equation 7.42 shows the transformation law for the components of a one-form is

$$\alpha_j = \frac{\partial \bar{x}^i}{\partial x^j} \bar{\alpha}_i \quad \Rightarrow \quad \bar{\alpha}_i = \frac{\partial x^j}{\partial \bar{x}^i} \alpha_j, \tag{7.43}$$

which is why (in contrast to Eq. 7.40) we refer to tensors with lower indices as "*co*variant" tensors—a one-form changes in the "same" sense, or "with," the change in coordinates.

Fig. 7.17 Given ϕ, the pullback ϕ^* must preserve the differential structure in the two manifolds

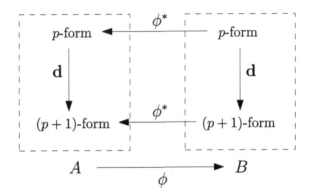

We need to verify that Eqs. 7.41 and 7.42 satisfy Eq. 7.38. That is, we ask whether the lefthand side of the equation (to be evaluated in A) and the righthand side of the equation (to be evaluated in B) have the same functional form in their respective spaces and/or coordinate systems—a hallmark of a valid tensor equation.

If in A we let the vector $\mathbf{v} = v^k \mathbf{e}_k$, then the lefthand side of Eq. 7.38 becomes

$$\phi^* \boldsymbol{\alpha}(\mathbf{v}) = \bar{\alpha}_i \frac{\partial \bar{x}^i}{\partial x^j} \mathbf{dx}^j \, v^k \mathbf{e}_k = \bar{\alpha}_i \frac{\partial \bar{x}^i}{\partial x^j} v^k \delta_k^j = \bar{\alpha}_i v^j \frac{\partial \bar{x}^i}{\partial x^j}.$$

Similarly, if in B we let the one-form $\bar{\alpha} = \bar{\alpha}_k \mathbf{d\bar{x}}^k$, then the righthand side of Eq. 7.38 yields the same result:

$$\bar{\alpha} \phi_*(\mathbf{v}) = \bar{\alpha}_k \mathbf{d\bar{x}}^k \frac{\partial \bar{x}^i}{\partial x^j} v^j \bar{\mathbf{e}}_i = \bar{\alpha}_k \frac{\partial \bar{x}^i}{\partial x^j} v^j \delta_i^k = \bar{\alpha}_i v^j \frac{\partial \bar{x}^i}{\partial x^j}.$$

Therefore, Eq. 7.38 is satisfied by the transformations laws given in Eqs. 7.41 and 7.42. An application of these results is left as an end-of-chapter problem.

Equation 7.38 can serve as a definition of the pullback transformation, but another consideration is that the ϕ-ϕ^* pair of operations must preserve the differential structures between the two manifolds. This means these two maps must be consistent with $\mathbf{d}$, the exterior differential operator, as illustrated in Fig. 7.17. The substance of these relationships is captured in Definition 7.6, which serves as a more formal and complete definition of the pullback transformation and its properties[26].

Definition 7.6 Consider the case where: $\mathbf{v}$ is a vector; α and β are forms (possibly of different order); $\phi : \mathbb{R}^m \to \mathbb{R}^n$; $\psi : \mathbb{R}^n \to \mathbb{R}^p$; and $f : \mathbb{R}^n \to \mathbb{R}$. The properties of the pullback map ϕ^* satisfy the following conditions:

1. $\phi^* \boldsymbol{\alpha}(\mathbf{v}) = \alpha \phi_*(\mathbf{v})$, where $\phi_*(\mathbf{v}) = D\phi(\mathbf{v})$ (see Eqs. 7.39–7.41);
2. $\phi^*(\alpha + \beta) = \phi^* \alpha + \phi^* \beta$ (linearity over +);

---

[26]These properties may be considered propositions subject to proof. See, for example, [3–5] and [17] for various approaches to proofs and derivations.

3. $\phi^*(f\alpha) = (\phi^* f)(\phi^*\alpha)$; (reverses the arrows in Fig. 7.16);
4. $\phi^*(\alpha \wedge \beta) = \phi^*\alpha \wedge \phi^*\beta$ (linearity over $\wedge$);
5. $(\phi\psi)^*\alpha = \psi^*\phi^*\alpha$ (the pullback of a composition is the composition of pullbacks);
6. $d(\phi^*\alpha) = \phi^* d\alpha$ (the differential structure is preserved under ϕ^* (Fig. 7.17)). ∎

Definition 7.6(6) will have particular relevance, and indeed a central role to play, when we discuss the integration of differential forms in Sect. 8.4. At that time we will see how the integral theorems we learn in vector analysis are really just special cases of a more comprehensive structure called the *Generalized Stokes's Theorem*—a theorem which allows us to extend the reach of integration methods to manifolds beyond our familiar three-dimensional space, and where the methods of ordinary vector analysis no longer apply.

Problems

7.1 As a review of elementary calculus, directly apply Eq. 7.1 (and not the formulas you know from elementary calculus) to find the derivatives of the following functions:

$$(a)\phi(x) = x^2; \ (b)\phi(x) = \sin x; \ (c)\phi(x) = e^x.$$

7.2 Referring to Example 7.1, and letting $\mathbf{a} = (a^1, a^2) = (x_0, y_0)$:
(a) Find the directional derivative for the same $\phi(\mathbf{x}) = \phi(x, y) = x^2 y$, but now with $|e\rangle = (1/\sqrt{2})(1, 1)$. [*Hint*: The answer is given in Example 7.1.];
(b) Find $\phi'(\mathbf{a}, |e\rangle)$ for $\phi(\mathbf{x}) = \phi(x, y) = x^3 y^2$, with $|e\rangle = (1/\sqrt{2})(1, 1)$.

7.3 In Cartesian coordinates, find $[D\phi(\mathbf{x})]$ in each of the following cases:

(a) $\phi(\mathbf{x}) = x^2 - y^2 + 2z^2$;
(b) $\phi(\mathbf{x}) = e^{-2y} \cos 2x$;
(c) $\phi(\mathbf{x}) = e^{4x+3y} \sin 5z$;

7.4 Evaluate the Kronecker tensor (Eq. 7.10) in each of the following cases:

$$(a) \ \epsilon_{1342}^{1234}; \ (b) \ \epsilon_{2143}^{1234}; \ (c) \ \epsilon_{4231}^{1234}.$$

[*Hint*: If necessary, you may wish to review Example 2.5 in Sect. 2.3.3.]

7.5 Consider a four-dimensional space X on which we define second-rank tensors.

(a) What is the dimension of T^*, the covariant tensor space?
(b) What is the dimension of Λ^2, the subspace of two-forms?
(c) Letting $\hat{e}^i$ designate a basis one-form, write down the basis two-forms for the space in part (b).

[*Hint*: See Example 7.5 and the references therein for guidance on this problem. See also Eq. 7.18. The first step is to properly interpret what is meant when we say a tensor is defined "on" a space. If necessary, see Sect. 4.5.3 for a review.]

7.6 Derive Eq. 7.21. [*Note*: Hints are given in the text.]

7.7 Derive (or otherwise justify) the Leibnitz product rule for the exterior differential operator $\mathbf{d}$ as given in Definition 7.5(5). [*Note*: You may use the other properties of $\mathbf{d}$ as listed in Definition 7.5. See also Example 7.6.]

7.8 Consider a function $f(x, y, z)$ in Cartesian coordinates.

(a) Show that the one-form $\mathbf{df}$ is a gradient [*Ans*: See Eq. 7.25.];
(b) Show that $\star\mathbf{df}$ corresponds to the curl of a vector;
(c) Evaluate $\mathbf{d}(\star\mathbf{df})$ and interpret the result in terms of vector operators.

7.9 A differential vector displacement in spherical coordinates may be written as

$$\mathbf{dr} = (dr)\hat{\mathbf{e}}_1 + (r d\theta)\hat{\mathbf{e}}_2 + (r \sin\theta d\phi)\hat{\mathbf{e}}_3,$$

where $\hat{\mathbf{e}}_1 = \hat{\mathbf{r}}$, $\hat{\mathbf{e}}_2 = \hat{\boldsymbol{\theta}}$ and $\hat{\mathbf{e}}_3 = \hat{\boldsymbol{\phi}}$ are unit vectors along the directions of the spherical coordinate system. [*Note*: ϕ is the azimuthal angle from the x-axis in the x-y-plane, and θ is the zenith angle measured from the z-axis (Fig. 4.6a).] The corresponding p-form is the one-form

$$\alpha = \alpha_i \mathbf{dx}^i = (1)\mathbf{dr} + (r)d\theta + (r\sin\theta)d\phi.$$

Applying the methods of exterior differentiation as described in this chapter (i.e., without resorting to the methods of ordinary vector calculus), find an expression for the curl of a vector in spherical coordinates.

7.10 Given a vector field $\mathbf{v}(x, y, z)$ in Cartesian coordinates, and using the methods of exterior differentiation, find $\nabla \cdot \mathbf{v}$, the divergence of the vector field.

7.11 Derive Eq. 7.30, the expression for $\star \star \alpha$ in terms of α.

7.12 Regarding the Legendre transformation and Hamilton's equations for a single particle,

(a) Work through all the details of the derivations in the Eqs. 7.32–7.34, and show that the result is the set of equations in Eq. 7.35;
(b) Show that $\mathcal{H}$ is the total mechanical energy of the particle.

7.13 A particle of mass m moves without friction in one dimension along the x-axis under the influence of a potential energy

$$U(x) = kx^2 e^{-x},$$

where $k > 0$ is a constant.

(a) Applying what you know from elementary physics about the relationship between a force and a corresponding potential energy, find the equilibrium points and assess whether each point is stable or unstable to small perturbations;
(b) Make a qualitatively correct sketch of $U(x)$, labeling the equilibrium points;
(c) Indicate on your sketch in part (b) the region(s) where the motion is bound, and find an expression for the maximum total energy E the particle can have in the bound region(s) and still be bound;
(d) By a direct application of Newton's second law, find the equation of motion for this particle;
(e) By the Lagrangian method, find the equation of motion for this particle and show that you get the same result as in part (d);
(f) Is linear momentum conserved in this problem? Explain.

7.14 Consider the same one-dimensional dynamical system in Problem 7.13 from a Hamiltonian perspective.

(a) Write an expression for the Hamiltonian for this system (here, $q = x$);
(b) Write Hamilton's equations (the conjugate pair of equations for $\dot{q}$ and $\dot{p}$ in Eq. 7.35) for this system;
(c) Let $k = 2$ and $m = 1$, and assume the state of the system evolves in such a way that at some time t the state of the system is at the point $A = (x, p) = (1, 2)$ in phase space. Locate point A on a sketch of the phase space, and draw the Hamiltonian velocity vector representing the direction of the evolution of the system from that point. Do the same for the point $B = (1, 4)$, being careful to distinguish the relative directions of the vectors at A and B (for example, which vector is "steeper," the one at A or the one at B?);
(d) Which point, A or B, represents a higher energy state for the system?
(e) As a much more ambitious graphing exercise, plot a selection of orbits in phase space for this system. Include in your graph some representative orbits in both the bound and unbound regimes. Also be sure to include on each orbit the direction in which the system evolves.

7.15 One of the central results in this chapter is the definition of the pullback transformation as given by Eq. 7.38—an equation that was premised on the statement that "...scalar quantities are invariant under coordinate transformations...." One consequence of this definition was the derivation of the the transformation laws for contravariant and covariant tensors, as given in Eqs. 7.40 and 7.43, respectively.

In traditional tensor analysis the process is reversed (though with no mention of pullbacks or push-forwards), and Eqs. 7.40 and 7.43 are taken as definitions of the two different types of tensors. From this we can then show that a scalar quantity is invariant under coordinate transformations. That is what we want to do in this problem.

Given the scalar product $\alpha_i v^i$, show that it is invariant under transformation laws Eqs. 7.40 and 7.43 (work only with tensor components, as is the custom in tensor analysis).

[*Hint*: In order to work this problem properly, you will need to make liberal use of the two-index Kronecker δ symbol. Be careful not to have more than two occurrences of any one index, one index up and one down.]

7.16 Rewrite Eq. 7.39 for a covariant vector (another name for a one-form).

7.17 Consider a path $\mathbf{x}(t) \in A = \mathbb{R}^2$ such that $\mathbf{x}(t) = (x_1(t), x_2(t)) = (3t, 4t^2)$ for a parameter $t \in \mathbb{R} \geq 0$. Further, let $\phi : A \to B = \mathbb{R}^2$ such that $\phi : \mathbf{x}(t) \to \bar{\mathbf{x}}(t) = (\bar{x}_1, \bar{x}_2) = (x_1 x_2, 2x_2)$.

(a) Find $\phi_*(\mathbf{v}) = [D\phi_p(\mathbf{v})]$ for the push-forward transformation, where $\mathbf{v} = d\mathbf{x}/dt$;
(b) Find $\bar{\mathbf{v}} = d\bar{\mathbf{x}}/dt$;
(c) Given $\bar{\alpha} = (\bar{\alpha}_1, \bar{\alpha}_2) \in B$, show that the scalar product as evaluated in B is

$$\bar{\alpha}\phi_*(\mathbf{v}) = 36\bar{\alpha}_1 t^2 + 16\bar{\alpha}_2 t.$$

[*Note*: This expression also serves as a check on your answer in part (a).]
(d) Show that you get the same result for the scalar product if you do a pullback of $\bar{\alpha}$ to A and evaluate the scalar product $\phi^*\alpha(\mathbf{v})$ in A.

Guide to Further Study
There is a veritable library of sources to consider if you wish to engage in the further study of differential forms and differential geometry. I will start by listing (in alphabetical order) a few texts that I believe are appropriate next steps, depending on your interests and what you wish to emphasize. A few other sources are mentioned in Chap. 8, and of course there are still many others not mentioned.

Arnold [1] emphasizes classical mechanics, but is probably most appropriate as collateral reading with a graduate-level mechanics course; Bachman [2] emphasizes geometry and is more visual than most; do Carmo [4] emphasizes algebra and the differential geometry of surfaces; Flanders [5] emphasizes physics and is often cited in physics-oriented texts on differential geometry; Singer and Thorpe [15] is primarily a topology text but includes a concise treatment of differential forms aimed at mathematics majors.

Special mention should be made of the thin one-volume work of Spivak [17], which treats the content of the typical "advanced calculus" course in the context of manifolds. Spivak's five-volume work [18] is for the person dedicated to mastering the most modern perspectives on differential geometry, but Volume 1 would be readily recognizable to you after you have completed our text.

Choquet-Bruhat, et al. [3] is a comprehensive text that touches on many of the topics we have discussed in this book, but may not be the ideal next step immediately after our text. Still, you should explore it.

In the area of classical tensor analysis, one of the "originals" is Levi-Civita [13]; a later work that is short, wonderfully instructive and starts with the basics is Spain [16], but it may be difficult to find. A very good source is Lovelock and Rund [14].

As noted in Chap. 5, every physics professor seems to have their favorite text for classical mechanics; for upper division and beginning graduate students mine is Landau and Lifshitz [12], which begins with a concise account of those aspects of the calculus of variations most relevant to mechanics. A long-time graduate-level standard textbook in mechanics is Goldstein [6]. The text by Thornton and Marion [19] is an excellent place to start if you are new to classical mechanics.

If you seek a more complete account of the calculus of variations beyond the three pages(!) in Landau and Lifshitz [12], then the easily-accessible and wide-ranging text by Weinstock [21] is highly recommended. However, your reading in the subject would not be complete without perusing Lanczos [11].

With regard to statistical mechanics you have many choices, depending on where you wish to start and where you wish to go. Kittel and Kroemer [9] begins with counting states accessible to a simple system of spin-up and spin-down "magnets" and then builds up thermodynamics "from scratch." Tolman [20] (a classic) starts with Lagrange's equations before moving to statistical mechanics, quantum mechanics and then thermodynamics. Goodstein [7] starts with thermodynamics, then briefly outlines statistical mechanics before devoting the bulk of his text to the states of matter. Each approach has its strengths, and all three texts are outstanding.

References

1. Arnold, V.I.: Mathematical Methods of Classical Mechanics, 2nd edn. Springer, New York (1989)
2. Bachman, D.: A Geometric Approach to Differential Forms. Birkhäuser, Boston (2006)
3. Choquet-Bruhat, Y., DeWitt-Morette, C., Dillard-Bleick, M.: Analysis, Manifolds and Physics, Part I: Basics, 1996 Printing. Elsevier, Amsterdam (1982)
4. do Carmo, M.P.: Differential Forms and Applications. Springer, Berlin and Heidelberg (1994)
5. Flanders, H.: Differential Forms With Applications to the Physical Sciences. Dover (1989) with corrections, originally published by Academic Press, New York (1963)
6. Goldstein, H., Poole, C., Safko, J.: Classical Mechanics, 3rd edn. Addison-Wesley, San Francisco (2002)
7. Goodstein, D.L.: States of Matter. Unabridged, Corrected Republication, Dover, New York (1985) of the edition first published by Prentice-Hall, Inc., Englewood Cliffs, NJ (1975)
8. Kelley, J.L.: General Topology. Springer, Berlin (1975), Originally Published by Van Nostrand, New York (1955)
9. Kittel, C., Kroemer, H.: Thermal Physics, 2nd edn. W.H. Freeman and Co., New York (1980)
10. Kline, M.: Mathematical Thought from Ancient to Modern Times, Published in Three Volumes. Oxford University Press, Oxford (1990)
11. Lanczos, C.: The Variational Principles of Mechanics, 4th edn. Dover, New York (1986), an unabridged republication of the 4th edition published by University of Toronto Press, Toronto (1970)
12. Landau, L.D., Lifshitz, E.M.: Mechanics, 3rd edn. 1988 Printing (With Corrections). Pergamon Press, Oxford (1976)

13. Levi-Civita, T.: The Absolute Differential Calculus, (English trans.), 1961 printing. Blackie & Son Ltd., Glasgow (1926)
14. Lovelock, D., Rund H.: Tensors, Differential Forms, and Variational Principles, corrected and revised edition. Dover, New York (1988), originally published by Wiley and Sons, New York (1975)
15. Singer, I.M. and Thorpe, J.A.: Lecture Notes on Elementary Topology and Geometry. Scott, Foresman and Company, Glenview, IL (1967). Republished by Springer, New York (1976)
16. Spain, B.: Tensor Calculus, 3rd edn. Oliver and Boyd, Edinburgh and London (1960)
17. Spivak, M.: Calculus on Manifolds—A Modern Approach to Classical Theorems of Advanced Calculus. Addison-Wesley, Reading, MA (1965)
18. Spivak, M.: A Comprehensive Introduction to Differential Geometry, Volumes 1–5, 3rd edn. Publish or Perish Inc, Houston, TX (1999)
19. Thornton, S.T., Marion, J.B.: Classical Dynamics of Particles and Systems, 5th edn. Cengage Learning, Boston (2003)
20. Tolman, R.C.: The Principles of Statistical Mechanics. Oxford University Press, London (1938), Unabridged Republication by Dover, New York (1980)
21. Weinstock, R.: Calculus of Variations. Dover, New York (1974), an unabridged corrected republication of the edition first published by McGraw-Hill, New York (1952)

Chapter 8
Aspects of Integration and Elements of Lie Groups

8.1 The Generalized Stokes's Theorem

At least insofar as integration is concerned, the manner by which we study mathematical physics follows a fairly standard path. Typically, among the first things we study in calculus is the Fundamental Theorem of Calculus, which defines the definite integral, after which we gain experience with various methods of integration. We then quickly progress from constructing the integral of a single-variable function, to the integration of multivariable functions over areas and volumes.

From there we move on to vector calculus, where we integrate vectors along curves (as when calculating the work done by a force), across surfaces (a flux integral, such as when finding the radiative flux through a surface or the transport of circulation within a fluid) and within volumes (such as when calculating the divergence of fluid flow). The principal functional forms of these vector integrals are

$$\int_c \mathbf{F} \cdot d\mathbf{s}, \quad \int_A \mathbf{F} \cdot \mathbf{n}\, dA, \quad \int_A (\nabla \times \mathbf{F}) \cdot \mathbf{n}\, dA \quad \text{and} \quad \int_V (\nabla \cdot \mathbf{F}) dV, \tag{8.1}$$

where $\mathbf{F}$ is the relevant vector field (e.g., force, flux density or velocity). Various combinations of these integrals appear in the classical vector integral theorems, culminating in Stokes's theorem.

When developed in the usual way, these integral expressions appear to embody distinct relationships uniquely framed for three dimensional space. As it happens, all of the above-referenced vector integrals and theorems are special cases of a *Generalized Stokes's Theorem* (GST), an integral theorem that applies to manifolds in general. Further, the GST is a more direct descendant of the Fundamental Theorem of Calculus (FTC) than our accounting of the usual curriculum would suggest, and may be derived (or at least inferred) without reference to vector calculus at all.

© Springer Nature Switzerland AG 2021
S. P. Starkovich, *The Structures of Mathematical Physics*,
https://doi.org/10.1007/978-3-030-73449-7_8

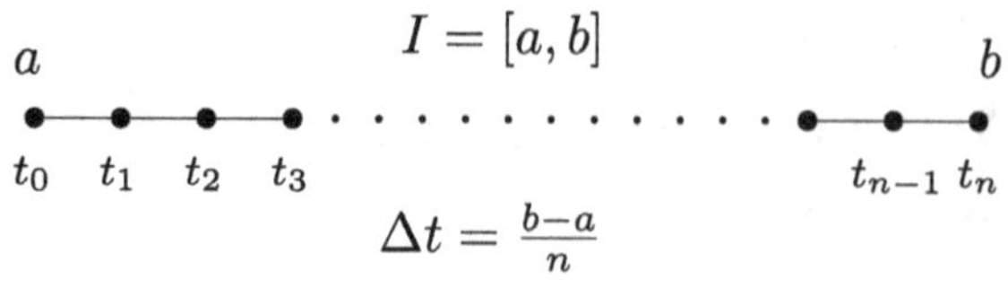

Fig. 8.1 The interval
$I = [a, b]$ with parameter t;
I is divided into n
equal-length subintervals

$$I = [a, b]$$

$$\Delta t = \frac{b-a}{n}$$

We start, therefore, with a short review of the Fundamental Theorem of Calculus.[1] Figure 8.1 shows a one-dimensional closed interval $I = [a, b]$, where points in the interval are identified by a parameter t. The interval I is divided into n equal-length subintervals such that $t_0 = a$ and $t_n = b$. We let $f(t)$ be a continuous function defined on I, and define a continuous function $F(t)$ (called the *primitive* of $f(t)$) such that $f(t) = F'(t)$, the derivative of $F(t)$. Finally, we let p_i be a unique point within each subinterval, where $t_{i-1} \le p_i \le t_i$ for $i = 1 \ldots n$.

We now form a sum S_n over all subintervals, defined as

$$S_n = \sum_{i=1}^{n} F'(p_i)\Delta t = \sum_{i=1}^{n} f(p_i)\Delta t. \tag{8.2}$$

In order to evaluate the sum we need to decide how to choose p_i in each subinterval, and there are different equivalent choices. The standard approach in Riemann integration is to apply the Mean Value Theorem at each subinterval and write

$$F(t_i) - F(t_{i-1}) = F'(p_i) \cdot (t_i - t_{i-1}) = F'(p_i)\Delta t = f(p_i)\Delta t, \tag{8.3}$$

for $i = 1, 2, \ldots n$. Evaluating Eq. 8.2 on each subinterval yields the sequence

$$F(t_1) - F(a) = F'(p_1) \cdot (t_1 - a) = f(p_1)\Delta t$$
$$F(t_2) - F(t_1) = F'(p_2) \cdot (t_2 - t_1) = f(p_1)\Delta t$$
$$F(t_3) - F(t_2) = F'(p_3) \cdot (t_3 - t_2) = f(p_1)\Delta t \tag{8.4}$$
$$\vdots \qquad \vdots$$
$$F(b) - F(t_{n-1}) = F'(p_n) \cdot (b - t_{n-1}) = f(p_1)\Delta t.$$

Summing the terms in Eq. 8.4 yields

$$F(b) - F(a) = \sum_{i=1}^{n} f(p_i)\Delta t = \lim_{n \to \infty} \sum_{i=1}^{n} f(p_i)\Delta t, \tag{8.5}$$

---

[1] It is highly recommended that you keep … forever … a good calculus text on hand for easy reference. The derivation of the FTC given here follows [16], but it is the standard development and is found in virtually all calculus texts.

where the last term in Eq. 8.5 arises because the middle term is independent of n; as n increases, Δt correspondingly decreases. Adopting the usual notation for the definite Riemann integral, we write Eq. 8.5 as

$$F(b) - F(a) = \int_a^b F'(t)\, dt. \qquad (8.6)$$

This is the *Fundamental Theorem of Calculus*. To the untrained eye, Eq. 8.6 might seem rather odd. The righthand side of the equation contains the continuous function $F'(t)$, and each subinterval or point in $I = [a, b]$ contributes something to the evaluation. Contrast this with the lefthand side, where only the two discrete endpoints of the interval contribute to the evaluation. Also note that because $I = [a, b]$ is a one-dimensional interval, the two endpoints form the zero-dimensional *boundary*[2] of I, which is denoted as $\partial I = \{a\} \cup \{b\}$.

Anticipating the time when we will be dealing with manifolds of more than one dimension (and thus with boundaries of dimension greater than zero), we would prefer to write each side of Eq. 8.6 as an integral. However, we would need to make it clear as to which manifold is the domain of integration—the boundary of the manifold, or the manifold itself. If we further let $F'(t)dt = dF$, then Eq. 8.6 may be written as

$$\int_{\partial I} F = \int_I dF, \qquad (8.7)$$

which thus far is just a stylized version of the Fundamental Theorem of Calculus.

We now make an inference, the justification for which will be borne out by the results. As written, F is a function, which we also know as a zero-form. If instead we let F be a p-form ω, then dF will become $d\omega$, a $(p + 1)$-form. Concurrently, if we denote the manifold as σ rather than I, then Eq. 8.7 becomes

$$\int_{\partial \sigma} \omega = \int_\sigma d\omega. \qquad (8.8)$$

This is the *Generalized Stokes's Theorem*, and as we have just shown it is a direct descendant of the Fundamental Theorem of Calculus. We need to justify the name, and this will be done in due course. We also need a means of evaluating it, and this is where the push-forwards and the pullbacks introduced in Chap. 7 play a central role. We will demonstrate the application of the GST in the following sections where, unless otherwise specified, we will adopt Cartesian coordinates.

---

[2]Note that the interval is one-dimensional, but the boundary is zero-dimensional, formally, a union of two zero-dimensional spaces. This is a general relationship; if a manifold is n-dimensional, then the boundary of the manifold (also a manifold) will be $(n - 1)$-dimensional.

Fig. 8.2 The integral of a
one-form is equivalent to an
oriented line integral

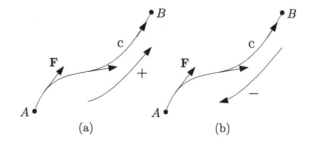

8.2   Line Integrals and the Integration of One-Forms

The integral of a one-form $\alpha = \alpha_i \mathbf{dx}^i$ has a direct correspondence to a line integral of the scalar product of a vector $\mathbf{F}$ with a differential displacement $\mathbf{ds}$. This is apparent when we write $\mathbf{F} \cdot d\mathbf{s} = F_x dx + F_y dy + F_z dz = F_i \, dx^i$. The corresponding integrals over a curve c are

$$\int_c \mathbf{F} \cdot d\mathbf{s} \quad \text{and} \quad \int_c \alpha, \tag{8.9}$$

the latter expression being disconcerting to the eye of most people who see it for the first time because it appears to be "missing" a differential.

Reversing the direction in which we traverse the curve (Fig. 8.2) reverses the sign of the integral so that

$$\int_c \alpha \;=\; -\int_{-c} \alpha, \tag{8.10}$$

thereby giving an *orientation* to the curve. This illustrates a more general consideration, namely, that integration on manifolds rests on the manifold being both orientable and compact.[3]

The procedure for evaluating Eq. 8.10 essentially involves integrating both sides of Eq. 7.38. First, we parameterize the curve c by the map[4] $\phi : [a, b] \subset \mathbb{R}^1 \to c \subset \mathbb{R}^n$. From our discussion in Sect. 7.7, the effect of ϕ is to assign a tangent vector $D\phi(\mathbf{v})$ to each point on the curve. We then evaluate the inner product $\alpha(D\phi(\mathbf{v}))$ at each point on c and integrate the resulting function along the curve.

Equivalently, we could apply a pullback transformation and map the one-form α from $c \subset \mathbb{R}^n$ to the parameter space $[a, b] \subset \mathbb{R}^1$ and then evaluate the integral of $\phi^*\alpha(\mathbf{v})$ by Riemann integration on $[a, b] \subset \mathbb{R}^1$. The equivalence of these two approaches leads us back to the Generalized Stokes's Theorem.

---

[3] See Sect. 6.5 for a discussion of compactness, a necessary condition here for the integral to be finite. A formal treatment of the topological necessities of manifolds such that integration may be defined on them is discussed in more advanced works. See, for example, the discussion in [4].

[4] This is the "push-forward" ϕ_* from Sect. 7.7, where $\phi_*(\mathbf{v}) = D\phi(\mathbf{v})$. See Definition 7.6(1).

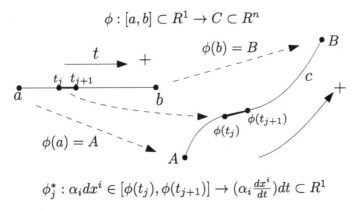

Fig. 8.3 The parameterization ϕ of a curve c from the parameter space $[a, b]$, and the pullback ϕ_j^* of a one-form α from an interval of the curve to an interval in the parameter space

This procedure is depicted in Fig. 8.3, where the interval $I = [a, b] \subset \mathbb{R}^1$ is mapped to the curve c between points A and B. Specifically, for the jth subinterval, $\phi_j : [t_j, t_{j+1}] \to [\phi(t_j), \phi(t_{j+1})]$ containing the vector $D\phi(\mathbf{v}_j)$. The one-form within each subinterval on c is $\alpha_j = (\alpha_i dx^i)_j$ with which we form $\alpha_j(D\phi(\mathbf{v}_j))$ and then integrate. Equivalently, the pullback ϕ_j^* of α_j to a subinterval on $I = [a, b]$ yields a function on $[t_j, t_{j+1}]$. Either way, summing over all infinitesimal subintervals yields a Riemann integral of α along the curve that is equivalent to the integral of the pullback on $I = [a, b]$. That is,

$$\int_c \alpha = \sum_j \int_{t_j}^{t_{j+1}} \phi_j^* \alpha = \int_a^b \phi^* \alpha = \int_a^b \left(\alpha_i \frac{dx^i}{dt} \right) dt. \qquad (8.11)$$

We note that the orientation of the parameter space in Fig. 8.3 determines the orientation of the curve. Any new parameterization of $I = [a, b]$ that maintains its orientation will likewise maintain the orientation of the curve. Again, knowing the orientation of the space on which the integration is being performed is essential when integrating p-forms in n-dimensional spaces.

Further, if α is an exact differential one-form, then there exists a zero-form (a function) f such that $\alpha = \mathbf{d}f$, and Eq. 8.11 becomes

$$\int_c \alpha = \int_a^b \left(\alpha_i \frac{dx^i}{dt} \right) dt = \int_a^b \frac{df}{dt} dt = \int_A^B df = f(B) - f(A), \qquad (8.12)$$

which is the Fundamental Theorem of Calculus expressed as the integration of an exact one-form. From there, we recover the Generalized Stokes's Theorem.

In elementary physics, Eq. 8.12 resembles the calculation of the work done by a *conservative force* **F**—one in which the work done in going between two *fixed* points is independent of the path between them, and where d**F** exact.

This example suggests that there is great value in being able to determine whether a *p*-form is exact. For this, we first recall the three relevant provisions in Definition 7.5(6)-(8): (i) $\mathbf{d}(d\alpha) \equiv \mathbf{d}^2\alpha = 0$; (ii) if $d\alpha = 0$, then α is *closed*; (iii) if $\alpha = d\beta$ for some form β, then α is *exact*.

Assessing whether a differential form α is closed is very straightforward; we simply determine whether $d\alpha = 0$. Further, if α is *not* closed, then we know immediately that it cannot be exact; if it were exact, then it would be closed.

However, if α *is* closed, then all we can say for sure is that α *may* be exact. The answer depends on whether the manifold is *simply connected* (Sect. 8.3); if α is closed *and* the manifold is simply connected, then α is exact. This is the substance of the *Poincare lemma*, the formal proof of which is done using the tools of cohomology, which are beyond the scope of this text. Still, if we are prepared to accept Poincare's lemma, then all we need to know for a closed *p*-form to be exact is whether the manifold is simply connected.

In the meantime we'll consider a "brute force" approach and endeavor to find (via integration) a form β such that $\alpha = d\beta$. This approach works well for one-forms and functions, but it is much more problematic for higher-order *p*-forms.

Example 8.1 Consider the one-form $\alpha = 6xy^3\mathbf{dx} + 9x^2y^2\mathbf{dy} \equiv A\mathbf{dx} + B\mathbf{dy}$ in $\mathbb{R}^2$. We first ask whether α is closed by assessing whether $d\alpha = 0$. If it is closed, then we will try to look for some function (zero-form) $f(x, y)$ such that $\alpha = \mathbf{df}$.

From Definition 7.5, and applying the methods of Sect. 7.4, we find

$$d\alpha = \left(\frac{\partial A}{\partial x}\mathbf{dx} + \frac{\partial A}{\partial y}\mathbf{dy}\right) \wedge \mathbf{dx} + \left(\frac{\partial B}{\partial x}\mathbf{dx} + \frac{\partial B}{\partial y}\mathbf{dy}\right) \wedge \mathbf{dy},$$

which reduces to

$$d\alpha = \left(\frac{\partial B}{\partial x} - \frac{\partial A}{\partial y}\right)(\mathbf{dx} \wedge \mathbf{dy}).$$

Evaluating the partial derivatives gives

$$d\alpha = (18xy^2 - 18xy^2)(\mathbf{dx} \wedge \mathbf{dy}) = 0,$$

a two-form that equals zero. Therefore, α is closed.

Now the question is whether we can integrate α in some consistent way to find $f(x, y)$. We proceed on the assumption that we can; if it turns out we're wrong, then we would know α is not exact (this is the "brute force" part of this process).

The integration is carried out by first assuming one of the independent variables (we'll choose x) is constant ($\mathbf{dx} = 0$). We find

$$\int \alpha = 9x^2 \int y^2 \mathbf{dy} = 3x^2y^3 + g(x) \overset{?}{=} f(x, y), \tag{8.13}$$

where $g(x)$ is an arbitrary function of x that serves as an integration "constant" when x is constant. Now we apply the condition that $\partial f / \partial x = A(x, y)$. From Eq. 8.13 we find

$$\frac{\partial f}{\partial x} = 6xy^3 + \frac{dg}{dx}. \tag{8.14}$$

This equals $A(x, y)$ provided $g(x)$ is a constant. Therefore, $f(x, y) = 3x^2 y^3 + C$, and from this we conclude that $\alpha (= \mathbf{df})$ is exact.   ▲

Example 8.2 For the one-form $\beta = 4xy^2 \mathbf{dx} + 2x^2 y\mathbf{dy}$, applying the same procedure as in Example 8.1 yields the non-zero two-form $\mathbf{d}\beta = (-4xy)(\mathbf{dx} \wedge \mathbf{dy}) \neq 0$, except identically at $(x, y) = (0, 0)$. Therefore, β is not closed, and we conclude immediately that β is not exact.

To illustrate one effect of "non-exactness," consider two continuous paths from the origin $(0, 0)$ to the point $(x, y) = (2, 4)$. Let path (i) be the parabola $y = x^2$, and path (ii) be the straight line $y = 2x$. Direct substitution into $\beta = 4xy^2 \mathbf{dx} + 2x^2 y\mathbf{dy}$ yields:

1. Path (i) : $\beta = 8x^5 \mathbf{dx}$. Integrating from $x = 0$ to $x = 2$ gives the result 256/3.
2. Path (ii) : $\beta = 24x^3 \mathbf{dx}$. Integrating from $x = 0$ to $x = 2$ gives a result of 96.

We mention in passing for now that the minus sign in $\mathbf{d}\beta = (-4xy)(\mathbf{dx} \wedge \mathbf{dy})$ is indicative of an orientation of an area—a topic we will discuss in Sect. 8.4.   ▲

In a physics context, if we were to interpret β in Example 8.2 as $F_i dx^i$, where the F_i are the components of a force, then we recognize this as a work-integral problem associated with a non-conservative force, i.e., where the work done by the force acting over a path between two points depends on the path taken. In such a circumstance, we would not be able to define a potential energy. Contrast this with Example 8.1 where the force associated with α would be a conservative force, and $f(x, y)$ would be the potential energy.

The conditions under which a closed one-form is necessarily exact (and how to know when it is not) are described as part of our discussion of homotopy..

8.3 Homotopy and the Cauchy Theorems of Complex Analysis

We now examine one of the many topological aspects of differential forms. Consider two continuous curves with the same endpoints A and B in $\mathbb{R}^2$. Taken together, the two curves form a closed loop, and in Fig. 8.4a.

one curve may be mapped to the other by a smooth transformation within $\mathbb{R}^2$. Two curves related in this way are said to be *homotopic*. If the curves remain homotopic as the endpoints in Fig. 8.4a move about the plane, then the curves are said to be *freely homotopic*.

Fig. 8.4 In $\mathbb{R}^2$ the two curves from A to B in (**a**) are homotopic; in (**b**) they are not homotopic in $\mathbb{R}^2$, but would be in $\mathbb{R}^3$. Curves 1 and 2 together form a closed curve

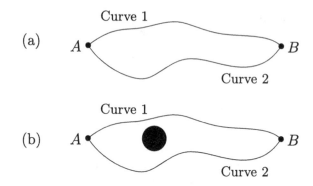

The presence of the "hole" in Fig. 8.4b makes a continuous transformation between the two curves in $\mathbb{R}^2$ impossible without leaving the space (conceivably somehow "lifting up" one curve so it can cross the hole). The dimension of $\mathbb{R}^n$ is relevant when assessing whether two curves are homotopic. For example, if Fig.8.4b were drawn for $\mathbb{R}^3$ and the planar hole became a spherical hole (a hollow sphere), then the two curves would be homotopic.

It follows from Eq. 8.12 that the integral of an exact one-form around the closed curve in Fig. 8.4a equals zero. In addition, if the closed curve in Fig. 8.4a is freely homotopic, then we can imagine a sequence of homotopic closed curves which converge to a point. In this instance we say that these closed curves are *freely homotopic to a point*, where the integral of an exact one-form around each of the curves in this sequence would be zero. As we will see shortly, this property has important consequences in complex analysis.

A connected topological space in which *every* continuous closed curve is freely homotopic to a point is said to be *simply connected*, of which $\mathbb{R}^2$ is an example. Simply connected spaces are special cases of *arcwise connected* spaces in which any two points may be joined by a curve, all of whose points are in the space.[5]

As we have seen in contrast, the space $\mathbb{R}^2 - \{0\}$ (the plane with a "hole" at the origin, as in Fig. 8.5) is not simply connected. From among all possible closed curves are those that encompass the origin, and these curves cannot be continuously contracted to a point without leaving the space.

Integration around closed loops and concepts related to homotopy have particular significance in the study of complex analysis.[6] Among the types of things we

---

[5]See also the definition of connected spaces in Definition 6.9. Topological spaces that are not simply connected appear in the study of the Dirac monopole and the Aharanov-Bohm effect. Homotopic transformations form groups (the homotopy groups) whose applications to physics include the study of gauge transformations. An introduction to these and related topics appears in [12]; a more advanced account is found in [11].

[6]Although I have assumed you are familiar with complex variables and functions, I have not assumed you are familiar with other features of complex analysis (e.g., the Cauchy integral theorem, Cauchy integral formula, or the Cauchy-Riemann equations). Consequently, we make only this one connection to complex analysis here, but see the Guide to Further Study at the end of the chapter.

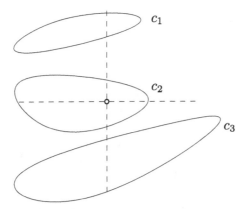

Fig. 8.5 The space $\mathbb{R}^2 - \{0\}$, where curves c_1 and c_3 are freely homotopic to a point, but c_2 is not. Therefore, even though the space is arcwise connected, it is not simply connected

Fig. 8.6 A circle of radius $|z|$ and center z_0 in the complex plane

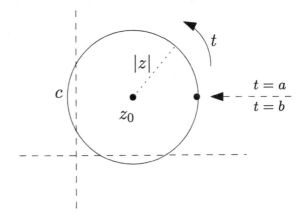

encounter in complex analysis (calculus in the complex plane) are closed-loop integrals of the form

$$\oint \frac{1}{z - z_0} dz, \tag{8.15}$$

where $(z - z_0)$ represents a circle of radius $|z|$ centered at z_0 in the complex plane. In this case the integration is taken around the closed circle c in Fig. 8.6, and the integrand in Eq. 8.15 is said to have a "pole" (becomes singular) at $z = z_0$. Consequently, the circle in Fig. 8.6 is not freely homotopic to z_0 or to any other point within the circle, and we would not expect the integral in Eq. 8.15 to be zero.

In fact, if we define

$$W(z_0) = \frac{1}{2\pi i} \oint \frac{1}{z - z_0} dz, \tag{8.16}$$

we find the result $W(z_0)$ to be a non-zero integer so long as z_0 is *inside* the circle. This integer is called the *winding number*[7] of the point z_0 with respect to the closed curve, and it is the foundation of one of the more important results from complex analysis called the *Cauchy integral formula*:

$$f(z_0) = \frac{1}{2\pi i} \oint \frac{f(z)}{z - z_0} dz. \tag{8.17}$$

As remarkable as it seems, the Cauchy integral formula may used to evaluate a functional value for $f(z)$ at a *point* within a closed curve in the complex plane by performing an integral along that curve!

There are various ways of showing $W(z_0)$ is an integer, including approaches that rely on the Cauchy-Riemann equations of complex analysis or on Green's Theorem. Our approach here[8] applies our understanding of homotopy and the GST. The idea is to recast the integral around the circle in Fig. 8.6, with $z = z(t)$ for $t \in [a, b]$.

We first define two functions, g and f, such that $g(z) \equiv z - z_0$ and $g'(t) \equiv g(t) f'(t)$, where the prime $(')$ denotes a derivative with respect to t. Noting that $z(b) = z(a)$ for a complete loop around the circle, we have

$$\oint \frac{1}{z - z_0} dz = \oint \frac{dg}{g} = \int_a^b \frac{1}{g} \frac{dg}{dt} dt = \int_a^b \frac{df}{dt} dt = \int_a^b f' dt = f(b) - f(a). \tag{8.18}$$

However, the definition $g' \equiv gf'$ means that

$$g(t) = z(t) - z_0 = Ce^{f(t)}, \tag{8.19}$$

where C is a constant of integration. Again, because $z(b) = z(a)$ we have

$$g(b) - g(a) = [z(b) - z_0] - [z(a) - z_0] = 0 = C[e^{f(b)} - e^{f(a)}]. \tag{8.20}$$

Because we are in the complex plane, this establishes that $f(b) = f(a) + 2\pi i k$ for an integer k—a result that should remind us of Euler's equation (Sect. 3.4.2). Therefore, Eq. 8.16 becomes

$$W(z_0) = \frac{1}{2\pi i} \oint \frac{1}{z - z_0} dz = k. \tag{8.21}$$

If z_0 were outside the circle in Fig. 8.6, then the circle would be freely homotopic to all interior points, and the integer k would equal zero. This is another way of stating the Poincare lemma, but it is also the substance of the *Cauchy integral theorem* (not to be confused with the Cauchy integral *formula* in Eq. 8.17). This theorem is usually

---

[7]The winding number is often referred to as the *index*.
[8]Our approach follows that in [10], pp. 134–5.

stated in terms of analytic functions by saying if $f(z)$ is analytic everywhere inside and on the closed curve in the complex plane, then

$$\oint f(z)dz = 0. \tag{8.22}$$

In this event, the closed curve is said to be *homologous to zero*. However, for the configuration described in Fig. 8.6 and Eq. 8.21, k is a non-zero integer and equals the number of complete trips taken around the circle when evaluating the integral.

The typical derivation of the Cauchy integral theorem (Eq. 8.22) and integral formula (Eq. 8.17) rely on much more analytic function machinery than we have used here, and these topics are part of a standard course in complex analysis. However, we have now shown that these theorems also arise from a consideration of homotopy and (in a fundamental sense) the Generalized Stokes's Theorem when both are applied to the complex plane.

8.4 Integration of *p*-Forms and the Vector Integral Theorems

The first step in extending the integration of one-forms to higher-order p-forms is to show that the central result of Eq. 8.11, namely,

$$\int_c \alpha = \int_a^b \phi^* \alpha, \tag{8.23}$$

is consistent with the Generalized Stokes's Theorem beyond one-dimension. That is, rather than a one-dimensional parameter space $I = [a, b]$ and a curve c (Fig. 8.3), we must now accommodate a higher-dimensional coordinate space and manifold, respectively. Figure 8.7 illustrates the conceptual picture.

This consistency is demonstrated[9] in Eq. 8.24:

$$\underset{\partial\phi(A)}{\int \alpha} \overset{1}{=} \underset{\phi(\partial A)}{\int \alpha} \overset{2}{=} \underset{\partial A}{\int \phi^*\alpha} \overset{3}{=} \underset{A}{\int d(\phi^*\alpha)} \overset{4}{=} \underset{A}{\int \phi^*(d\alpha)} \overset{5}{=} \underset{\phi(A)}{\int d\alpha}. \tag{8.24}$$

Starting with the integral of a p-form α on the boundary of $\phi(A)$:

- Step 1 recognizes that ϕ maps the boundary of A to the boundary of $\phi(A)$;
- Step 2 involves the pullback of α to the boundary of A and incorporates Eq. 8.23;
- Step 3 incorporates the Generalized Stokes's Theorem;

---

[9]Our approach here is a slightly embellished version of that in [3], p. 109.

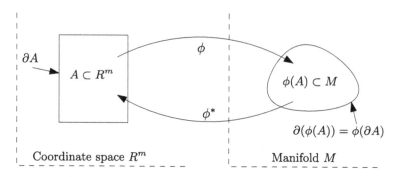

Fig. 8.7 The mapping ϕ of a coordinate system onto a manifold induces the pullback ϕ^*. This is a more general version of the one-dimensional configuration shown in Fig. 8.3

- Step 4 recognizes that ϕ and ϕ^* preserve the differential structures of A and $\phi(A)$ (Definition 7.6(6) and Fig. 7.17);
- Step 5 arises from a push forward of $d\alpha$ to $\phi(A)$.

The first and last terms in Eq. 8.24 constitute the customary formulation of the Generalized Stokes's Theorem on M. From the steps above, we see that this formulation is consistent with the pullback and push forward transformations between $A \subset \mathbb{R}^m$ and $\phi(A) \subset M$ as depicted in Fig. 8.7.

Thus far we have assumed very little about the manifold M beyond smoothness, the ability to identify subspaces and the continuity of the transformations. Consequently, the Generalized Stokes's Theorem, rewritten here again for convenience as

$$\int_\sigma d\omega = \int_{\partial\sigma} \omega, \tag{8.25}$$

is applicable to a wide range of manifolds—a range that includes essentially all the manifolds we regularly use in mathematical physics.

The actual evaluation of the GST on a manifold M requires that we map a coordinate patch (called a *chart*) $A \subset \mathbb{R}^m$ onto $\phi(A) \subset M$. Overlapping charts are mapped to overlapping portions of M (see, for example, Fig. 7.9). The collection of all charts that cover the manifold M is called an *atlas*; different charts and atlases (i.e., different coordinate systems) may used to cover any given manifold.

Recall that when we were considering a one-dimensional parameter space being mapped onto a curve, the orientation of the curve was linked directly to the orientation of the parameter space. In higher dimensions the natural choice is to map basis vectors in $\mathbb{R}^m$ onto $\phi(A)$, and as in the one-dimensional case whichever orientation we choose for $\mathbb{R}^m$ will carry over to the orientation of the manifold.

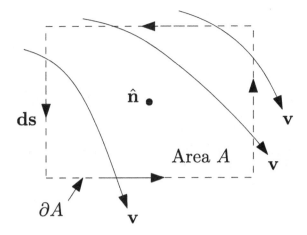

Fig. 8.8 A two-dimensional fluid flow with the velocity field passing across an enclosed area A with boundary ∂A. The integral of $\mathbf{v} \cdot d\mathbf{s}$ around ∂A is called the circulation

We now show that several important vector integral theorems—usually derived by other, more coordinate-specific, means—are really just special cases of the Generalized Stokes's Theorem.[10]

Example 8.3 *The work integral*: We covered this example thoroughly in Sects. 8.1 and 8.2, but we repeat it here for context. If we let the manifold σ be a curve c with endpoints A and B, then $\partial \sigma = \{A\} \cup \{B\}$. Further, if we let $d\boldsymbol{\omega}$ be a one-form, then $\boldsymbol{\omega}$ is a function.

Applying this to the work integral from elementary physics, let $\mathbf{F}$ represent a force acting on a particle and $d\mathbf{s}$ be an infinitesimal displacement. Then

$$\int_c \mathbf{F} \cdot d\mathbf{s} = \int_\sigma d\boldsymbol{\omega} = \int_{\partial\sigma} \boldsymbol{\omega} = \int_A^B f = f(B) - f(A).$$

In this context, the quantity $f(B) - f(A)$ is the work done by the force, which (by the *work-energy theorem*) is the change in the kinetic energy of the particle. If the force is conservative then the one-form $d\boldsymbol{\omega}$ is an exact differential. The work done is then path-independent, and we can define a mechanical potential energy.   ▲

Example 8.4 *The circulation of a fluid and Stokes's Theorem*: In two-dimensional fluid dynamics, the normal component of the curl of the velocity vector is defined as the circulation per area, as defined at a point. In Fig. 8.8 the unit vector $\hat{\mathbf{n}}$ points out of the page, and integrating $(\nabla \times \mathbf{v}) \cdot \mathbf{n}$ over the area gives the total circulation. The *circulation theorem* states that this integral equals the line integral of the velocity vector around the boundary of the enclosed area. In a more generic setting, the circulation theorem is the traditional form of Stokes's Theorem.

---

[10]See also the correspondences between exterior and vector calculus in $\mathbb{R}^3$ in Sects. 7.4 and 7.5.

We can frame this in terms of the Generalized Stokes's Theorem by letting the manifold σ be an enclosed area A in $\mathbb{R}^2$ with ∂A as the boundary. Further, let $d\omega$ be a two-form (see below) so that ω is a one-form. Applying this to the circulation within a fluid gives

$$\int_A (\nabla \times \mathbf{v}) \cdot \mathbf{n} dA = \int_\sigma d\omega = \int_{\partial\sigma} \omega = \int_{\partial A} \mathbf{v} \cdot \mathbf{ds},$$

where the equality of the first and last terms constitute the circulation theorem.[11] The interpretation of $\nabla \times \mathbf{v}$ as "circulation per area" follows from writing the circulation theorem as

$$(\nabla \times \mathbf{v}) \cdot \mathbf{n} = \lim_{\Delta A \to 0} \int_{\partial A} \mathbf{v} \cdot \mathbf{ds}.$$

The assertion that $(\nabla \times \mathbf{v}) \cdot \mathbf{n} dA$ corresponds to a two-form $d\omega$ is borne out by Exs. 7.11 and 7.13. If we let $\omega = A(x, y, z)\mathbf{dx} + B(x, y, z)\mathbf{dy}$, then

$$d\omega = \left(\frac{\partial B}{\partial x} - \frac{\partial A}{\partial y} \right) (\mathbf{dx} \wedge \mathbf{dy}),$$

where $(\mathbf{dx} \wedge \mathbf{dy})$ corresponds to an oriented differential area with a unit normal vector perpendicular to the x-y plane (Example 7.13 and Problem 8.5).

For a non-zero circulation, ω is not a closed one-form. Compare this example with Example 8.1, where the two-form vanished. In fluid dynamics a closed one-form would correspond (as one example) to a uniform fluid flow (parallel velocity vectors of constant magnitude with no circulation) and an exact one-form $\omega = \mathbf{v} \cdot \mathbf{ds}$. ▲

Example 8.5 *The divergence theorem*[12]: In three-dimensional fluid dynamics, the divergence of the velocity vector is defined as the integrated surface flux per volume as defined at a point. We may write this as

$$(\nabla \cdot \mathbf{v}) = \lim_{\Delta V \to 0} \int_{\partial V} \mathbf{v} \cdot \hat{\mathbf{n}} dA.$$

Figure 8.9 depicts a volume V, where the unit vector $\hat{\mathbf{n}}$ points radially outward at the surface.

Integrating the divergence of the fluid over the volume gives the net rate fluid flow into or out of the volume. This is equal to the integrated surface flux through the surface (think of this as conservation of mass), and this equality is known as the *divergence theorem*.

---

[11] In three-dimensional fluid flow the area A is that which is projected onto some plane.

[12] We saw the divergence theorem in our derivation of Liouville's theorem vis-a-vis Hamiltonian mechanics in Sect. 7.6. There, it referred to the conservation of phase space volume.

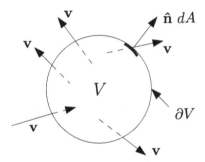

Fig. 8.9 A three-dimensional fluid flow with velocity vectors passing into and out of an enclosed volume V with boundary (the surface) ∂V. The integral of $\mathbf{v} \cdot \hat{n} dA$ is the integrated flux through the surface

In the language of differential forms, let the manifold σ be an enclosed volume V in $\mathbb{R}^3$ with ∂V as the boundary. Further, let $d\omega$ be a three-form so that ω is a two-form. The rational for asserting that ω is a two-form in this context is given in Example 7.12.

Applying this to the divergence of fluid flow gives

$$\int_V (\nabla \cdot \mathbf{v}) dV = \int_\sigma d\omega = \int_{\partial\sigma} \omega = \int_{\partial V} \mathbf{v} \cdot \hat{n} dA,$$

where the equality of the first and last terms constitute the divergence theorem. ▲

The usual manner by which we derive the vector integral theorems shown in the previous examples is so dependent on the coordinate system that we fail to see the general underlying structure behind them. That structure is the Generalized Stokes's Theorem, of which these vector theorems are special cases. However, the Generalized Stokes's Theorem is applicable to bounded manifolds generally, and it is one of the more important results in differential geometry and mathematical physics.

8.5 Lie Groups as Manifolds

A principal theme of this text has been how any one set may host different algebraic or differential structures, depending on the topology of the set and the operations defined among its elements. One example is the real line. In various places we have shown $\mathbb{R}^1$ to be a group, a field, a vector space, an algebra or a manifold, each the result of appropriately-selected operations having been defined on the space.

Given two or more structures, each defined on a given set in accordance with its own criteria, the question arises as to whether it is possible to unify them so as to create a hybrid structure that exhibits the characteristics of its constituent parts simultaneously. Such is the essence of a *Lie group*, a continuous group that is also a manifold. The "glue" that unifies the manifold and group aspects of a Lie group is an algebra—a Lie algebra, whose elements are the group generators.[13]

Lie groups arose in the late 19th-century as a product of Sophus Lie's efforts to solve differential equations using their symmetries. His work at the time was motivated in part by the earlier work of Galois in describing the symmetries associated with the roots of algebraic equations, but it was also shaped by the influence of his contemporary, Felix Klein. It is fair to say that topology "began" with Klein, whose work focused on the classification of topological spaces according to those properties that remain invariant under transformations.[14]

Continuous groups are typically represented by matrices, but we have already seen (Problem 5.5) how some matrices, such as the 2×2 rotation matrix in the plane, can be represented as an exponential of an operator. For Lie groups, this exponential map is applied within a tangent plane at a particular point in the manifold—a point that corresponds to the identity element of the group. Further, it is the group generators that appear in the argument of the exponential function (see Sect. 5.8).

The premise of a Lie group, therefore, is that the group parameters may be associated with a manifold, and that the *local* properties of a Lie group may be determined by its *local* behavior near its identity element. All of this will become more clear once we develop a particular example of a Lie group. The group $SO(2)$, which we associate with rotations in the two-dimensional plane, suffices for us to show the essential concepts, and we will devote our attention there. The groups $SO(3)$ and $SU(2)$ are left as (guided) end-of-chapter problems.

Although our treatment of Lie groups in this text is all-too-brief, you will at least get an overall sense of how to construct more elaborate groups. From here, you can progress to one or more of the specialized references mentioned in the Guide to Further Study at the end of this chapter. Also left for further study is the question of *global* group properties, a topic that would take us into representation theory, which we are not covering in this text.

In Sect. 2.3.5 we established that rotations in a plane form a continuous group, and in Problem 5.5(a) you were asked to show that the matrix representation for the group elements,

$$R(\theta) = \begin{pmatrix} \cos\theta & -\sin\theta \\ \sin\theta & \cos\theta \end{pmatrix},$$

is the matrix representation of $e^{i\theta} = \cos\theta + i\sin\theta$. The method in solving that problem was to apply the matrix representation for i and show that $R(\theta) = Ie^{i\theta}$, where I is the identity matrix. The group is $SO(2)$, the elements of which are 2×2 orthogonal matrices with unit determinant.

---

[13] See Sect. 2.3.1 for our first mention of group generators (there, in the context of cyclic groups), and Sect. 5.3.1 for a discussion of Lie algebras and a brief biographical footnote on Sophus Lie. You should also take a look at Problems 5.5, 5.7 and 5.8, where we presaged the discussion we are about to have on several Lie groups.

[14] A comprehensive history of the origin of Lie groups is [9]. An introductory treatment of Lie group methods in solving differential equations is [15]; a more advanced account is [13].

Fig. 8.10 Rotations in a
plane are elements of the
continuous group $SO(2)$

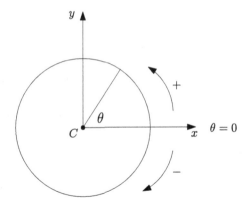

This same result may be found by a different method, and the advantage of this second approach is that it applies more broadly to other Lie groups. We will first outline the steps, and then apply it to $SO(2)$.

- Identify a parameter that we hypothesize is also a group parameter. For rotations in the plane, the choice is fairly obvious (an angle of rotation);
- Identify the zero point associated with the parameter, and associate this with the identity element of the group;
- Associate the identity element of the group with some aspect of the manifold;
- Write the group element for a differential displacement from the identity. This is the step where "group meets differentiable manifold," and was Lie's principal innovation. This displacement occurs in a tangent plane at the group identity;
- Write the group element for a differential displacement from some arbitrary value of the group parameter (again, this occurs in the tangent plane at that point);
- Combining these expressions, find a differential equation for a group element;
- Solve the differential equation to obtain a general expression for a group element.

We now apply this procedure to find the group elements of $SO(2)$:

- Let the group parameter be a rotation angle, and write a group element as $R(\theta)$;
- The zero point is $\theta = 0$, so $R(0) = I$;
- Associate the group identity element with the x-axis in the manifold (Fig. 8.10);
- The group element for a differential displacement in the tangent plane at the identity is

$$R(d\theta) = I - iJ(d\theta)$$

where we use $(-iJ)$ as a proportionality constant, knowing that the coefficient of $d\theta$ must be a matrix. The inclusion of $(-i)$ is for later convenience (try leaving it out!). The matrix J is independent of $d\theta$. It will eventually be identified as a group generator and as an element in the Lie algebra associated with the group;

- The group element for a differential displacement in the tangent plane at some arbitrary θ is $R(\theta + d\theta)$. This can be expressed in two different ways: first as a consequence of the group multiplication property,

$$R(\theta + d\theta) = R(\theta)R(d\theta) = R(\theta)[I - iJ(d\theta)],$$

but also in a form that would apply to any differentiable manifold:

$$R(\theta + d\theta) = R(\theta) + \frac{dR(\theta)}{d\theta}d\theta.$$

- From these two expressions we get the differential equation

$$\frac{dR(\theta)}{d\theta} = -iJR(\theta).$$

- The solution to this first-order differential equation (with boundary condition $R(0) = I$) is

$$R(\theta) = e^{-iJ\theta}, \tag{8.26}$$

where J is the group generator. Note that $R(\theta) = R(\theta + 2\pi)$, as it must.

We leave it as an exercise to show that

$$J = \begin{pmatrix} 0 & -i \\ i & 0 \end{pmatrix} = \sigma_2, \tag{8.27}$$

one of the Pauli matrices (see Problem 5.7). You should also verify by direct matrix multiplication that

$$e^{-iJ\theta} = I\cos\theta - iJ\sin\theta = R(\theta). \tag{8.28}$$

There is a three-fold significance of this second approach to finding $R(\theta)$ for $SO(2)$. First, the differentiability of the manifold plays a central role in the derivation. This is the essence of Lie groups, i.e., that a continuous group is also a manifold. This aspect of $SO(2)$ does not show through when finding $R(\theta)$ by the more strictly algebraic method in Problem 5.5, but it is essential for developing other Lie groups.

Second, in the previous method there was no hint of a group generator. Now we can find the generators and see that they are elements of a tangent space. If there were more than one generator they would form a Lie algebra, but for $SO(2)$ there is only one generator so the algebra is trivial.

Third, the result for $R(\theta)$ in Eq. 8.26 is representative of a general form. If we let R designate a generic group operator (i.e., not necessarily a rotation) and ψ_n represent the nth group parameter (not necessarily an angle), then it is fair to describe the general form

$$R(\psi_n) = e^{-iJ_n\psi_n} \tag{8.29}$$

as perhaps the single most succinct description of Lie group theory.

$$\mathcal{L} : f(\tau) \in D(\tau) \to \bar{f}(\omega) \in \bar{D}(\omega)$$

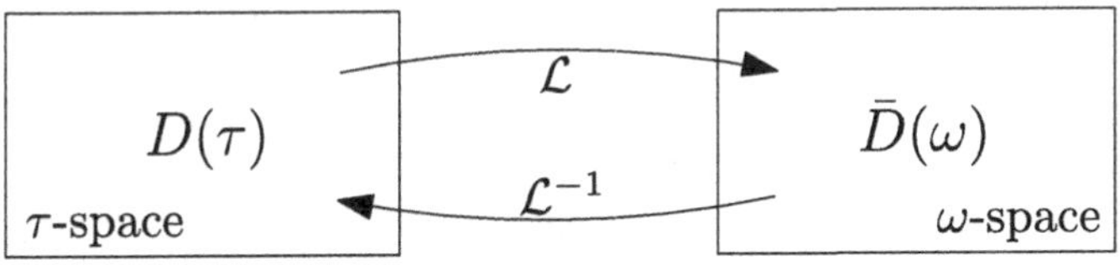

$$\mathcal{L}^{-1} : \bar{f}(\omega) \in \bar{D}(\omega) \to f(\tau) \in D(\tau)$$

Fig. 8.11 Integral transforms as linear functionals; $\mathcal{L}$ is the transform, and $\mathcal{L}^{-1}$ is its inverse

In $SO(3)$ there will be three generators, each representing a rotation around one of the axes in three-dimensional space. The group $SU(2)$ operates in "internal" complex spaces such as those used to describe spin and other inherent properties of elementary particles. The generators for $SU(2)$ will be the three Pauli matrices.

For both groups the generators are the elements of the *Lie algebras* associated with the corresponding group. The algebras and generators were given in Problems 5.7 and 5.8, and will be repeated in some problems at the end of this chapter.

8.6 Integral Transforms as Maps

When reduced to its fundamentals an integral transform $\mathcal{L}$ is a nonsingular linear map—a linear functional that maps elements of one function space to those in another and which has an inverse (Definition 4.5 and Fig. 8.11). Everything else we must know in order to apply transforms to problems in physics and engineering revolves around methodology and techniques for solution, which frequently involve methods of complex integration and the properties of special functions.

A transform and its inverse form a pair whose respective functional forms are

$$\mathcal{L}(f(\tau)) = A \int_D [f(\tau)][K(\tau, \omega)]d\tau \Leftrightarrow \mathcal{L}^{-1}(\bar{f}(\omega)) = \bar{A} \int_{\bar{D}} [\bar{f}(\omega)][\bar{K}(\tau, \omega)]d\omega,$$

(8.30)

where K ($\bar{K}$) is the *kernel* (*inverse kernel*) that defines the transform.[15] The constants A and $\bar{A}$ occasionally depend on the convention in use.

The fact that these functionals form a pair may cause you to wonder as to the rationale for designating one over the other as the "transform," while the other is the "inverse transform." In large part that terminology is rooted in how these functionals are applied.

---

[15]This use of the word "kernel" differs from its algebraic usage earlier in the text.

In physics, the usual reasons you would apply a transform include (a) changing your original problem into one that highlights particular features; or (b) converting your original problem into one that you can more easily solve. Among the transforms you are most likely to encounter, the Fourier transform is an example of the former application, and the Laplace transform is an example of the latter.

Probably in widest use in physics and engineering is the *Fourier transform,*

$$\bar{f}(\omega) = \frac{1}{\sqrt{2\pi}} \int_{-\infty}^{\infty} f(\tau) e^{-i\omega\tau} d\tau; \; f(\tau) = \frac{1}{\sqrt{2\pi}} \int_{-\infty}^{\infty} f(\omega) e^{i\omega\tau} d\omega. \qquad (8.31)$$

Among its many applications are those in signal and image processing, quantum mechanics and optics, but they appear in virtually all subdisciplines.

For example, we may observe a time signal $f(t)$, which then serves as input to the transform; the output $\bar{f}(\omega)$ would be a frequency spectrum. It may be that we are looking for a characteristic frequency that describes the dynamics of the system under observation, or perhaps we wish to filter out noise or other features at certain frequencies. In this latter case, these filtering operations would be carried out in the frequency space, and the inverse transform would then give as output a signal which is "cleaner" than the original.

The same concept holds for two-dimensional images or three-dimensional spatial distributions, where τ is now a length and ω is a wavenumber. For these applications, we would need to convert Eq. 8.31 into their spatial equivalents.

Of particular note is that the Fourier transform of a Gaussian distribution is another Gaussian distribution, and the respective widths of these distributions are inversely proportional to each other. A direct consequence of this in quantum mechanics is the Heisenberg uncertainty principle, which follows from taking the Fourier transform of the position- and momentum-space probability distribution functions.[16]

Laplace transforms[17] appear in the study of ordinary differential equations generally, and circuit analysis in particular. These transforms, defined as

$$\bar{f}(\omega) = \int_{0}^{\infty} f(\tau) e^{-\omega\tau} d\tau, \qquad (8.32)$$

can be used to convert a differential equation into an algebraic equation. Solving the algebra problem, and then applying an inverse transform, yields solutions to the differential equation. We should note that the inverse Laplace transform is an integral over a line in the complex plane that requires a deft touch with complex analysis. For that reason, you may often employ tables of Laplace transforms and their inverses in solving problems if you decide not to evaluate the integral directly.

---

[16]See [14], pp. 57–8.

[17]Here and in the transforms that follow, ω and τ are just parameters.

Another transform that appears on occasion is the *Hankel transform*,

$$\bar{f}(\omega) = \int_0^\infty f(\tau)[\tau J_n(\omega\tau)]d\tau, \tag{8.33}$$

where J_n is a Bessel function. Applications of this transform tend to be found in problems with cylindrical symmetry.

The *Mellin transform* appears in association with several special functions, and is given as

$$\bar{f}(\omega) = \int_0^\infty f(\tau)\tau^{\omega-1}d\tau. \tag{8.34}$$

The gamma function $\Gamma(\omega)$ is a special case of the Mellin transform, with $f(\tau) = e^{-\tau}$.

In addition to particular physical applications these transforms may be applied toward solving integral equations more generally, and most of the standard texts in mathematical methods cover integral transforms and the nuances of their methods of evaluation in great depth,[18]. Instead, our purpose in this section has been to emphasize that behind those methods and applications lies the fundamental structure of the integral transform—that of a linear map.

Problems

8.1 Consider Example 8.3, which refers to the work-energy theorem.
(a) Work this one-dimensional example with $\mathbf{F} = -kx\hat{\mathbf{i}}$, the linear restoring force that we usually assume applies to the "mass-on-a-spring" problem;
(b) Verify that the $d\omega$ associated with this problem is an exact differential by applying the methods of exterior differentiation;
(c) Evaluate the integral of $d\omega$ from $x = 1$ to $x = 3$;
(d) Applying the methods of this chapter, show that the work done by the force equals the change in the kinetic energy.

[*Hint*: For part (d) you will need to apply Newton's Second Law, and the definition of velocity as dx/dt. Then see if you can form the quantity $mvdv$. Once you do that, you will have another problem involving a one-form.]

8.2 In Cartesian coordinates, consider the one-form

$$\alpha = F_i \mathbf{dx}^i = (3x^2yz - 3y)\mathbf{dx} + (x^3z - 3x)\mathbf{dy} + (x^3y + 2z)\mathbf{dz}.$$

(a) Is α closed?
(b) Is α exact?

---

[18] See, for example [1] Chap. 20 whose conventions we have followed in the expressions for the transforms mentioned in this section.

(c) If the F_i represent the components of a force $\mathbf{F}$, can a potential energy $U(x, y, z)$ be defined? If so, find it. If not, explain.

8.3 Consider the one-form

$$\alpha = v_i \mathbf{dx}^i = xy \mathbf{dx} - y^2 \mathbf{dy},$$

where the v_i are the components of a velocity vector field in the Cartesian plane. Is there a non-zero circulation associated with this field? If so, find it. If not, explain.

8.4 In this problem, let the path of integration be $|z| = 5$, a closed circle that is centered at the origin in the complex plane and with $r = 5$. Apply the Cauchy integral formula (Eq. 8.17) and what we know about homotopy to evaluate the integral

$$\oint \frac{f(z)}{z - z_0} dz$$

and the winding number in each of the following cases:
(a) $f(z) = 1$ and $z_0 = 3$;
(b) $f(z) = z$ and $z_0 = 3$; (*Ans:* $6\pi i$)
(c) $f(z) = z$ and $z_0 = 4 + 5i$;
(d) $f(z) = \exp 3z$ and $z_0 = \ln 2$.

8.5 Example 7.13 showed us how a two-form maps two vectors to a scalar, where the scalar represented the signed magnitude of the area of the parallelogram formed by the two vectors. That particular example was limited by the fact that there was only one basis two-form in the two dimensional space (which is all there can be).

A more precise interpretation of this example would have been to say that the area of this parallelogram was projected into the planes containing the basis two-forms, but because we were dealing with just one basis two-form ($\mathbf{dx} \wedge \mathbf{dy}$), the projection was of a parallelogram in the x-y-plane into the x-y-plane itself.

A more interesting problem is when we have a two-form in a three-dimensional space. In this case, the parallelogram is projected onto each of the three coordinate planes. As a general example, let

$$\omega = P(x, y, z)(\mathbf{dy} \wedge \mathbf{dz}) + Q(x, y, z)(\mathbf{dz} \wedge \mathbf{dx}) + R(x, y, z)(\mathbf{dx} \wedge \mathbf{dy}),$$

and let the two vectors be

$$\mathbf{u} = u^1 \hat{\mathbf{i}} + u^2 \hat{\mathbf{j}} + u^3 \hat{\mathbf{k}} \text{ and } \mathbf{v} = v^1 \hat{\mathbf{i}} + v^2 \hat{\mathbf{j}} + v^3 \hat{\mathbf{k}}.$$

Then, following Eq. 7.15 and Example 7.13, we have

$$
\omega(\mathbf{u}, \mathbf{v}) = P \begin{vmatrix} \langle \mathbf{dy}|\mathbf{u} \rangle & \langle \mathbf{dy}|\mathbf{v} \rangle \\ \langle \mathbf{dz}|\mathbf{u} \rangle & \langle \mathbf{dz}|\mathbf{v} \rangle \end{vmatrix} + Q \begin{vmatrix} \langle \mathbf{dz}|\mathbf{u} \rangle & \langle \mathbf{dz}|\mathbf{v} \rangle \\ \langle \mathbf{dx}|\mathbf{u} \rangle & \langle \mathbf{dx}|\mathbf{v} \rangle \end{vmatrix} + R \begin{vmatrix} \langle \mathbf{dx}|\mathbf{u} \rangle & \langle \mathbf{dx}|\mathbf{v} \rangle \\ \langle \mathbf{dy}|\mathbf{u} \rangle & \langle \mathbf{dy}|\mathbf{v} \rangle \end{vmatrix}
$$

$$
= P \begin{vmatrix} u^2 & v^2 \\ u^3 & v^3 \end{vmatrix} + Q \begin{vmatrix} u^3 & v^3 \\ u^1 & v^1 \end{vmatrix} + R \begin{vmatrix} u^1 & v^1 \\ u^2 & v^2 \end{vmatrix} = A_{yz} + A_{zx} + A_{xy},
$$

where, for example, A_{yz} is the projection of the parallelogram into the y-z-plane. As a specific example, let $P = Q = R = 1$, and let the two vectors be

$$
\mathbf{u} = \hat{\mathbf{i}} + 2\hat{\mathbf{j}} - \hat{\mathbf{k}} \text{ and } \mathbf{v} = -2\hat{\mathbf{i}} + 2\hat{\mathbf{j}} + 2\hat{\mathbf{k}}.
$$

(a) Show that $A_{yz} = A_{xy} = 6$ and $A_{zx} = 0$.

It is possible from this information about the projections to find the area of the parallelogram, but we won't take that approach. Instead, we'll apply what we know about vector cross products and normal vectors.
(b) Evaluating $\mathbf{u} \times \mathbf{v}$, find the normal vector $\mathbf{N}$ and the area of the parallelogram formed by the two vectors;
(c) Sketch the edge-on view of the plane of the parallelogram in the z-x plane. [*Note*: We know this view is edge-on because of your answers in part (a).];
(d) What angle does the plane of the parallelogram make with y-z and x-y planes? [*Hint*: You will want to draw a sketch. *Ans*: 45 degrees in both cases.];
(e) Referring to your sketch, find the projection onto each of the other two planes using elementary geometry, and show that you get the same answers as in part (a).

8.6 Regarding the $SO(2)$ group:
(a) Complete the details of the calculations that led to Eqs. 8.26 and 8.27;
(b) Verify the result in Eq. 8.28.

8.7 The group $SO(3)$ describes rotations in three-dimensional space around successive axes, and the generators associated with it also serve as angular momentum operators in quantum mechanics. These are just two reasons we should take a look at this group more closely.

We start by considering a one-parameter subgroup of $SO(3)$ whose generators satisfy the Lie algebra $[J_i, J_j] = i\epsilon_{ij}^k J_k$ given in Problem 5.8(b). The generators are

$$
J_1 = \begin{pmatrix} 0 & 0 & 0 \\ 0 & 0 & -i \\ 0 & i & 0 \end{pmatrix}, \ J_2 = \begin{pmatrix} 0 & 0 & i \\ 0 & 0 & 0 \\ -i & 0 & 0 \end{pmatrix} \text{ and } J_3 = \begin{pmatrix} 0 & -i & 0 \\ i & 0 & 0 \\ 0 & 0 & 0 \end{pmatrix},
$$

where we can associate the axis labels $(1, 2, 3)$ with (x, y, z), respectively. For rotations around any one of the three axes we apply Eq. 8.29:

$$
R_n(\theta_n) = e^{-i J_n \theta_n}.
$$

Show that the rotation matrices $R_n(\theta)$ corresponding to the generators J_n are

$$R_1 = \begin{pmatrix} 1 & 0 & 0 \\ 0 & \cos\theta & -\sin\theta \\ 0 & \sin\theta & \cos\theta \end{pmatrix} ; \quad R_2 = \begin{pmatrix} \cos\theta & 0 & \sin\theta \\ 0 & 1 & 0 \\ -\sin\theta & 0 & \cos\theta \end{pmatrix} ; \quad R_3 = \begin{pmatrix} \cos\theta & -\sin\theta & 0 \\ \sin\theta & \cos\theta & 0 \\ 0 & 0 & 1 \end{pmatrix} .$$

Each generator and matrix forms a one-parameter abelian subgroup of $SO(3)$ describing a rotation around a single axis. However, these same generators can be used to describe a general rotation in the full non-abelian group (Problem 8.8).

8.8 The first task in describing the full non-abelian $SO(3)$ group is to decide how to describe the rotations around axes that don't stay in fixed positions. A system of angles called the *Euler angles* solves this issue very nicely; alas there are several different conventions for them.[19] One consideration is to distinguish between active and passive rotations.

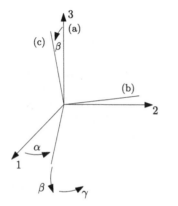

(a) α around the 3-axis

(b) β around the new 2-axis

(c) γ around the new 3-axis

Shown in the figure is a sequence of *passive rotations*, where we rotate the coordinate axes in a sequence. When we are finished, we will reverse the order to indicate the sequence associated with *active rotations*. In physics, active rotations are used to describe the transformations of operators and vectors, and to form rotation groups. This was the approach taken in our treatment of $SO(2)$.

Proceeding, the passive rotations are as follows:

- The first rotation (a) is around the 3-axis through an angle α. This moves the 1-axis and the 2-axis to new positions, but both are still in the 1-2 plane;
- The second rotation (b) is around the new 2-axis through an angle β in the direction shown. The original 3-axis now tips "toward you" to a new 3-axis position. The 1-axis now tips "down below" the 1-2 plane as seen from our perspective;
- The third rotation (c) is around the new 3-axis thru an angle γ. The effect of this rotation is not shown in the figure, but the new 1-2 plane now rotates around the new 3-axis at an angle β to its original orientation.

---

[19]We are using the convention in [17]; our figure is a rudimentary version of Tung's Fig. 7.3.

Applying $R_n(\theta_n) = e^{-iJ_n\theta_n}$ to these three rotations in sequence, and then *reversing the order* to reflect an active rotation sequence, the cummulative effect is

$$R(\alpha, \beta, \gamma) = e^{-i\alpha J_3} e^{-i\beta J_2} e^{-i\gamma J_3}.$$

As a final step, derive the general expression for the 3×3 matrix $R(\alpha, \beta, \gamma)$.

8.9 In quantum mechanics the generators of $SO(3)$ are angular momentum operators (more precisely, $\hbar J_i$ is the operator), and their Lie algebra (Problem 5.8(b))

$$[J_i, J_j] = i\epsilon_{ij}^k J_k$$

tells us that two different components of angular momentum are incompatible observables (cf. Problem 5.13 on the Heisenberg Uncertainty Principle). Although a full study of operator combinations would take us too far afield, one consequence concerns the total angular momentum.

The square of the total angular momentum operator is

$$J^2 \equiv J_1^2 + J_2^2 + J_3^2.$$

Show that $[J^2, J_i] = 0$, which means that the total angular momentum and any *one* of its components *are* compatible observables. [*Note:* You need only show this for one J_i, as all would be equivalent in this respect.]

8.10 We note that other operator combinations, such as the "raising" and "lowering" operators

$$J_\pm \equiv J_1 \pm i J_2,$$

can be used in combination with others to find the eigenvalues of angular momentum. If you have studied chemistry to the point of learning that angular momentum is quantized in units of $j(j + 1)$, then you should know that this result comes directly from the properties of the angular momentum operators—the $SO(3)$ generators—and their various combinations.[20]
(a) Show that $[J^2, J_\pm] = 0$;
(b) Show that $[J_3, J_\pm] = \pm J_\pm$.

8.11 There are many parallel constructions between the generators of $SO(3)$ and those of $SU(2)$, the latter being the Pauli matrices. This should not surprise us since (within a factor of two) they obey the same Lie algebra (Problem 5.8):

$$[\sigma_i, \sigma_j] = 2i\epsilon_{ij}^k \sigma_k,$$

where

$$\sigma_1 = \begin{pmatrix} 0 & 1 \\ 1 & 0 \end{pmatrix}, \sigma_2 = \begin{pmatrix} 0 & -i \\ i & 0 \end{pmatrix} \text{ and } \sigma_3 = \begin{pmatrix} 1 & 0 \\ 0 & -1 \end{pmatrix}.$$

---

[20] Any standard quantum mechanics text derives these results in detail. See, for example, [8].

Show that one consequence of this factor of two difference in the Lie algebras is that the arguments of the trigonometric functions in $R_n(\theta)$ (Problem 8.7) are halved, i.e., $\theta \to \theta/2$. [*Note*: The effect of this is that for a vector to "return to its initial position" under active rotations now requires a rotation through an angle of 4π rather than 2π. For this reason, there is a 2-to-1 map of the elements of $SU(2)$ to those of $SO(3)$, a map that is often (imprecisely) described as a 2-to-1 "isomorphism;" as we know, isomorphisms are 1-to-1.]

8.12 In quantum mechanics, the spin and orbital angular momentum operators yield parallel constructions. Among these are the raising and lowering operators defined for J_i (Problem 8.10) that are now written in terms of $S_i \equiv \sigma_i/2$; this accommodates the factor of 2 difference in the Lie algebras of J_i and σ_i. Specifically,

$$S_\pm \equiv S_1 \pm i S_2 = \frac{1}{2}(\sigma_1 \pm \sigma_2).$$

Rather than showing their effect on spin angular momentum, we take this opportunity to introduce the *isospin* transformation between a neutron and a proton. This transformation is motivated by the similar (though not identical) masses of the two particles.[21]

The $S_\pm$ transformations act on a (p, n) doublet ψ, in some instances switching the wave function of one particle for that of the other. Let

$$\psi_p = \begin{pmatrix} 1 \\ 0 \end{pmatrix} \text{ and } \psi_n = \begin{pmatrix} 0 \\ 1 \end{pmatrix}$$

denote the eigenfunctions for each particle in the context of these operators.
(a) Show that the eigenvalues for S_3 are $\pm 1/2$, so that

$$S_3 \psi_p = +\frac{1}{2} \psi_p \text{ and } S_3 \psi_n = -\frac{1}{2} \psi_p.$$

Show each of the following:
(b) $S_+ \psi_p = S_- \psi_n = 0$;
(c) $S_+ \psi_n = \psi_p$;
(d) $S_- \psi_p = \psi_n$.
    [*Note*: The action of the J_i and S_i operators on their respective angular momentum wavefunctions follows this same pattern. In that setting, these operators serve to either raise or lower the momentum state in each case.]

8.13 In this problem we consider the Fourier transform of a single-pulse square-wave function, which may be described by

$$f(t) = 1 \text{ for } -1 \le t \le +1,$$

---

[21] For a physical rationale that motivated the consideration of isospin, see [18], Sect. 4.6.

with $f(t) = 0$ for $|t| > 1$.

(a) Make a sketch if this function;

(b) Show that its Fourier transform is

$$\bar{f}(\omega) = \sqrt{\frac{2}{\pi}} \frac{\sin \omega}{\omega};$$

(c) Find $|\bar{f}(\omega)|^2$ as $\omega \to 0$, where $|\bar{f}(\omega)|^2$ is called the *power spectrum* of $f(t)$;

(d) Plot the power spectrum.

8.14 The normalized *Gaussian distribution function* centered at $x = \mu$ is given as

$$f(x) = \frac{1}{\sigma_x \sqrt{2\pi}} \exp\left[-(x - \mu)^2/(2\sigma_x^2)\right],$$

where σ_x is the standard deviation of the distribution. (The distribution is "normalized" in the sense that the area under its "bell-shaped curve" equals one.) Let $\bar{f}(p)$ denote the Fourier transform of $f(x)$, and for this problem let $\mu = 0$.

(a) Show that $\bar{f}(p)$ is a Gaussian distribution;

(b) Find σ_p, the standard deviation of $\bar{f}(p)$;

(c) Evaluate the product $\sigma_x \sigma_p$. How is your answer relevant to the Heisenberg Uncertainty Principle?

[*Note*: The method of integration in part (a) starts by completing the square in the exponential function and then performing an integral in the complex plane. An alternative is to use a table of integrals or software.]

8.15 Given a twice-differentiable function $f(t)$ and using p rather than ω in Eq. 8.32, the Laplace transforms $\mathcal{L}$ of the first and second derivatives of $f(t)$ are given as

$$\mathcal{L}[f'(t)] = p\mathcal{L}[f(t)] - f(0)$$

and

$$\mathcal{L}[f''(t)] = p^2\mathcal{L}[f(t)] - pf(0) - f'(0),$$

where the initial conditions appear in the transform. Suppose you are given

$$\mathcal{L}[\sin at] = \frac{a}{p^2 + a^2}; \quad \mathcal{L}[\cos at] = \frac{p}{p^2 + a^2}$$

$$\mathcal{L}[t \sin at] = \frac{2ap}{(p^2 + a^2)^2}; \quad \mathcal{L}[t \cos at] = \frac{p^2 - a^2}{(p^2 + a^2)^2}$$

and the differential equation for $y(t) = f(t)$ to be

$$y'' + 16y = 8\cos 4t,$$

with initial conditions $y(0) = 0$ and $y'(0) = 8$.

Apply the method of Laplace transforms to show that the solution to this differential equation is

$$y(t) = (t + 2) \sin 4t,$$

and verify your solution by direct substitution into the differential equation.

[*Note*: An alternative method of solution is to assume a general solution whose form is "motivated" by the appearance of the righthand side of the differential equation. In this case, you might think to let the general solution be some linear combination of sines and cosines.]

Guide to Further Study

The first part of this chapter is a continuation of Chap. 7, and for the integration of differential forms I draw your attention to Arnold [2], Bachman [3], do Carmo [5], Choquet-Bruhat, et al. [4] and Flanders [6] for many of the same reasons cited in the Guide to Further Study in the previous chapter.

For the study of Lie groups in physics, a very good place to start is Gilmore [7], but your choices are virtually endless. I have found Tung [17] to be thorough, but densely written. The emphasis there is on field theory and would be considered an advanced work.

Over the course of this chapter and the text I have made reference to various texts in quantum mechanics. Beyond the undergraduate-level standards, such as the work by Griffiths and Schroeter [8], the relatively recent text by Weinberg [18] offers many valuable insights that would be most appreciated by someone with an understanding of the basics beforehand.

Special functions and integral transforms are the "bread and butter" of most college-level courses and textbooks covering the methods of mathematical physics. Arfken, et al. [1] is very appropriate for upper-division undergraduates in physics and engineering, but others such as those cited in the Guide to Further Study in Chap. 4 are recommended for your consideration as well.

References

1. Arfken, G.B., Weber, H.J., Harris, F.E.: Mathematical Methods for Physicists—A Comprehensive Guide, 7th edn. Academic Press, Waltham, MA (2013)
2. Arnold, V.I.: Mathematical Methods of Classical Mechanics, 2nd edn. Springer, New York (1989)
3. Bachman, D.: A Geometric Approach to Differential Forms. Birkhäuser, Boston (2006)
4. Choquet-Bruhat, Y., DeWitt-Morette, C., Dillard-Bleick, M.: Analysis, Manifolds and Physics, Part I: Basics, 1996 Printing. Elsevier, Amsterdam (1982)
5. do Carmo, M.P.: Differential Forms and Applications. Springer, Berlin (1994)
6. Flanders, H.: Differential Forms With Applications to the Physical Sciences. Dover (1989) with corrections, originally published by Academic Press, New York (1963)
7. Gilmore, R.: Lie Groups, Physics and Geometry. Cambridge University Press, Cambridge (2008)

8. Griffiths, D.J., Schroeter, D.F.: Introduction to Quantum Mechanics, 3rd edn. Cambridge University Press, Cambridge (2018)
9. Hawkins, T.: Emergence of the Theory of Lie Groups: An Essay in the History of Mathematics, 1869–1926. Springer, New York (2000)
10. Lang, S.: Complex Analysis, 4th edn. Springer, New York (1999)
11. Naber, G.L.: Topology, Geometry and Gauge Fields—Foundations, 2nd Edn. Springer, New York (2011). A companion volume is Topology, Geometry and Gauge Fields—Interactions, 2nd Edn. Springer, New York (2011)
12. Nash C., Sen S.: Topology and Geometry for Physicists. Academic Press, New York (1983). Republished by Dover, New York (2011)
13. Olver, P.J.: Applications of Lie Groups to Differential Equations, 2nd edn. Springer, New York (1993)
14. Sakurai, J.J.: Modern Quantum Mechanics. Addison-Wesley, Redwood City, CA (1985)
15. Stefani, H.: Differential Equations: Their Solution Using Symmetries. Cambridge University Press, New York (1989)
16. Thomas Jr., G.B.: Calculus and Analytic Geometry, 4th edn. Addison-Wesley, Reading, MA (1968)
17. Tung, W.K.: Group Theory in Physics. World Scientific, Philadelphia and Singapore (1985)
18. Weinberg, S.: Lectures on Quantum Mechanics, 2nd edn. Cambridge University Press, Cambridge (2017)

Index

A

Abelian
  binary operation, 26
  group, 28
Accumulation point, 159
Action principle, 206
Algebraic structure, or system, 25
Algebraic topology, 39
Algebras, 117
  algebra ideal, 127
  associative, viii, 117
  basis, 119
  basis operators, 120
  commutative, 118
  commutator, 123, 124
  direct sum, 129
  Grassman, 193
  identity
    additive (zero vector), 118
  inverse
    additive, 118
  Lie, 123, 241
  non-associative, viii, 117
  order (dimension), 119
  Poisson, 125
  quotient, 129
  real, 118
  structure constants, 119
  with (multiplicative) inverse, 118
  with unity (identity), 118
  zero element, 118
Argand diagram, 64
Associated tensors, 107
Atlas (on a manifold); compare chart, 234
Axioms of countability (in topology), 191

B

Banach space, 97
Basis, 79
  antisymmetric (alternating), 195
  skew, 90
  standard coordinate, 79
Bivector, 193
Boundary (of a set), 161
Bracket
  Lie, 123
  Poisson, 125
Bra vector, 76

C

Calculus of variations, 206
Canonical transformations, 209
Cartesian product, 16
Category theory, 39
  arrows, 39
  category, 39
  functors, 39
  objects, 39
Cauchy integral theorem, 232
Cauchy sequence, 63, 172
Chain, 5
Characteristic, 56
  of $\mathbb{Z}_n$, 56
  zero, 56
Chart (on a manifold); compare atlas, 234
Circulation theorem, 235
Closure
  binary operation, 25
Closure (of a set), 161
Cluster point, 172, 174
Codomain, 9
Commutator algebra, 123, 124

© Springer Nature Switzerland AG 2021
S. P. Starkovich, *The Structures of Mathematical Physics*,
https://doi.org/10.1007/978-3-030-73449-7

Printed in the United States
by Baker & Taylor Publisher Services